Teresa Krakauer (Ed.)

Enterotoxins: Microbial Proteins and Host Cell Dysregulation

MDPI

This book is a reprint of the Special Issue that appeared in the online, open access journal, *Toxins* (ISSN 2072-6651) from 2013 – 2016 (available at: http://www.mdpi.com/journal/toxins/special_issues/enterotoxins_2013).

Guest Editor
Teresa Krakauer
U.S.Army Medical Research
Institute of Infectious Diseases
USA

Editorial Office
MDPI AG
Klybeckstrasse 64
Basel, Switzerland

Publisher
Shu-Kun Lin

Managing Editor
Chao Xiao

1. Edition 2016

MDPI • Basel • Beijing • Wuhan • Barcelona

ISBN 978-3-03842-163-4 (Hbk)
ISBN 978-3-03842-164-1 (PDF)

Table of Contents

List of Contributors VII

Preface XI

Teresa Krakauer
Update on Staphylococcal Superantigen-Induced Signaling Pathways and
Therapeutic Interventions
Reprinted from: *Toxins* **2013**, *5*(9), 1629-1654
http://www.mdpi.com/2072-6651/5/9/1629 1

Preeti Sharma, Ningyan Wang and David M. Kranz
Soluble T Cell Receptor Vβ Domains Engineered for High-Affinity Binding to
Staphylococcal or Streptococcal Superantigens
Reprinted from: *Toxins* **2014**, *6*(2), 556-574
http://www.mdpi.com/2072-6651/6/2/556 27

MaryAnn Principato and Bi-Feng Qian
Staphylococcal enterotoxins in the Etiopathogenesis of Mucosal Autoimmunity within the
Gastrointestinal Tract
Reprinted from: *Toxins* **2014**, *6*(5), 1471-1489
http://www.mdpi.com/2072-6651/6/5/1471 47

Robert J. McKallip, Harriet F. Hagele and Olga N. Uchakina
Treatment with the Hyaluronic Acid Synthesis Inhibitor 4-Methylumbelliferone Suppresses
SEB-Induced Lung Inflammation
Reprinted from: *Toxins* **2013**, *5*(10), 1814-1826
http://www.mdpi.com/2072-6651/5/10/1814 67

Teresa Krakauer
Sulfasalazine Attenuates Staphylococcal Enterotoxin B-Induced Immune Responses
Reprinted from: *Toxins* **2015**, *7*(2), 553-559
http://www.mdpi.com/2072-6651/7/2/553 81

IV

Lily Zhang and Thomas J. Rogers
Assessment of the Functional Regions of the Superantigen Staphylococcal Enterotoxin B
Reprinted from: *Toxins* **2013**, *5*(10), 1859-1871
http://www.mdpi.com/2072-6651/5/10/1859 88

Norbert Stich, Nina Model, Aysen Samstag, Corina S. Gruener, Hermann M. Wolf and Martha M. Eibl
Toxic Shock Syndrome Toxin-1-Mediated Toxicity Inhibited by Neutralizing Antibodies Late in the Course of Continual *in Vivo* and *in Vitro* Exposure
Reprinted from: *Toxins* **2014**, *6*(6), 1724-1741
http://www.mdpi.com/2072-6651/6/6/1724 101

Stacey X. Xu, Katherine J. Kasper, Joseph J. Zeppa and John K. McCormick
Superantigens Modulate Bacterial Density during *Staphylococcus aureus* Nasal Colonization
Reprinted from: *Toxins* **2015**, *7*(5), 1821-1836
http://www.mdpi.com/2072-6651/7/5/1821 120

Bradley G. Stiles, Gillian Barth, Holger Barth and Michel R. Popoff
Clostridium perfringens Epsilon Toxin: A Malevolent Molecule for Animals and Man?
Reprinted from: *Toxins* **2013**, *5*(11), 2138-2160
http://www.mdpi.com/2072-6651/5/11/2138 137

Masahiro Nagahama, Sadayuki Ochi, Masataka Oda, Kazuaki Miyamoto, Masaya Takehara and Keiko Kobayashi
Recent Insights into *Clostridium perfringens* Beta-Toxin
Reprinted from: *Toxins* **2015**, *7*(2), 396-406
http://www.mdpi.com/2072-6651/7/2/396 161

Simone Roos, Marianne Wyder, Ahmet Candi, Nadine Regenscheit, Christina Nathues, Filip van Immerseel and Horst Posthaus
Binding Studies on Isolated Porcine Small Intestinal Mucosa and *in vitro* Toxicity Studies Reveal Lack of Effect of *C. perfringens* Beta-Toxin on the Porcine Intestinal Epithelium
Reprinted from: *Toxins* **2015**, *7*(4), 1235-1252
http://www.mdpi.com/2072-6651/7/4/1235 172

Bradley G. Stiles, Kisha Pradhan, Jodie M. Fleming, Ramar Perumal Samy, Holger Barth and Michel R. Popoff
Clostridium and *Bacillus* Binary Enterotoxins: Bad for the Bowels, and Eukaryotic Being
Reprinted from: *Toxins* **2014**, *6*(9), 2626-2656
http://www.mdpi.com/2072-6651/6/9/2626 191

Alexandra Olling, Corinna Hüls, Sebastian Goy, Mirco Müller, Simon Krooss, Isa Rudolf, Helma Tatge and Ralf Gerhard
The Combined Repetitive Oligopeptides of *Clostridium difficile* Toxin A Counteract
Premature Cleavage of the Glucosyl-Transferase Domain by Stabilizing Protein Conformation
Reprinted from: *Toxins* **2014**, *6*(7), 2162-2176
http://www.mdpi.com/2072-6651/6/7/2162 223

Jonathan D. Black, Salvatore Lopez, Emiliano Cocco, Carlton L. Schwab, Diana P. English and Alessandro D. Santin
Clostridium Perfringens Enterotoxin (CPE) and CPE-Binding Domain (*c*-CPE) for the
Detection and Treatment of Gynecologic Cancers
Reprinted from: *Toxins* **2015**, *7*(4), 1116-1125
http://www.mdpi.com/2072-6651/7/4/1116 238

Keegan J. Baldauf, Joshua M. Royal, Krystal Teasley Hamorsky and Nobuyuki Matoba
Cholera Toxin B: One Subunit with Many Pharmaceutical Applications
Reprinted from: *Toxins* **2015**, *7*(3), 974-996
http://www.mdpi.com/2072-6651/7/3/974 249

Debaleena Basu and Nilgun E. Tumer
Do the A Subunits Contribute to the Differences in the Toxicity of Shiga Toxin 1 and
Shiga Toxin 2?
Reprinted from: *Toxins* **2015**, *7*(5), 1467-1485
http://www.mdpi.com/2072-6651/7/5/1467 273

List of Contributors

Keegan J. Baldauf: Department of Pharmacology and Toxicology, University of Louisville School of Medicine, Louisville, KY 40202, USA.

Gillian Barth: Veterinary Medical Technology Department, Wilson College, 1015 Philadelphia Avenue, Chambersburg, PA 17201, USA.

Holger Barth: Institute of Pharmacology and Toxicology, University of Ulm Medical Center, Albert-Einstein-Allee 11, Ulm D-89081, Germany.

Debaleena Basu: Department of Plant Biology and Pathology, School of Environmental and Biological Sciences, Rutgers University, New Brunswick, NJ 08901-8520, USA.

Jonathan D. Black: Department of Obstetrics, Gynecology and Reproductive Sciences, Yale University School of Medicine, 333 Cedar Street, PO Box 208063, New Haven, CT 06520-8063, USA.

Ahmet Candi: Department of Infectious Diseases and Pathobiology, Institute of Animal Pathology, Vetsuisse Faculty, University of Bern, Bern 3012, Switzerland.

Emiliano Cocco: Department of Obstetrics, Gynecology and Reproductive Sciences, Yale University School of Medicine, 333 Cedar Street, PO Box 208063, New Haven, CT 06520-8063, USA.

Martha M. Eibl: Biomedizinische ForschungsgmbH Lazarettgasse 19/2, Vienna A-1090, Austria; Immunology Outpatient Clinic, Schwarzspanierstrasse 15, Vienna A-1090, Austria.

Diana P. English: Department of Obstetrics, Gynecology and Reproductive Sciences, Yale University School of Medicine, 333 Cedar Street, PO Box 208063, New Haven, CT 06520-8063, USA.

Jodie M. Fleming: Department of Biology, North Carolina Central University, 1801 Fayetteville Street, Durham, NC 27707, USA.

Ralf Gerhard: Institute of Toxicology, Hannover Medical School, Carl-Neuberg-Str. 1, 30625 Hannover, Germany.

Sebastian Goy: Institute of Toxicology, Hannover Medical School, Carl-Neuberg-Str. 1, 30625 Hannover, Germany.

Corina S. Gruener: Biomedizinische ForschungsgmbH Lazarettgasse 19/2, Vienna A-1090, Austria.

Harriet F. Hagele: Division of Basic Medical Sciences, Mercer University School of Medicine, 1550 College St, Macon, GA 31207, USA.

Krystal Teasley Hamorsk: Owensboro Cancer Research Program of James Graham Brown Cancer Center at University of Louisville School of Medicine, Owensboro, KY 42303, USA; Department of Medicine, University of Louisville School of Medicine, Louisville, KY 40202, USA.

Corinna Hüls: Institute of Toxicology, Hannover Medical School, Carl-Neuberg-Str. 1, 30625 Hannover, Germany.

Katherine J. Kasper: Department of Microbiology and Immunology, Schulich School of Medicine and Dentistry, Western University, London, ON N6A 5C1, Canada.

Keiko Kobayashi: Department of Microbiology, Faculty of Pharmaceutical Sciences, Tokushima Bunri University, Yamashiro-cho 770-8514, Tokushima, Japan.

Teresa Krakauer: Department of Immunology, Molecular Translational Sciences Division, United States Army Medical Research Institute of Infectious Diseases, Fort Detrick, Frederick, MD 21702-5011, USA.

David M. Kranz: Department of Biochemistry, University of Illinois, Urbana, IL 61801, USA.

Simon Krooss: Institute of Toxicology, Hannover Medical School, Carl-Neuberg-Str. 1, 30625 Hannover, Germany.

Salvatore Lopez: Division of Gynecologic Oncology, University Campus Bio-Medico of Rome, Via Alvaro del Portillo 21, 00128 Rome, Italy.

Nobuyuki Matoba: Department of Pharmacology and Toxicology, University of Louisville School of Medicine, Louisville, KY 40202, USA; Owensboro Cancer Research Program of James Graham Brown Cancer Center at University of Louisville School of Medicine, Owensboro, KY 42303, USA.

John K. McCormick: Department of Microbiology and Immunology, Schulich School of Medicine and Dentistry, Western University, London, ON N6A 5C1, Canada; Lawson Health Research Institute, London, ON N6A 5C1, Canada.

Robert J. McKallip: Division of Basic Medical Sciences, Mercer University School of Medicine, 1550 College St, Macon, GA 31207, USA.

Kazuaki Miyamoto: Department of Microbiology, Faculty of Pharmaceutical Sciences, Tokushima Bunri University, Yamashiro-cho 770-8514, Tokushima, Japan.

Nina Model: Biomedizinische ForschungsgmbH Lazarettgasse 19/2, Vienna A-1090, Austria.

Mirco Müller: Institute for Biophysical Chemistry, Hannover Medical School, Carl-Neuberg-Str. 1, 30625 Hannover, Germany.

Masahiro Nagahama: Department of Microbiology, Faculty of Pharmaceutical Sciences, Tokushima Bunri University, Yamashiro-cho 770-8514, Tokushima, Japan.

Christina Nathues: Veterinary Public Health Institute, Vetsuisse Faculty, University of Bern, Bern 3012, Switzerland.

Sadayuki Ochi: Department of Microbiology, Fujita Health University School of Medicine, Toyoake 470-1192, Aichi, Japan.

Masataka Oda: Division of Microbiology and Infectious Diseases, Niigata University Graduate School of Medical and Dental Sciences, Gakkocho-dori, Chuo-ku 951-8514, Niigata, Japan.

Alexandra Olling: Institute of Toxicology, Hannover Medical School, Carl-Neuberg-Str. 1, 30625 Hannover, Germany.

Michel R. Popoff: Bacteries Anaerobies et Toxines, Institut Pasteur, 28 rue du Docteur Roux, Paris 75724, France.

Horst Posthaus: Department of Infectious Diseases and Pathobiology, Institute of Animal Pathology, Vetsuisse Faculty, University of Bern, Bern 3012, Switzerland.

Kisha Pradhan: Environmental Science Department, Wilson College, 1015 Philadelphia Avenue, Chambersburg, PA 17201, USA.

MaryAnn Principato: Division of Toxicology, Office of Applied Research and Safety Assessment, Center for Food Safety and Applied Nutrition, US Food and Drug Administration, 8301 Muirkirk Road, Laurel, MD 20708, USA.

Bi-Feng Qian: Commissioner's Fellowship Program, Division of Toxicology, Office of Applied Research and Safety Assessment, Center for Food Safety and Applied Nutrition, US Food and Drug Administration, 8301 Muirkirk Road, Laurel, MD 20708, USA.

Nadine Regenscheit: Department of Infectious Diseases and Pathobiology, Institute of Animal Pathology, Vetsuisse Faculty, University of Bern, Bern 3012, Switzerland.

Thomas J. Rogers: Center for Inflammation, Translational and Clinical Lung Research, Temple University School of Medicine, 3500 N. Broad Street, Philadelphia, PA 19140, USA.

Simone Roos: Department of Infectious Diseases and Pathobiology, Institute of Animal Pathology, Vetsuisse Faculty, University of Bern, Bern 3012, Switzerland.

Joshua M. Royal: Owensboro Cancer Research Program of James Graham Brown Cancer Center at University of Louisville School of Medicine, Owensboro, KY 42303, USA.

Isa Rudolf: Institute of Toxicology, Hannover Medical School, Carl-Neuberg-Str. 1, 30625 Hannover, Germany.

Aysen Samstag: Immunology Outpatient Clinic, Schwarzspanierstrasse 15, Vienna A-1090, Austria.

Ramar Perumal Samy: Venom and Toxin Research Programme, Department of Anatomy, Yong Loo Lin School of Medicine, National University Health System, National University of Singapore, Kent Ridge 117597, Singapore; Infectious Diseases Programme, Department of Microbiology, Yong Loo Lin School of Medicine, National University Health System, National University of Singapore, Kent Ridge 117597, Singapore.

Alessandro D. Santin: Department of Obstetrics, Gynecology and Reproductive Sciences, Yale University School of Medicine, 333 Cedar Street, PO Box 208063, New Haven, CT 06520-8063, USA.

Carlton L. Schwab: Department of Obstetrics, Gynecology and Reproductive Sciences, Yale University School of Medicine, 333 Cedar Street, PO Box 208063, New Haven, CT 06520-8063, USA.

Preeti Sharma: Department of Biochemistry, University of Illinois, Urbana, IL 61801, USA.

Norbert Stich: Biomedizinische ForschungsgmbH Lazarettgasse 19/2, Vienna A-1090, Austria.

Bradley G. Stiles: Biology Department, Wilson College, 1015 Philadelphia Avenue, Chambersburg, PA 17201, USA.

Masaya Takehara: Department of Microbiology, Faculty of Pharmaceutical Sciences, Tokushima Bunri University, Yamashiro-cho 770-8514, Tokushima, Japan.

Helma Tatge: Institute of Toxicology, Hannover Medical School, Carl-Neuberg-Str. 1, 30625 Hannover, Germany.

Nilgun E. Tumer: Department of Plant Biology and Pathology, School of Environmental and Biological Sciences, Rutgers University, New Brunswick, NJ 08901-8520, USA.

Olga N. Uchakina: Division of Basic Medical Sciences, Mercer University School of Medicine, 1550 College St, Macon, GA 31207, USA.

Filip van Immerseel: Department of Pathology, Bacteriology and Avian Medicine, Ghent University, Ghent 9000, Belgium.

Ningyan Wang: Department of Biochemistry, University of Illinois, Urbana, IL 61801, USA.

Hermann M. Wolf: Immunology Outpatient Clinic, Schwarzspanierstrasse 15, Vienna A-1090, Austria.

Marianne Wyder: Department of Infectious Diseases and Pathobiology, Institute of Animal Pathology, Vetsuisse Faculty, University of Bern, Bern 3012, Switzerland.

Stacey X. Xu: Department of Microbiology and Immunology, Schulich School of Medicine and Dentistry, Western University, London, ON N6A 5C1, Canada.

Joseph J. Zeppa: Department of Microbiology and Immunology, Schulich School of Medicine and Dentistry, Western University, London, ON N6A 5C1, Canada.

Lily Zhang: Center for Inflammation, Translational and Clinical Lung Research, Temple University School of Medicine, 3500 N. Broad Street, Philadelphia, PA 19140, USA.

Preface

Enterotoxins encompass a diverse group of microbial toxins affecting the gut, and are major contributors to bacterial food borne illness, gastrointestinal and systemic diseases, for which limited therapeutics are available. Although the pathogenic effects arise from mucosal perturbation, dysregulation of immune cells through mediator release, cell activation or damage are major factors disrupting homeostasis in gut mucosa. Whereas proinflammatory cytokines and chemokines mediate toxic shock induced by staphylococcal enterotoxins, apoptosis and cytotoxic events are responsible for Clostridium perfringens enterotoxin and cholera toxin. Elucidation of cell receptors, signaling pathways and the communication between cells of the gastrointestinal tract, immune and neuroendocrine system will facilitate the development of new therapeutics.

Teresa Krakauer
Guest Editor

Update on Staphylococcal Superantigen-Induced Signaling Pathways and Therapeutic Interventions

Teresa Krakauer

Abstract: Staphylococcal enterotoxin B (SEB) and related bacterial toxins cause diseases in humans and laboratory animals ranging from food poisoning, acute lung injury to toxic shock. These superantigens bind directly to the major histocompatibility complex class II molecules on antigen-presenting cells and specific Vβ regions of T-cell receptors (TCR), resulting in rapid hyper-activation of the host immune system. In addition to TCR and co-stimulatory signals, proinflammatory mediators activate signaling pathways culminating in cell-stress response, activation of NFκB and mammalian target of rapamycin (mTOR). This article presents a concise review of superantigen-activated signaling pathways and focuses on the therapeutic challenges against bacterial superantigens.

Reprinted from *Toxins*. Cite as: Krakauer, T. Update on Staphylococcal Superantigen-Induced Signaling Pathways and Therapeutic Interventions. *Toxins* **2013**, *5*, 1629-1654.

1. Overview

Staphylococcus aureus produces several exotoxins, staphylococcal enterotoxins A through U (SEA-SEU), and toxic shock syndrome toxin 1 (TSST-1), with potent immunostimulating activities that cause a variety of diseases in humans, including food poisoning, acute lung injury, autoimmune diseases, and toxic shock [1–15]. These bacterial toxins were originally known for their enterotoxicity and pyrogenicity. A considerable effort was directed early on at defining their structure and cellular receptors to understand how these toxins exert their biological effects. Staphylococcal exotoxins bind to the major histocompatibility complex (MHC) class II on antigen-presenting cells (APC) and specific regions of Vβ chains of the T-cell receptor (TCR), leading to activation of both APC and T-cells [7,11,14–17]. The term "superantigen" was coined by Kappler and colleagues in 1989 to describe the novel hyper-stimulatory properties of these bacterial toxins [16]. A decade of crystallographic and structural studies revealed their common molecular structure and binding motifs [18], paving the way for investigations of their signaling mechanisms and the way in which these superantigens exert their potent immunological effects. Unlike conventional antigens, superantigens bypass normal "processing" by APC and induce a large proportion (5%–30%) of T-cells to proliferate at picomolar concentrations [7,16]. The excessive release of proinflammatory cytokines and chemokines from APC, T-cells, and other cell types mediate the toxic effects of staphylococcal superantigens [19–25]. The proinflammatory cytokines, tumor necrosis factor α (TNFα), interleukin 1 (IL-1) and gamma interferon (IFNγ) have tissue damaging effects [26] and together with matrix metalloproteinases (MMPs) and tissue factor produced by superantigen-activated host cells [27], activate both the inflammatory and coagulation pathways. The increased expression of adhesion molecules and chemokine gradient changes direct leukocyte migration to sites of tissue injury [28]. IL-2 from superantigen-activated T-cells causes vasodilation, vascular leak, and edema [29]. Toxic reactive oxygen species (ROS) from activated neutrophils increase vascular permeability and cause acute lung injury [28]. These molecular changes occur rapidly upon superantigen exposure and progress to hypotension, multi-organ failure and death. In addition to inflammatory pathways activated by staphylococcal superantigens, *S. aureus* also produces numerous virulence factors that aid in its survival and subsequent dissemination in the host. For example, staphylococcal extracellular adherence protein [30] and superantigen-like protein 5 [31] as well as two other staphylococcal surface proteins (the clumping factors A and B) [32] stimulate platelet aggregation which leads to disseminated intravascular coagulation. Targeting the inflammatory and coagulation pathways/molecules represent widely diverse strategies to prevent toxic shock and organ damage resulting from superantigens and various virulence factors [33].

SEB is considered a Category B select agent by the Centers for Disease Control and Prevention (CDC) as it is extremely toxic to humans and can be used as an air-borne, food-borne, and water-borne toxicant. The biodefense objective of mitigation of SEB toxicity in the absence of staphylococcal infection seems simpler when compared to the scenario of replicating pathogens with other virulence factors they produced. Recent efforts have been directed at preventing superantigenic shock, acute lung injury and organ damage resulting from the cumulative biological effects elicited by proinflammatory cytokines. Many reviews and books on superantigens have been published and I will present a concise review on the signaling pathways and give a perspective on the therapeutic modalities for counteracting superantigen-induced shock.

2. Staphylococcal Superantigen Structure and Binding to Host Cells

Staphylococcal superantigens are stable, single-chain proteins of 22- to 30-kD that are highly resistant to proteases and denaturation. Despite differences in sequence homology among staphylococcal enterotoxins (SEs) and the streptococcal pyrogenic exotoxins, they have similar protein folds and conserved receptor binding sites [5,15]. These bacterial toxins are classified into five distinct homology groups based on amino acid sequence and similarities in modes of binding to MHC class II molecules [13,15]. Among the different SE "serotypes", SEA, SED, and SEE share the highest amino acid sequence homology, ranging from 53%–81%, whereas SEB is 50%–66% homologous with SECs. TSST-1 has only a limited sequence homology with other SEs. It has a shorter primary sequence of 194 amino acids with no cysteines, and binds TCR Vβ differently than other SEs [17]. TSST-1 lacks enterotoxicity in non-human primates [34] and has a missing "disulfide loop", which may be responsible for the emetic activity of SEs, as mutation of residues in this loop abolishes the emetic activity of SEC2 [35]. There is a separation of the emetic and superantigenic domains of SEs since carboxymethylation of histidine residues of SEB resulted in the loss of emetic activity but not superantigenity [36]. Despite varying sequences, structural and crystallographic analysis of SEA, SEB, and TSST-1 show a conserved conformation with two tightly packed domains containing β-sheets and α-helices [18], separated by a shallow groove representing the TCR-binding site [37,38]. The C-terminal domain has a β-grasp motif found in other unrelated proteins. The N-terminal domain contains an oligosaccharide/oligonucleotide-binding (OB) fold, characterized by the presence of hydrophobic residues in the solvent-exposed regions [18].

Superantigens bind to common, conserved elements outside the peptide-binding groove on MHC class II molecules with relatively high affinity ($K_d = 10^{-8}$–10^{-7} M) [3,39]. Structural analysis shows at least two distinct binding sites on MHC class II molecules for superantigen. A common, low-affinity binding site involving the invariant α-chain of MHC class II and a high-affinity, zinc-dependent binding site on the polymorphic β-chain [39–46]. SEA can cross-link MHC class II molecules on APC by binding to both sites, and persists longer on the cell surface of APC, prolonging its biological effects [47].

The groove formed between the conserved *N*- and *C*-terminal domains of staphylococcal superantigens represents an important interaction site for the TCR Vβ chain [48–51]. Each superantigen binds to a distinct repertoire of Vβ-bearing T-cells, revealing a unique biological "fingerprint" which might be useful for diagnosing toxin exposure [51,52]. The binding of superantigens to the Vβ chain of TCR is of low affinity ($K_d = 10^{-4}$–10^{-6} M), with contacts mostly between the side-chain atoms of the superantigen and the complementarity-determining regions 1 and 2 and the hypervariable region 4 of the Vβ chain. Studies with mutants of SEB and SEC3 indicate that a small increase in the affinity of a superantigen for MHC can overcome a large decrease in their affinity for the TCR [48]. Thus, the multiple modes of superantigen binding to MHC and TCR indicate a cooperative effect of interactions in the formation of the trimolecular complex, hyper-activating the host immune system. The superantigen/MHC interactions strengthen their binding to TCR such that they mimic TCR binding to a conventional MHC-peptide complex [49]. Other co-stimulatory receptors on both cells also interact to further stabilize superantigen binding to many cell types [53,54]. A direct binding of SEB to the T-cell co-stimulatory receptor CD28 was reported recently [55]. Peptides derived from the CD28 binding region protected mice from SEB-induced lethality and reduced TNFα, IL-2 and IFNγ expression [55]. This correlates with previous reports of the resistance of CD28-deficient mice to superantigen-induced shock and the lack of serum TNFα and IFNγ after toxin challenge in these mice [56,57].

3. Three Signals Synergize to Sustain Cell Activation

The three signals required for T-cell activation by superantigens and conventional antigens are similar even though superantigens bind outside the peptide-binding groove of MHC class II molecules. The first signal is induced upon the binding of superantigen with TCR-CD3 complex, which activates the Src family of protein tyrosine kinases (PTKs) [58–60]. The engagement of co-stimulatory molecules on APC and T-cells, subsequent to superantigen binding, results in a second signal that optimizes and sustains T-cell activation [61–63]. The interactions between the adhesion molecules LFA-1 with intercellular adhesion molecule 1 (ICAM-1), and the co-stimulatory molecules CD28 with CD80 on T-cells and APC, respectively, promotes stable cell conjugates. Co-ligation of receptors results in extensive cytoskeletal remodeling and the formation of immunological synapse, initiating signaling cascades [61,64]. PTKs, including Lck and ZAP-70, phosphorylate tyrosine-based motifs of the TCR intracellular components and other adaptors [58,59,65]. The TCR-induced kinases activate phospholipase C gamma (PLCγ) resulting in the generation of second messengers and increase in intracellular calcium levels. One specific second messenger, diacylglycerol (DAG), subsequently activates protein kinase C (PKC) and the proto-oncogene Ras [64,66]. PKC activates downstream signaling pathways including the mitogen-activated protein kinase (MAPK) and the NFκB cascade [67]. Many proinflammatory cytokine genes contain NFκB binding sites in the promoter region and are activated by NFκB [68]. The cytokines IL-1, TNFα, IFNγ, IL-2 and IL-6, and chemokines, in particular, MCP-1, are induced directly by superantigens *in vitro* and *in vivo*. The inflammatory environment provided by these proinflammatory mediators represents the third signal for T-cell activation. IL-1 and TNFα activate many other cell types including fibroblasts, epithelial, and endothelial cells to produce other mediators, cell adhesion molecules, tissue protease MMPs, and ROS. IFNγ from superantigen-activated T-cells activates expression of MHC class II and adhesion molecules, and synergizes with IL-1 and TNFα to promote tissue injury, specifically in the gut [10]. Collectively and individually, these mediators from superantigen-activated cells exert damaging effects on the immune and cardiovascular systems, culminating in multi-organ failure and lethal shock.

4. Cross-Talk among Key Signaling Pathways

The three signals of T-cell activation exert their potent effects by activating the phosphoinositide 3 kinase (PI3K)/mammalian target of rapamycin (mTOR), NFκB and MAPK pathways [67–69]. A description of these signal transduction pathways upon superantigen binding to host cell receptors was presented recently (Figure 1) [70]. Phosphorylation and dephosphorylation events modulate all three cascades with specific kinases and phosphatases. PTKs and lipid molecules from PLCγ activation trigger the PI3K pathway upon specific ligand binding to a number of receptors besides the TCR. Co-stimulatory receptor CD28, IL-2 receptor (IL-2R), IFNγR, growth factor receptor, and G-protein-coupled receptor (GPCR) all activate PI3K [69]. Different PTK inhibitors including genistein, tyrphostin, and herbimycin A, reduced IL-1 levels in TSST-1-stimulated cells [65]. PI3K activates Akt (also known as PKB) and mTOR downstream and modulates many biological processes including cell growth, differentiation, proliferation, survival, migration and metabolism [71–76]. The importance of the PI3K/mTOR pathway is shown by the efficacy of rapamycin, a specific inhibitor of mTOR complex 1 (mTORC1), in protecting mice from SEB-induced lethal shock [77]. Rapamycin inhibited SEB-stimulated T-cell proliferation and reduced SEB-induced IL-2 and IFNγ *in vitro* and *in vivo*. An alternative pathway of T-cell activation by SEE bypasses PTK tyrosine phosphorylation and uses PLCβ to activate PKC, ultimately activating extracellular signal-regulated kinase 1 and 2 (ERK1/2), NF-AT and NFκB [78].

The MAPK pathway is induced by mitogens, superantigens, cytokines, chemokines, growth factors, as well as environmental stress, and comprises of three major kinase cascades, ERK1/2, c-Jun-*N*-terminal kinase (JNK), and p38 MAPK. These MAP kinases control fundamental cellular processes to signal cell stress, culminating in the activation of transcription factors NFκB, NF-AT and AP-1 [79], affecting proliferation, differentiation and apoptosis. One common upstream activator of the MAPK pathway is PKC which is activated by TCR, co-stimulatory receptors and GPCR. MAPK promotes inflammation by targeting NFκB to promoters of inflammatory genes [80]. IL-1 and TNFα are both activators and effectors of MAPKs, as these mediators both activate

MAPK via various intracellular TNF receptor-associated factors (TRAFs), and are themselves induced by MAPK activation.

The proinflammatory cytokines IL-1 and TNFα can directly activate the transcriptional factor NFκB in many cell types that include epithelial and endothelial cells. IL-1 interacts with IL-1 receptor 1 (IL-1R1) and receptor accessory protein, uses signaling molecules, the adaptor myeloid differentiation factor 88 (MyD88), IL-1R-associated protein kinase 1 (IRAK1), and TRAF-6 to activate IκB kinases (IKK), leading to NFκB activation [81]. Phosphorylation of the inhibitory protein IκBα by IKK leads to IκBα degradation and release from cytoplasmic NFκB. This allows NFκB to translocate to the nucleus where it binds to promoter regions of various inflammatory genes [82]. Activation of NFκB leads to induction of many proinflammatory and anti-apoptotic genes. IL-1R1 has structural homology to toll-like receptors (TLRs) which use similar intracellular adaptors and molecules as those used for IL-1R1 for signaling (Figure 1). TLRs are receptors used by the host to sense pathogen associated molecules such as lipoprotein, peptidoglycan, lipopolysaccharide, flagellin, dsRNA and viral RNA to activate a rapid innate response [83]. Recently, SEB was shown to upregulate the expression of TLR2 and TLR4, thereby enhancing the host response to other microbial products [84–86]. This might partially account for the synergistic effects of LPS and SEB in mouse models of SEB-induced shock [87–89].

TNFα binds to TNF receptor 1 and 2 (TNFR1, TNFR2) and signals with different intracellular TRAFs, ultimately activating MAPK and NFκB, and results in the expression of other cytokines, adhesion and co-stimulatory molecules [26,90]. An important and damaging component of signaling by the TNFR superfamily which includes various death receptors is caspase activation via the intracellular death domains of these TNFRs. Receptors in this superfamily use intracellular adaptors, TNFR-associated death domain (TRADD) and Fas-associated death domain (FADD) to activate the caspase 8 cascade, JNK, and NFκB. These multiple pathways account for the pleiotropic effects of TNFα including apoptosis, cell activation, coagulation, inflammation, and host defense [90]. The synergistic effects of TNFα and IFNγ on epithelial cells increase ion transport, leading to cell damage and epithelial leakage [10]. The critical role of TNFα in mediating lethality was shown by anti-TNFα antibodies protecting mice from SEB-induced shock in a D-galactosamine (Dgal)-sensitized model [91].

Chemokines, and T-cell cytokines, IFNγ and IL-2, bind to their respective receptors and activate the PI3K/mTOR and MAPK pathways with diverse signal transducers. IFNγ binds to IFNγR, uses Janus kinase 1 and 2 (JAK1, JAK2) to phosphorylate the signal transducer and activator of transcription 1 (STAT1) [92,93]. The main function of IFNγ is in antimicrobial defense as it activates antiviral genes, adhesion molecules, immunoproteasome, and E3 ligase. The IFNγ-activated JAKs also activate PI3K/mTOR independent of STAT1 [94]. Additionally, IFNγ induces the expression and activation of death receptors including Fas (CD95), leading to cell apoptosis [95]. Thus, IFNγ-induced immunoproteasome and CD95 death signaling pathways contribute to vascular cell apoptosis and cardiovascular inflammation [95]. The death receptors use intracellular death domains to activate FADD and caspase 8, resulting in mitochondrial cytochrome c release and DNA fragmentation. IFNγ disrupts ion transport and barrier function in superantigen-activated epithelial cells and these biological effects are amplified by TNFα [96]. However, anti-IFNγ had no effect on mortality and only reduced SEB-induced weight loss and hypoglycemia in the Dgal-sensitized mouse model of lethal shock [97]. A recent study suggests that IFNγ from SEB-stimulated cells plays an important role in autoimmunity in HLA-DQ8 transgenic mice [98].

Figure 1. Cell receptors, intracellular signaling molecules, and signal transduction pathways used by superantigens and mediators induced by superantigens. Potential targets of inhibition are represented by stop signs 1–14, numbered in order of their description in the text. 1. Major histocompatibility complex (MHC) class II (not shown); 2. T-cell receptor (TCR) Vβ; 3. CD28; 4. Tyrosine kinases; 5. Phospholipase C (PLC); 6. Mammalian target of rapamycin (mTOR); 7. Protein kinase C (PKC); 8. Extracellular signal-regulated kinase (ERK1/2); 9. NFκB; 10. p38 MAPK; 11. Myeloid differentiation factor 88 (MyD88); 12. Proteasomes; 13. Caspases; 14. Signal transducer and activator of transcription (STAT).

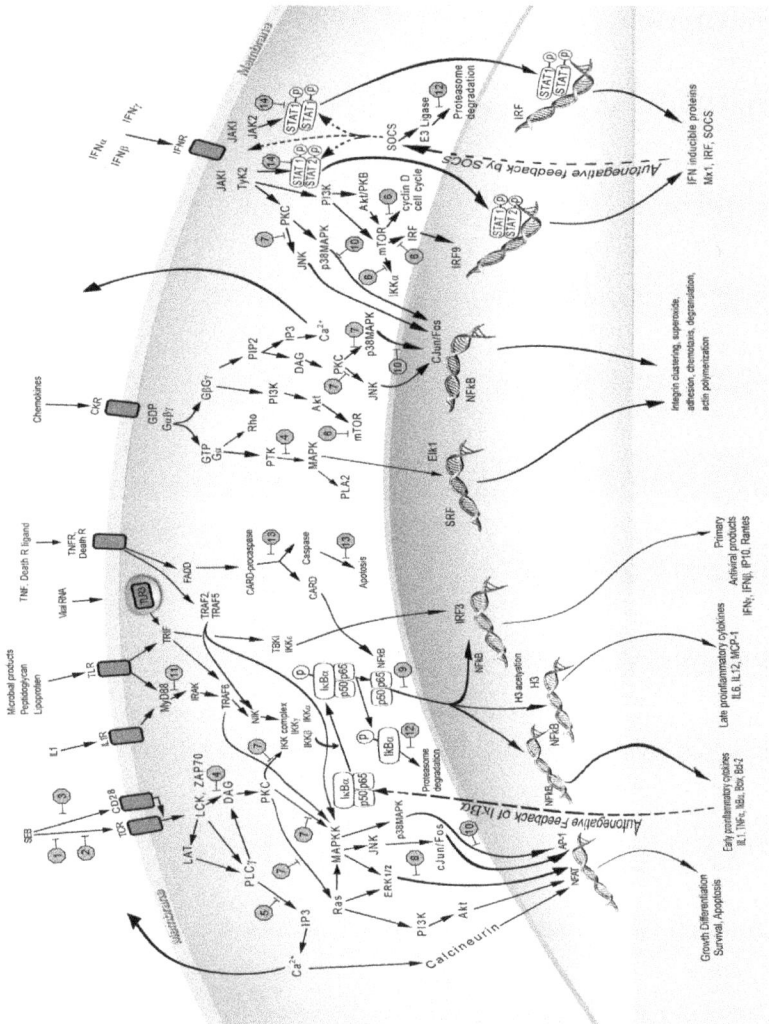

IL-2 binds to the IL-2R and signals through JAK1 and JAK3 to activate PI3K and Ras, affecting proliferation, growth, and differentiation of many cell types [99]. Ras signals through the MAPK pathway to activate AP-1, cJun/Fos and NFAT. IL-2 increases microvascular permeability and induces vasodilation, resulting in perivascular edema in SEB-induced lung injury [100,101]. IL-2-deficient mice are resistant to SEB-induced toxic shock [102].

IL-6, from both macrophages and activated T-cells, has some overlapping activities with IL-1 and TNFα, and activates JAK3 and Ras upon binding to IL6R [103]. Additionally, IL-6R also signals through PI3K/mTOR to promote cell survival. The Ras pathway used by IL6, IL2, IFNγ, TCR and co-stimulatory receptors results in MAPK activation whereas the alternate PI3K pathway activates mTOR.

The chemokines IL-8, MCP-1, MIP-1α, and MIP-1β, are induced directly by SEB or TSST-1 and are potent chemoattractants and leukocytes activators [22,26,104]. Chemokines bind to seven-transmembrane GPCR, induce early calcium flux, activate PLC and signal via the PI3K/mTOR pathway [26,104,105]. Chemokines orchestrate leukocyte migration to promote inflammation and increase tissue injury. Exudates from superantigen-injected air pouches contained predominantly neutrophils with few macrophages [22]. Recruited- and activated-neutrophils produce cytotoxic superoxide and MMPs, contributing to organ damage. Systemic or intranasal exposure to SEB resulted in acute lung injury characterized by increased expression of adhesion molecules ICAM-1 and VCAM, increased neutrophil and mononuclear cell infiltrate, endothelial cell injury, and increased vascular permeability [28,106].

TCR, co-stimulatory receptors and cytokines signal with diverse intracellular molecules to activate PI3K/mTOR, MAPK, and IKK/NFκB cascades. There is cross-talk among these pathways as the MAPKs cascade downstream from TCR, co-stimulatory receptors and T-cell cytokines all activate NFκB, whereas TRAFs from IL-1 and TNFα signaling activate MAPK and NFκB independently. There is some overlap and redundancy of these activation pathways as multiple receptors activate PI3K/mTOR, MAPK and NFκB. However, specificity exists as illustrated by the different classes of MAPKs and their targets. JNK regulates c-Jun and AP-1, and has detrimental effects in the liver whereas p38 MAPK has an additional effect on the phosphorylation of eukaryotic initiation factor (eIF-4E) and promotes translation [79]. The cellular responses to individual cytokines are also different and specific with IFNγ increasing cellular permeability in activated epithelial and endothelial cells whereas IL-1 has prothrombotic effects on the endothelium through the increased production of tissue factor and prostaglandins.

5. Mouse Models of Superantigen-Induced Shock

Superantigens from *S. aureus* and *Streptococcus pyogenes* are the causative agents of serious life threatening toxic shock syndrome (TSS) and the excessive release of cytokines contributes to the pathogenesis of TSS [1–3,33]. SEB has historically been used as a prototype superantigen in biological and biodefense research investigations, as humans are extremely sensitive to SEB especially by inhalation. An obvious step in developing new therapeutic approaches for SEB-induced toxic shock is finding relevant models that mimic human disease.

Mice are often used as a model to study the immunological mechanisms of superantigen mediated shock [21,22,25,28,55–57,87–89,101]. Although these animals lack an emetic response, they are ideal to work with as immunological reagents are available, the strains and genetic backgrounds including specific MHC class II are well-defined, and the low cost of maintenance allows more animals to be used in experimental protocols. However, mice are naturally less susceptible to SEs when compared to humans because of the lower toxin affinity to murine MHC class II [88,107]. As a result, mice do not develop lethal SEB shock and potentiating agents such as Dgal, actinomycin D, LPS, or viruses are used together with toxin to induce toxic shock [88,91,108–111]. Depending on the injury model, sensitizing agents and route of delivery, the severity of disease may involve different organs and distinct profile of mediators. Both Dgal and actinomycin D induced TNFα-dependent hepatotoxicity, and SEB-induced shock models using these potentiators showed much higher serum levels of TNFα not present when SEB was used alone and liver injury was a key feature in these models. Both IL-2 and IFNγ are also critical in Dgal-sensitized models of superantigen-induced shock as IL-2 deficient mice were resistant to SEB-induced shock and antibodies to IFNγ inhibited SEB-induced weight loss and hypoglycemia [97,102].

LPS, a cell wall component of gram negative bacteria, is the most frequently used synergistic agent in mouse model of SEB-induced shock [25,87–89,111]. Relatively high doses of SEB or LPS are used together in these models, usually with Balb/c mice. The shock syndrome induced by superantigens in these models results from the culmination of the biological effects of elevated levels of IL-1, TNFα, and IFNγ, not seen in the absence of LPS [88]. An analysis of the interdependent effects of various doses of SEB used alone and together with LPS in different dose combinations indicated that prolonged levels of certain cytokines were necessary to induce lethal shock in Balb/c mice [25]. Non-survivors in SEB plus LPS groups have significantly higher levels of TNFα, IL-6, MIP-2, and MCP-1 early (eight hours) after SEB exposure [25]. In addition to these mediators, non-survivors showed higher levels of IFNγ and IL-2 later at 24 h. In this LPS-sensitized shock model, lethality and cytokine response were both influenced mostly by the LPS dose and not by SEB. The early TNFα release and sustained levels of IFNγ, IL-2, IL-6, MIP-2 and MCP-1 later correlated with acute lethal shock at 48 h. Since LPS and SEB activate similar cytokines and cells, although using different receptors, it is difficult to compare the molecular mechanisms of shock in this traditional model with human TSS.

To avoid the confounding effects of potentiating agents, a "double-hit" low dose SEB model was developed in a LPS resistant mouse strain C3H/HeJ to simulate human TSS [101]. This model mimics human TSS as intranasal delivery of SEB triggers lung inflammation, systemic release of cytokines, and hypothermia that culminate in death at later time points unlike the various potentiated models with much earlier lethal end points. An alternative model using transgenic mice expressing human HLA class II molecules was established to recapitulate human TSS, with different susceptibility to various superantigens dependent on the inserted human HLA class II type, DQ or DR [112–116]. Transgenic mice expressing HLA-DQ6 still required Dgal to potentiate the effects of SEB [112] whereas mice with HLA-DR3 were sensitive to SEB alone [116]. A recent study revealed multiple organ inflammation in lung, liver, kidney, heart and the small intestine that accompanied lethal shock in HLA-DR3 transgenics [116]. Moreover, intestinal absorptive functions were also interrupted in this transgenic model of SEB-induced shock.

6. Vaccines and Therapeutic Antibodies

There is currently no available treatment for superantigen-induced shock except for the use of intravenous human immunoglobulin [117]. Antibodies to superantigens can provide broad spectrum protection as neutralizing antibodies against one superantigen cross-react and block the biological effects of a different superantigen [118]. Naturally protective antibodies against superantigens can be found in *S. aureus* bacteremia and increase in neutralizing titers during infection correlated with recovery [119]. Other studies showed that there is a correlation of lower serum antibodies to TSST-1 in patients with recurring TSS [120,121]. Both monoclonal and human-mouse chimeric antibodies against SEB with high affinities in the picomolar range have been used effectively to target SEB-induced host responses [122–124]. The use of antibodies has certain limitations since neutralization of toxins is effective only at the early stages of exposure as it blocks the first step of host receptor interaction before cell activation. Vaccination is a proven method to prevent SEB-induced shock and attenuated mutants of SEB with defective MHC class II binding which lack superantigenicity were efficacious against toxin challenge in mice, piglets and monkeys [125,126].

7. Inhibitors of Cell Receptor-Toxin Interaction

A number of small overlapping peptides, encompassing a conserved region of SEB (residues 150–161), bind to host cell receptors and have been tested to block superantigen-induced effects both *in vitro* and *in vivo* with contradictory results using the same peptide [111,113,127]. Although the dodecapeptide prevented transcytosis of various SEs across human intestinal epithelial cell monolayer and may block co-stimulatory signals [128], this and other "SEB peptide antagonists" failed to block SEB-induced T-cell proliferation in human peripheral blood mononuclear cells (PBMC) and had no effect on SEB-induced lethal shock in HLA-DR3 transgenic mice [113]. Blockade of the CD28 co-stimulatory receptor by its synthetic ligand, CTLA4Ig (also known as abatacept) prevented TSST-1-induced shock, and reduced the serums TNFα, IL-2, and IFNγ [129]. Peptide mimetics of CD28 also prevented co-stimulatory receptor interaction with SEB and inhibited SEB effects *in vitro* and *in vivo* [55]. Furthermore, a specific CD28-peptide mimetic blocked superantigen-binding to CD28 and attenuated toxic shock and necrotizing soft-tissue infection induced by *Streptococcus pyogenes* [130]. Anti-CD44 reduced the binding of SEB-activated leukocytes to lung epithelial cells and prevented acute lung injury [131]. Bi-specific chimeric inhibitors composed of MHC class II and TCR Vβ domains competitively blocked SEB binding and cell activation in human PBMC [132]. A soluble TCR Vβ mutant also neutralized the effects of superantigens *in vitro* [133]. Blocking receptor interaction has many limitations as blockade of host receptors might adversely affect immune function and the inhibitor has to be administered pre-exposure or soon after exposure to toxins for it to be effective. Aside from cross-reactive antibodies, receptor blockade inhibitors are usually specific and have to be tailor-made for a specific superantigen [132]. A list of the therapeutics used in mouse SEB-induced toxic shock models is shown in Table 1, arranged in order of their description in the text.

8. Inhibitors of Signal Transduction and Cytokines

Targeting host responses after superantigen exposure is an attractive strategy as these events occur later and may be more amenable to interruption. An important class of therapeutic compounds to be considered is inhibitors that can block signal transduction pathways activated by superantigens. Inhibitors of signal transduction are often cytokine inhibitors as cytokines are the best known "signalers", acting both as activators and effectors. An obvious advantage of signal transduction inhibitors is that they can be administered post-exposure and are likely broad spectrum, inhibiting many different superantigens or even pathogens that trigger similar host responses and signaling pathways. The pathways central to superantigen activation are PI3K/mTOR, MAPK, and NFκB, which are also used by other pathogens, TLRs, and cytokines.

Various animal models indicated TSS results from the cytokine signals activating host pathways and inducing damage in multiple organs. The convergence of multiple receptor signaling allows activation of innate host pathway signals to persist and dominate, and these "signals" are likely potential therapeutic targets. An example is NFκB, which binds to the promoter regions of many inflammatory genes implicated in TSS including cell adhesion molecules, cytokines, chemokines, acute phase proteins, and inducible nitric oxide synthase [82]. The downstream activation of NFκB leads to the inducible expression of mediators involved in inflammation and tissue injury seen in SEB-induced lung injury and toxic shock models. A cell-permeable cyclic peptide targeting NFκB nuclear transport prevented lethal shock in Dgal-sensitized mice accompanied by attenuation in liver apoptosis and hemorrhagic necrosis [100,134]. This NFκB inhibitor also reduced SEB-induced inflammatory cytokines and T-cell responses [100].

There are other NFκB inhibitors which are FDA-approved for treatment of inflammatory diseases and cancers [148,149]. Dexamethasone, an immunosuppressive corticosteroid, potently attenuated superantigen-induced T-cell proliferation, cytokine release, and cell activation marker expression in human PBMC [150]. Dexamethasone also prevented lethal shock accompanied by attenuation of the hypothermic response, weight loss and serum cytokines in the LPS-potentiated SEB model and the SEB "double-hit" model of toxic shock [106,135]. The pulmonary lesions were reduced by dexamethasone treatment only at later time points (96 to 168 h) and resolution of lung inflammation lagged behind the reduction in cytokines such that a long course of steroid treatment was necessary to rescue mice from lethal shock [106]. Bortezomib, another inhibitor of NFκB, and a proteasome inhibitor, blocked SEB-induced cytokine release but had no effect on lethality or liver necrosis in transgenic mice [136]. Natural products such as epigallocatechin gallate (EGCG) from green tea, and resveratrol (RES) from red wine, are also NFκB inhibitors that separately reduced superantigen-induced T-cell proliferation and cytokine release from human PBMC [151]. EGCG attenuated IFNγ-induced epithelial permeability increases and suppressed T-cell activation and cytokines from SEB-stimulated human PBMC and murine lymph node cells [152]. RES reduced lung injury by blocking SEB-induced T-cell activation, pulmonary permeability increases, and caspase 8-dependent apoptosis [153]. Another upstream inhibitor of NFκB, a synthetic mimetic (EM-163) to the BB-loop of MyD88, reduced multiple cytokines in superantigen-stimulated human PBMC and protected mice from lethal shock in the LPS-sensitized model [137,138]. However, the complete or long-term blockade of NFκB would likely produce adverse side effects as NFκB is essential in maintaining normal host defense and homeostasis [68,154].

Table 1. Therapeutics tested for efficacy in murine models of staphylococcal enterotoxins (SEB)-induced toxic shock. * indicates drug is FDA-approved.

Pharmacologic agent	Target	Biological effects against SEB
Anti-SEB monoclonal antibodies	SEB	Neutralized mitogenic activity of SEB *in vitro*. Prevented SEB-induced shock in HLA-DR3 transgenic mice [124].
SEB-peptide antagonists	MHC	Blocked superantigen binding to MHC class II in human PBMC and inhibited T-cell proliferation [111]. Afforded 83% protection in mouse model of SEB + LPS-induced shock [111]. Failed to block SEB-induced T-cell proliferation in human PBMC [113]. No protective effect against SEB-induced shock in HLA-DR3 transgenic mice [113]. Decreased SEB-induced IL-2, IFNγ and TNFβ gene expression [127]. Protected 80% of Dgal-sensitized mice against SEB-induced shock [127].
Mimetic peptides of CD28	CD28	Blocked superantigen binding to CD28 and attenuated SEB-induced IL-2, TNFα, and IFNγ [55]. Protected mice from SEB-induced shock [55,130].
Cell-permeable peptide targeting NFκB	NFκB nuclear translocation	Attenuated serum TNFα, IFNγ and IL-6. Protected mice from liver injury and SEB-induced shock in Dgal-sensitized mice [134].
Dexamethasone *	NFκB	Inhibited SEB-induced proinflammatory cytokines and chemokines in human PBMC. Reduced serum levels of cytokines, attenuated hypothermia due to SEB, and protected mice 100% in both SEB-induced and SEB + LPS-induced shock models [106,135].
Bortezomib *	NFκB, proteasome	Decreased serum cytokine but no effect on lethality in HLA-DR3 transgenic mice challenged with SEB [136].
Mimetic peptides of BB loop of MyD88	MyD88	Reduced SEB-induced IL-1β, TNFα and IFNγ. Provided 83% protection in SEB + LPS-induced shock model [137,138].
D609	PLC	Blocked SEB-induced cytokines and chemokines [139]. Protected 90% of mice from SEB-induced lethal shock [140].
Cell-permeable SOCS3	STAT1	Inhibited cytokine production, attenuated liver necrosis, and prevented SEB + LPS-induced lethal shock [141].
Rapamycin *	mTORC1	Blocked SEB-induced cytokines and chemokines. Protected mice 100% from lethality even when administered 24 h after SEB [77,142].
Tacrolimus *	Calcineurin phosphatase	Suppressed serum cytokines but no protection against SEB-induced shock in HLA-DR3 transgenic mice [143].
N-acetyl cysteine *	NFκB, ROS	Suppressed NFκB activation but protected only 30% of mice from SEB-induced lethal shock [144,145].
Dexamethasone * + *N*-acetyl cysteine *	NFκB, ROS	When used in tandem, reduced SEB-induced cytokines, hypothermia, and protected 75% of mice from lethal shock [145].
Niacinamide	Nitric oxide	Reduced SEB-induced cytokines and lethality in SEB + LPS-induced shock model [146].
Pentoxifylline *	Phospho-diesterase	Attenuated SEB-induced cytokines *in vitro* and *in vivo*. Prevented lethality in SEB + LPS-induced shock model [147].

Other pathway inhibitors include those directed against the various kinases, PKC, MAPK and PTK. Genistein, a tyrosine kinase inhibitor, and H7, a PKC inhibitor, separately reduced TNFα but not IL-1 from TSST-1-stimulated PBMC [155]. A selective inhibitor of p38 MAPK (SB203580) and an inhibitor of ERK (PD098059) each partially blocked TNFα production from SEB-stimulated human T cell clones [156]. D609, an inhibitor of PLC, which is downstream from superantigen binding to TCR and CD28, blocked SEB-induced effects both *in vitro* and *in vivo* [139,140]. SOCS3, an intracellular feedback inhibitor of the various STATs used by IFNγ and IL-2 signaling, reduced the effects of these two cytokines [92]. A cell-permeable form of SOCS3 reduced the lethal effects of SEB and LPS by inhibiting the production of inflammatory cytokines and attenuating liver apoptosis and hemorrhagic necrosis [141].

Immunosuppressive drugs are also good candidates to block superantigen-induced immune responses as they are potent inhibitors against many cell types including T-cells and macrophages. Three FDA-approved drugs for preventing transplant rejection have been used in three different animal models of SEB-induced toxic shock. Cyclosporine A (CsA) inhibited SEB-induced T-cell proliferation *in vitro*, reduced serum cytokines, and attenuated pulmonary inflammation, but has no effect on lethality in monkeys [157]. Rapamycin, a specific inhibitor of mTORC1, was efficacious even when given 24 h after SEB in the SEB "double-hit" model [77]. Rapamycin blocked SEB-induced T-cell proliferation, reduced serum cytokines, and prevented hypothermia and weight loss induced by SEB. Intranasal rapamycin also protected mice against SEB-induced shock when administered as late as 17 h after toxin exposure, providing a practical route of drug delivery against SEB [142]. Another structural analog of rapamycin, tacrolimus, suppressed superantigen-induced T-cell activation *in vitro* but did not reduce lethality in HLA-DR3 transgenic mice [143].

Another hallmark of SEB-intoxication is acute lung injury which is most likely a result of oxidative stress inducing damage in the lung. Acute lung injury arises as SEB-, cytokine- and chemokine-activated neutrophils migrate into lung areas producing high levels of superoxide, which is capable of inducing vascular permeability and apoptosis [28]. The anti-oxidants *N*-acetyl cysteine (NAC) and pyrrolidine dithiocarbamate (PDTC) each mitigated NFκB signaling and T-cell proliferation, and blocked cytokine production in superantigen-activated human PBMC [144]. However, NAC has only a minor effect *in vivo*, reducing lethality by 30% in the SEB "double-hit" model [145]. Dexamethasone, although effective against SEB-induced shock, required a prolonged dosing of up to four days, which might not be ideal in a clinical setting as dexamethasone is immunosuppressive. Treatment with a short course of dexamethasone (up to five hours post-SEB) provided only 20% protection. Importantly, the combined effects of a short treatment course of intranasal dexamethasone followed by NAC prevented SEB-induced shock, hypothermia and weight loss [145]. Both dexamethasone and NAC are FDA-approved drugs that act distal to toxin binding. Another combination treatment, using a human-mouse chimeric anti-SEB antibody and lovastatin concomitantly and immediately after toxin exposure, also protected transgenic mice from SEB-induced shock [158].

Most therapeutic testing in animal models of SEB-induced shock have targeted proinflammatory cytokines as there is a strong correlation between toxicity and elevated serum levels of these mediators [21–25]. Inhibitors aimed at blocking proinflammatory mediator release overlap with inhibitors of signal transduction triggered by superantigens. The critical role of TNFα in lethal shock was established by the prevention of SEB-induced lethality with neutralizing antibodies against TNFα in Dgal-sensitized mice [91]. IL-10, an anti-inflammatory cytokine, prevented superantigen-induced toxic shock by reducing the production of the proinflammatory mediators IL-1, TNFα and IFNγ [159,160]. The nitric oxide inhibitor niacinamide improved the survival of mice given LPS plus SEB by attenuating serum IL-2 and IFNγ [146]. Doxycycline, an antibiotic, inhibited SEB-induced T-cell proliferation, proinflammatory cytokines, and chemokines in human PBMC [161]. Recently, a panel of different antibiotics was tested for inhibitory effects on cytokine release from SEA- and TSST-1-stimulated human PBMC [162]. Tigecycline decreased IL-6 and IFNγ whereas trimethoprim increased IL-8 and TNFα from superantigen-stimulated cells. Clindamycin, daptomycin, vanomycin and azithromycin had no effect on cytokine release in these stimulated cells. Another study showed that azithromycin suppressed TSST-1-induced T-cell proliferation by blocking ERK and JNK activity [163]. Pentoxyfylline, a phophodiesterase inhibitor used clinically to treat peripheral vascular disease, reduced cytokines and T-cell proliferation in SEB- or TSST-1-stimulated cells [147]. Pentoxyfylline prevented lethal shock accompanied by a reduction in serum cytokines in the LPS plus SEB mouse model [147].

Caspase inhibitors have also been used to attenuate the toxic effects of superantigens as caspases initiate cellular apoptosis and the release of certain cytokines from inactive precursors. The release of IL-1β is dependent on caspase 1, a proteolytic enzyme that cleaves pro-IL-1 into active IL-1β [26]. The caspase 1 specific inhibitor, Ac-YVAD-cmk, attenuated both IL-1β and MCP-1 production in SEB- and TSST-1-stimulated PBMC cultures but had no effect on other cytokines or T-cell proliferation [164]. Caspase 3 and caspase 8 are enzymes involved in SEB-induced cell apoptosis but inhibitors of these two caspases were ineffective in reducing superantigen-induced cytokines or T-cell proliferation [164]. In contrast, a pan-caspase inhibitor, Z-D-CH$_2$-DCB, blocked the production of IL-1β, TNFα, IL-6, IFNγ, MCP-1, MIP-1α, MIP-1β, and inhibited T-cell proliferation in SEB- and TSST-1-stimulated PBMC [164].

9. Repurposing of FDA-Approved Drugs for Biodefense Agents

As seen from the above studies, FDA-approved drugs currently used for other indications including dexamethasone, rapamycin, cyclosporine A, tacrolimus, bortezomib, doxycycline, pentoxyfylline, NAC, PDTC, have been tested as therapeutics against superantigens with varying degree of success since the 1990s. The testing of FDA-approved drugs for preventing superantigen-induced shock should speed up the approval process for biodefense use in case of exposure. However, as seen from the various FDA-approved drugs tested, even knowing the mechanism of action of these drugs is no guarantee for success as *in vivo* dosages, dosing routes and schedules as well as animal models all affect the outcome. Rapamycin, by decreasing the levels and effects of IL-2 and IFNγ through mTOR inhibition, is proven to be the most effective single agent to counter both intranasal and systemic exposure to SEB [77,142].

Repurposing of FDA-approved therapeutics makes sense for biodefense use as the therapeutics approval process for human use requires resources and time that might not work for biodefense-related agents. Currently, the approval rate for therapeutics through the FDA is low with 90% of drugs rejected due to safety concerns, inadequate bioavailability or lack of efficacy [165,166]. Intuitively, drug repurposing makes use of the drug's mechanism of action and applics it to diseases or bioterror agents with known or putative pathogenic effects. It bypasses the usual time and resource consuming process of target discovery, optimization, preclinical development and clinical safety testing, and might possibly obtain faster regulatory review by the FDA. This fast track method of repurposing FDA-approved drugs is especially suited for biodefense agents as clinical evaluation of efficacy is usually not possible or ethical. The development of animal models that simulate human diseases by bioterror agents is of critical importance in this non-traditional route of drug repurposing for biodefense use. New avenues to be considered include the use of FDA-approved drugs singularly, in combination, or in tandem. In the case of simultaneous dosing, a lower dose of each drug with different mechanisms of action might produce synergistic beneficial effects and limit individual drug toxicity. Drugs used in tandem will likely be cooperative with the first drug muting out the early host inflammatory response and the second drug acting on secondary signals. Systematic identification of novel synergistic drug combinations will be beneficial to treat a multi-system and complex disease such as TSS.

10. Summary

Significant advances have been made in cell activation signals and pathways induced by staphylococcal superantigens. The superantigenic properties of SEB make it an "ideal" toxin to study the cellular interactions, biological effects and therapeutic interventions. Newer mouse models of toxic shock using human HLA class II transgenic mice or SEB un-potentiated mice can better define the systemic effects of SEB and aid in the therapeutic discovery to prevent TSS. Targeting proinflammatory mediators and T-cell cytokines appears to be most beneficial yet not all anti-inflammatory drugs are effective in preventing shock. The use of FDA-approved drugs, rational combinations of FDA-approved drugs, and changing treatment modality are avenues to fast-track and repurpose old drugs for biodefense use. Immunosuppressants, combinations of an immunosuppressant with an anti-oxidant and other carefully tailored combinations hold promise as treatment options for TSS.

Disclaimer

The views expressed in this publication are those of the author and do not reflect the official policy or position of the Department of the Army, the Department of Defense, or the U.S. Government.

Acknowledgments

I thank the Defense Threat Reduction Agency for generous support.

Conflicts of Interest

The author declares no conflict of interest.

References

1. DeVries, A.S.; Lesher, L.; Schlievert, P.M.; Rogers, T.; Villaume, L.G.; Danila, R.; Ruth, L. Staphylococcal toxic shock syndrome 2000–2006: Epidemiology, clinical features, and molecular characteristics. *PLoS One* **2011**, *6*, e22997.

2. Brosnahan, A.J.; Schlievert, P.M. Gram-positive bacterial superantigen outside-in signaling causes toxic shock syndrome. *FEBS J.* **2011**, *278*, 4649–4667.

3. Langley, R.; Patel, D.; Jackson, N.; Clow, F.; Fraser, J.D. Staphylococcal superantigen super-domains in immune evasion. *Crit. Rev. Immunol.* **2010**, *30*, 149–165.

4. Argudin, M.A.; Mendoza, M.C.; Rodicio, M.R. Food poisoning and *Staphylococcus aureus* enterotoxins. *Toxins* **2010**, *2*, 1751–1773.

5. Schlievert, P.M.; Bohach, G.A. Staphylococcal and Streptococcal Superantigens: An Update. In *Superantigens: Molecular Basis for Their Role in Human Diseases*; Kotb, M.A., Fraser, J.D., Eds.; ASM Press: Washington, DC, USA, 2007; pp. 21–36.

6. Uchiyama, T.; Imanishi, K.; Miyoshi-Akiyama, T.; Kata, H. Staphylococcal Superantigens and the Diseases They Cause. In *The Comprehensive Sourcebook of Bacterial Protein Toxins*, 3rd ed.; Alouf, J.E., Popoff, M.R., Eds.; Academic Press: London, UK, 2006; pp. 830–843.

7. Kotzin, B.L.; Leung, D.Y.M.; Kappler, J.; Marrack, P. Superantigens and their potential role in human disease. *Adv. Immunol.* **1993**, *54*, 99–166.

8. Brocke, S.; Hausmann, S.; Steinmam, L.; Wucherpfennig, K.W. Microbial peptides and superantigens in the pathogenesis of autoimmune diseases of the central nervous system. *Semin. Immunol.* **1998**, *10*, 57–67.

9. Yarwood, J.M.; Leung, D.Y.; Schlievert, P.M. Evidence for the involvement of bacterial superantigens in psoriasis, atopic dermatitis, and Kawasaki syndrome. *FEMS Microbiol. Lett.* **2000**, *192*, 1–7.

10. McKay, D.M. Bacterial superantigens: Provocateurs of gut dysfunction and inflammation? *Trends Immunol.* **2001**, *22*, 497–501.

11. Marrack, P.; Kappler, J. The staphylococcal enterotoxins and their relatives. *Science* **1990**, *248*, 705–709.

12. Kotb, M. Bacterial pyrogenic exotoxins as superantigens. *Clin. Microbiol. Rev.* **1995**, *8*, 411–426.

13. McCormick, J.K.; Yarwood, J.M.; Schlievert, P.M. Toxic shock syndrome and bacterial superantigens: An update. *Annu. Rev. Microbiol.* **2001**, *55*, 77–104.

14. Proft, T.; Fraser, J.D. Bacterial superantigens. *Clin. Exp. Immunol.* **2003**, *133*, 299–306.

15. Fraser, J.D.; Proft, T. The bacterial superantigen and superantigen-like proteins. *Immunol. Rev.* **2008**, *225*, 226–243.

16. Choi, Y.; Kotzin, B.; Hernon, L.; Callahan, J.; Marrack, P.; Kappler, J. Interaction of *Staphylococcus aureus* toxin "superantigens" with human T cells. *Proc. Natl. Acad. Sci. USA* **1989**, *86*, 8941–8945.

17. McCormick, J.K.; Tripp, T.J.; Llera, A.S.; Sundberg, E.J.; Dinges, M.M.; Mariuzza, R.A.; Schlievert, P.M. Functional analysis of the TCR binding domain of toxic shock syndrome toxin-1 predicts further diversity in MHC class II/superantigen/TCR ternary complexes. *J. Immunol.* **2003**, *171*, 1385–1392.

18. Papageorgiou, A.C.; Acharya, K.R. Microbial superantigens: From structure to function. *Trends Microbiol.* **2000**, *8*, 369–375.

19. Jupin, C.; Anderson, S.; Damais, C.; Alouf, J.E.; Parant, M. Toxic shock syndrome toxin 1 as an inducer of human tumor necrosis factors and gamma interferon. *J. Exp. Med.* **1988**, *167*, 752–761.

20. Trede, N.S.; Geha, R.S.; Chatila, T. Transcriptional activation of IL-1 beta and tumor necrosis factor-alpha genes by MHC class II ligands. *J. Immunol.* **1991**, *146*, 2310–2315.

21. Miethke, T.; Wahl, C.; Heeg, K.; Echtenacher, B.; Krammer, P.H.; Wagner, H. Superantigen mediated shock: A cytokine release syndrome. *Immunobiology* **1993**, *189*, 270–284.

22. Tessier, P.A.; Naccache, P.H.; Diener, K.R.; Gladue, R.P.; Neotem, K.S.; Clark-Lewis, I.; McColl, S.R. Induction of acute inflammation *in vivo* by staphylococcal superantigens. II. Critical role for chemokines, ICAM-1, and TNF-alpha. *J. Immunol.* **1998**, *161*, 1204–1211.

23. Krakauer, T. The induction of CC chemokines in human peripheral blood mononuclear cells by staphylococcal exotoxins and its prevention by pentoxifylline. *J. Leukco. Biol.* **1999**, *66*, 158–164.

24. Faulkner, L.; Cooper, A.; Fantino, C.; Altmann, D.M.; Sriskandan, S. The mechanism of superantigen-mediated toxic shock: Not a simple Th1 cytokine storm. *J. Immunol.* **2005**, *175*, 6870–6877.

25. Krakauer, T.; Buckley, M.; Fisher, D. Proinflammatory mediators of toxic shock and their correlation to lethality. *Mediat. Inflamm.* **2010**, doi:10.1155/2010/517594.

26. Krakauer, T.; Vilcek, J.; Oppenheim, J.J. Proinflammatory Cytokines: TNF and IL-1 Families, Chemokines, TGFß and Others. In *Fundamental Immunology*, 4th ed.; Paul, W., Ed.; Lippincott-Raven: Philadelphia, PA, USA, 1998; pp. 775–811.

27. Mattsson, E.; Herwald, H.; Egsten, A. Superantigen from *Staphylococcus aureus* induce procoagulant activity and monocyte tissue factor expression in whole blood and mononuclear cells via IL-1β. *J. Thromb. Haemost.* **2003**, *1*, 2569–2575.

28. Neumann, B.; Engelhardt, B.; Wagner, H.; Holzmann, B. Induction of acute inflammatory lung injury by staphylococcal enterotoxin B. *J. Immunol.* **1997**, *158*, 1862–1871.

29. Vial, T.; Descotes, J. Immune-mediated side-effects of cytokines in human. *Toxicology* **1995**, *105*, 31–57.

30. Bertling, A.; Niemann, S.; Hussain, M.; Holbrook, L.; Stanley, R.G.; Brodde, M.F.; Pohl, S.; Schifferdecker, T.; Roth, J.; Jurk, K.; *et al.* Staphylococcal extracellular adherence protein induces platelet activation by stimulation of thiol isomerases. *Arterioscler. Thromb. Vasc. Biol.* **2012**, *32*, 1979–1990.

31. Armstrong, P.C.J.; Hu, H.; Rivera, J.; Rigby, S.; Chen, Y.-C.; Howden, B.P.; Gardiner, E.; Peter, K. Staphylococcal superantigen-like protein 5 induces thrombotic and bleeding complications *in vivo*: Inhibition by an anti-SSL5 antibody and the glycan Bimosiamose. *J. Thromb. Haemost.* **2012**, *10*, 2607–2609.

32. O'Brien, L.; Kerrigan, S.W.; Kaw, G.; Hogan, M.; Penadés, J.; Litt, D.; Fitzgerald, D.J.; Foster, T.J.; Cox, D. Multiple mechanisms for the activation of human platelet aggregation by *Staphylococcus aureus*: Roles for the clumping factors ClfA and ClfB, the serine-aspartate repeat protein SdrE and protein A. *Mol. Microbiol.* **2002**, *44*, 1033–1044.

33. Lappin, E.; Ferguson, A.J. Gram-positive toxic shock syndromes. *Lancet Infect. Dis.* **2009**, *9*, 281–290.

34. Reiser, R.F.; Robbins, R.N.; Khoe, G.P.; Bergdoll, M.S. Purification and some physicochemical properties of toxic-shock toxin. *Biochemistry* **1983**, *22*, 3907–3912.

35. Wang, X.; Xu, M.; Cai, Y.; Yang, H.; Zhang, H.; Zhang, C. Functional analysis of the disulphide loop mutant of staphylococcal enterotoxin C2. *Appl. Microbiol. Biotechnol.* **2009**, *82*, 861–871.

36. Alber, G.; Hammer, D.K.; Fleischer, B. Relationship between enterotoxic- and T lymphocyte stimulating activity of staphylococcal enterotoxin B. *J. Immunol.* **1990**, *144*, 4501–4506.

37. Kappler, J.W.; Herman, A.; Clements, J.; Marrack, P. Mutations defining functional regions of the superantigen staphylococcal enterotoxin B. *J. Exp. Med.* **1992**, *175*, 387–396.

38. Li, H.; Llera, A.; Tsuchiya, D.; Leder, L.; Ysern, X.; Schlievert, P.M.; Karjalainen, K.; Mariuzza, R.A. Three-dimensional structure of the complex between a T cell receptor beta chain and the superantigen staphylococcal enterotoxin B. *Immunity* **1998**, *9*, 807–816.

39. Mollick, J.A.; Chintagumpala, M.; Cook, R.G.; Rich, R.R. Staphylococcal exotoxin activation of T cells. Role of exotoxin-MHC class II binding affinity and class II isotype. *J. Immunol.* **1991**, *146*, 463–468.

40. Chintagumpala, M.M.; Mollick, J.A.; Rich, R.R. Staphylococcal toxins bind to different sites on HLA-DR. *J. Immunol.* **1991**, *147*, 3876–3882.

41. Ulrich, R.G.; Bavari, B.; Olson, M.A. Staphylococcal enterotoxins A and B share a common structural motif for binding class II major histocompatibility complex molecules. *Nat. Struct. Biol.* **1995**, *2*, 554–560.

42. Hudson, K.R.; Tiedemann, R.E.; Urban, R.G.; Lowe, S.C.; Strominger, J.L.; Fraser, J.D. Staphylococcal enterotoxin A has two cooperative binding sites on major histocompatibility complex class II. *J. Exp. Med.* **1995**, *182*, 711–720.

43. Tiedemann, R.E; Urban, R.J.; Strominger, J.L.; Fraser, J.D. Isolation of HLA-DR1 (staphylococcal enterotoxins A)2 trimers in solution. *Proc. Natl. Acad. Sci. USA* **1995**, *92*, 12156–12159.

44. Thibodeau, J.; Cloutier, I.; Lavoie, P.M.; Labrecque, N.; Mourad, W.; Jardetzky, T.; Sekaly, R.P. Subsets of HLA-DR1 molecules defined by SEB and TSST-1 binding. *Science* **1994**, *266*, 1874–1878.

45. Herrmann, T.; Acolla, R.S.; MacDonald, H.R. Different staphylococcal enterotoxins bind preferentially to distinct MHC class II isotypes. *Eur. J. Immunol.* **1989**, *19*, 2171–2174.

46. Herman, A.; Croteau, G.; Sekaly, R.P.; Kappler, J.; Marrack, P. HLA-DR alleles differ in their ability to present staphylococcal enterotoxins to T cells. *J. Exp. Med.* **1990**, *172*, 709–712.

47. Pless, D.D.; Ruthel, G.; Reinke, E.K.; Ulrich, R.G.; Bavari, S. Persistence of zinc-binding bacterial superantigens at the surface of antigen-presenting cells contributes to the extreme potency of these superantigens as T-cell activators. *Infect. Immun.* **2005**, *73*, 5358–5366.

48. Leder, L.; Llera, A.; Lavoie, P.M.; Lebedeva, M.I.; Li, H.; Sékaly, R.P.; Bohach, G.A.; Gahr, P.J.; Schlievert, P.M.; Karjalainen, K.; Mariuzza, R.A. A mutational analysis of the binding of staphylococcal enterotoxins B and C3 to the T cell receptor beta chain and major histocompatibility complex class II. *J. Exp. Med.* **1998**, *187*, 823–833.

49. Seth, A.; Stern, L.J.; Ottenhoff, T.H.; Engel, I.; Owen, M.J.; Lamb, J.R.; Klausner, R.D.; Wiley, D.C. Binary and ternary complexes between T-cell receptor, class II MHC and superantigen *in vitro*. *Nature* **1994**, *369*, 324–327.

50. Moza, B.; Varma, A.K.; Buonpane, R.A.; Zhu, P.; Herfst, C.A.; Nicholson, M.J.; Wilbuer, A.K.; Seth, N.P.; Wucherpfennig, K.W.; McCormick, J.K.; *et al.* Structural basis of T-cell specificity and activation by the bacterial superantigen TSST-1. *EMBO J.* **2007**, *26*, 1187–1197.

51. Ferry, T.; Thomas, D.; Perpoint, T.; Lina, G.; Monneret, G.; Mohammedi, I.; Chidiac, C.; Peyramond, D.; Vandenesch, F.; Etienne, J. Analysis of superantigenic toxin Vbeta T-cell signatures produced during cases of staphylococcal toxic shock syndrome and septic shock. *Clin. Microbiol. Infect.* **2008**, *14*, 546–554.

52. Seo, K.S.; Park, J.Y.; Terman, D.S.; Bohach, G.A. A quantitative real time PCR method to analyze T cell receptor Vb subgroup expansion by staphylococcal superantigens. *J. Transl. Med.* **2010**, *8*, 2–9.

53. Linsley, P.S.; Ledbetter, J.A. The role of the CD28 receptor during T cell responses to antigen. *Annu. Rev. Immunol.* **1993**, *11*, 191–212.

54. Krakauer, T. Co-stimulatory receptors for the superantigen staphyloccocal enterotoxin B on human vascular endothelial cells and T cells. *J. Leukco. Biol.* **1994**, *56*, 458–463.

55. Arad, G.; Levy, R.; Nasie, I.; Hillman, D.; Rotfogel, Z.; Barash, U.; Supper, E.; Shpilka, T.; Minis, A.; Kaempfer, R. Binding of superantigen toxins into CD28 homodimer interface is essential for induction of cytokine genes that mediate lethal shock. *PLoS Biol.* **2012**, *9*, e1001149.

56. Saha, B.; Harlan, D.M.; Lee, K.P.; June, C.H.; Abe, R. Protection against lethal toxic shock by targeted disruption of the CD28 gene. *J. Exp. Med.* **1996**, *183*, 2675–2680.

57. Mittrücker, H.W.; Shahinian, A.; Bouchard, D.; Kündig, T.M.; Tak, T.W. Induction of unresponsiveness and impaired T cell expansion by staphylococcal enterotoxin B in CD28-deficient mice. *J. Exp. Med.* **1996**, *183*, 2481–2488.

58. Weiss, A. T Lymphocyte Activation. In *Fundamental Immunology*, 4th ed.; Paul, W., Ed.; Lippincott-Raven: Philadelphia, PA, USA, 1998; pp. 411–447.

59. Van Leeuwen, J.E.; Samelson, L.E. T cell-antigen receptor signal transduction. *Curr. Opin. Immunol.* **1999**, *11*, 242–248.

60. Smith-Garvin, J.E.; Koretzky, G.A.; Jordan, M.S. T cell activation. *Annu. Rev. Immunol.* **2009**, *27*, 591–619.

61. Cemerski, S.; Shaw, A. Immune synapses in T-cell activation. *Curr. Opin. Immunol.* **2006**, *18*, 298–304.

62. Fraser, J.; Newton, M.; Weiss, A. CD28 and T-cell antigen receptor signal transduction coordinately regulates interleukin 2 gene expression in response to superantigen stimulation. *J. Exp. Med.* **1992**, *175*, 1131–1134.

63. Isakov, N.; Altman, A. PKC-theta-mediated signal delivery from the TCR/CD28 surface receptors. *Front. Immunol.* **2012**, *3*, 273–284.

64. Cartwright, N.G.; Kashyap, A.K.; Schaefer, B.C. An active kinase domain is required for retention of PKCθ at the immunological synapse. *Mol. Biol. Cell* **2011**, *22*, 3491–3497.

65. Scholl, P.R.; Trede, N.; Chatila, T.A.; Geha, R.S. Role of protein tyrosine phosphorylation in monokine induction by the staphylococcal superantigen toxic shock syndrome toxin-1. *J. Immunol.* **1992**, *148*, 2237–2241.

66. Chatila, T.; Wood, N.; Parsonnet, J.; Geha, R.S. Toxic shock syndrome toxin-1 induces inositol phospholipid turnover, protein kinase C translocation, and calcium mobilization in human T cells. *J. Immunol.* **1988**, *140*, 1250–1255.

67. Park, S.G.; Schulze-Luehrman, J.; Hayden, M.S.; Hashimoto, N.; Ogawa, W.; Kasuga, M.; Ghosh, S.P. Phosphoinositide-dependent kinase 1 integrates T cell receptor and CD28 co-receptor signaling to effect NFκB induction and T cell activation. *Nat. Immunol.* **2009**, *10*, 158–166.

68. DiDonato, J.A.; Mercurio, F.; Karin, M. NFκB and the link between inflammation and cancer. *Immunol. Rev.* **2012**, *246*, 379–400.

69. Deane, J.A.; Fruman, D.A. Phosphoinositide 3-kinase: Diverse roles in immune cell activation. *Annu. Rev. Immunol.* **2004**, *22*, 563–598.

70. Krakauer, T. PI3K/Akt/mTOR, a pathway less recognized for staphylococcal superantigen-induced toxicity. *Toxins* **2012**, *4*, 1343–1366.

71. Manning, B.D.; Cantley, L.C. AKT/PBK signaling: Navigating downstream. *Cell* **2007**, *129*, 1261–1274.

72. Memmott, R.M.; Dennis, P.A. Akt-dependent and independent mechanisms of mTOR regulation in cancer. *Cell. Signal.* **2009**, *21*, 656–664.

73. Thomson, A.W.; Turnquist, H.R.; Raimondi, G. Immunoregulatory functions of mTOR inhibition. *Nat. Rev. Immunol.* **2009**, *9*, 324–337.

74. Laplante, M.; Sabatini, D.M. mTOR signaling at a glance. *J. Cell Sci.* **2009**, *122*, 3389–3394.

75. Wullschleger, S.; Loewith, R.; Hall, M.N. TOR signaling in growth and metabolism. *Cell* **2006**, *124*, 471–484.

76. Abraham, R.T.; Wiederrecht, O.J. Immunopharmacology of rapamycin. *Annu. Rev. Immunol.* **1996**, *14*, 483–510.

77. Krakauer, T.; Buckley, M.; Issaq, H.J.; Fox, S.D. Rapamycin protects mice from staphylococcal enterotoxin B-induced toxic shock and blocks cytokine release *in vitro* and *in vivo*. *Antimicrob. Agents Chemother.* **2010**, *54*, 1125–1131.

78. Bueno, C.; Lemke, C.D.; Criado, G.; Baroja, M.L.; Ferguson, S.S.; Rahman, A.K.; Tsoukas, C.D.; McCormick, J.K.; Madrenas, J. Bacterial superantigens bypass Lck-dependent T cell receptor signaling by activating a Galpha11-dependent, PLC-beta-mediated pathway. *Immunity* **2006**, *25*, 67–78.

79. Kyriakis, J.M.; Avruch, J. Mammalian MAPK signal transduction pathways activated by stress and inflammation. *Physiol. Rev.* **2012**, *92*, 689–737.

80. Saccani, S.; Pantano, S.; Natoli, G. p38-Dependent marking of inflammatory genes for increased NF-kappa B recruitment. *Nat. Immunol.* **2002**, *3*, 69–75.

81. Sims, J.E.; Smith, D.E. The IL-1 family: Regulators of immunity. *Nat. Rev. Immunol.* **2010**, *10*, 89–102.

82. Vallabhapurapu, S.; Karin, M. Regulation and function of NFκB transcription factors in the immune system. *Annu. Rev. Immunol.* **2009**, *27*, 693–733.

83. Takeuchi, O.; Akira, S. Pattern recognition receptors and inflammation. *Cell* **2010**, *140*, 805–820.

84. Mele, T.; Madrenas, J. TLR2 signalling: At the crossroads of commensalism, invasive infections and toxic shock syndrome by *Staphylococcus aureus*. *Int. J. Biochem. Cell. Biol.* **2010**, *42*, 1066–1071.

85. Hopkins, P.A.; Fraser, J.D.; Pridmore, A.C.; Russell, H.H.; Read, R.C.; Sriskandan, S. Superantigen recognition by HLA class II on monocytes up-regulates toll-like receptor 4 and enhances proinflammatory responses to endotoxin. *Blood* **2005**, *105*, 3655–3662.

86. Hopkins, P.A.; Pridmore, A.C.; Ellmerich, S.; Fraser, J.D.; Russell, H.H.; Read, R.C.; Sriskandan, S. Increased surface toll-like receptor 2 expression in superantigen shock. *Crit. Care Med.* **2008**, *36*, 1267–1276.

87. Sugiyama, H.; McKissic, E.M.; Bergdoll, M.S.; Heller, B. Enhancement of bacterial endotoxin lethality by staphylococcal enterotoxin. *J. Infect. Dis.* **1964**, *4*, 111–118.

88. Stiles, B.G.; Bavari, S.; Krakauer, T.; Ulrich, R.G. Toxicity of staphylococcal enterotoxins potentiated by lipopolysaccharide: Major histocompatibility complex class II molecule dependency and cytokine release. *Infect. Immun.* **1993**, *61*, 5333–5338.

89. Blank, C.; Luz, A.; Bendigs, S.; Erdmann, A.; Wagner, H.; Heeg, K. Superantigen and endotoxin synergize in the induction of lethal shock. *Eur. J. Immunol.* **1997**, *27*, 825–833.

90. Keystone, E.C.; Ware, C.F. Tumor necrosis factor and anti-tumor necrosis factor therapies. *J. Rheumatol.* **2010**, *85*, 27–39.

91. Miethke, T.; Wahl, C.; Heeg, K.; Echtenacher, B.; Krammer, P.H.; Wagner, H. T cell-mediated lethal shock triggered in mice by the superantigen staphylococcal enterotoxin B: Critical role of tumor necrosis factor. *J. Exp. Med.* **1992**, *175*, 91–98.

92. Ghoreschi, K.; Laurence, A.; O'Shea, J.J. Janus kinases in immune cell signaling. *Immunol. Rev.* **2009**, *228*, 273–287.

93. Murray, P.J. The JAK-STAT signaling pathway: Input and output integration. *J. Immunol.* **2007**, *178*, 2623.

94. Ramana, C.V.; Gil, M.P.; Schreiber, R.D.; Stark, G.R. Stat-1-dependent and -independent pathways in IFN-dependent signaling. *Trends Immunol.* **2002**, *23*, 96–101.

95. Yang, Z.; Gagarin, D.; St Laurent, G., 3rd; Hammell, N.; Toma, I.; Hu, C.A.; Iwasa, A.; McCaffrey, T.A. Cardiovascular inflammation and lesion cell apoptosis: A novel connection via the interferon-inducible immunoproteasome. *Arterioscler. Thromb. Vasc. Biol.* **2009**, *29*, 1213–1219.

96. Lu, J.; Philpott, D.J.; Saunders, P.R.; Perdue, M.H.; Yang, P.C.; McKay, D.M. Epithelial ion transport and barrier abnormalities evoked by superantigen-activated immune cells are inhibited by interleukin-10 but not interleukin-4. *J. Pharmacol. Exp. Ther.* **1998**, *287*, 128–136.

97. Matthys, P.; Mitera, T.; Heremans, H.; van Damme, J.; Billiau, A. Anti-gamma interferon and anti-interleukin-6 antibodies affect staphylococcal enterotoxin B-induced weight loss, hypoglycemia, and cytokine release in D-galactosamine-sensitized and unsensitized mice. *Infect. Immun.* **1995**, *63*, 1158–1164.

98. Chowdhary, V.R.; Tilahun, A.Y.; Clark, C.R.; Grande, J.P.; Rajagopalan, G. Chronic exposure to staphylococcal superantigen elicts a systemic inflammatory disease mimicking lupus. *J. Immunol.* **2012**, *189*, 2054–2062.

99. Malek, T.R.; Castro, I. Interleukin-2 receptor signaling: At the interface between tolerance and immunity. *Immunity* **2010**, *33*, 153–165.

100. Liu, D.; Zienkiewicz, J.; DiGiandomenico, A.; Hawiger, J. Suppression of acute lung inflammation by intracellular peptide delivery of a nuclear import inhibitor. *Mol. Ther.* **2009**, *17*, 796–802.

101. Huzella, L.M.; Buckley, M.J.; Alves, D.A.; Stiles, B.G.; Krakauer, T. Central roles for IL-2 and MCP-1 following intranasal exposure to SEB: A new mouse model. *Vet. Res. Sci.* **2009**, *86*, 241–247.

102. Khan, A.A.; Priya, S.; Saha, B. IL-2 regulates SEB induced toxic shock syndrome in BALB/c mice. *PLoS One* **2009**, *4*, e8473.

103. Wang, X.; Lupardus, P.; LaPorte, S.L.; Garcia, K.C. Structural biology of shared cytokine receptors. *Annu. Rev. Immunol.* **2009**, *27*, 27–60.

104. Sadik, C.D.; Kim, N.D.; Luster, A.D. Neutrophils cascading their way to inflammation. *Trends Immunol.* **2011**, *32*, 452–460.

105. Zlotnik, A.; Yoshie, D. The chemokine superfamily revisited. *Immunity* **2012**, *36*, 705–716.

106. Krakauer, T.; Buckley, M.; Huzella, L.M.; Alves, D. Critical timing, location and duration of glucocorticoid administration rescues mice from superantigen-induced shock and attenuates lung injury. *Int. Immunopharmacol.* **2009**, *9*, 1168–1174.

107. Scholl, P.; Sekaly, R.; Diez, A.; Glimcher, L.; Geha, R. Binding of toxic shock syndrome toxin-1 to murine major histocompatibility complex class II molecules. *Eur. J. Immunol.* **1990**, *20*, 1911–1916.

108. Chen, J.Y.; Qiao, Y.; Komisar, J.L.; Baze, W.B.; Hsu, I.C.; Tseng, J. Increased susceptibility to staphylococcal enterotoxin B intoxication in mice primed with actinomycin D. *Infect. Immun.* **1994**, *62*, 4626–4631.

109. Sarawar, S.R.; Blackman, B.A.; Doherty, P.C. Superantigen shock in mice with an inapparent viral infection. *J. Infect. Dis.* **1994**, *170*, 1189–1194.

110. Zhang, W.J.; Sarawar, S.; Nguyen, P.; Daly, K.; Rehig, J.E.; Doherty, P.C.; Woodland, D.L.; Blackman, M.A. Lethal synergism between influenza infection and staphylococcal enterotoxin B in mice. *J. Immunol.* **1996**, *157*, 5049–5060.

111. Visvanathan, K.; Charles, A.; Bannan, J.; Pugach, P.; Kashfi, K.; Zabriskie, J.B. Inhibition of bacterial superantigens by peptides and antibodies. *Infect. Immun.* **2001**, *69*, 875–884.

112. Yeung, R.S.; Penninger, J.M.; Kundig, J.; Khoo, W.; Ohashi, P.S.; Kroemer, G.; Mak, T.W. Human CD4 and human major histocompatibility complex class II (DQ6) transgenic mice: Supersensitivity to superantigen-induced septic shock. *Eur. J. Immunol.* **1996**, *26*, 1074–1082.

113. Rajagopalan, G.; Sen, M.M.; David, C.S. *In vitro* and *in vivo* evaluation of staphylococcal superantigen peptide antagonists. *Infect. Immun.* **2004**, *72*, 6733–6737.

114. DaSilva, L.; Welcher, B.; Ulrich, R.; Aman, J.; David, C.S.; Bavari, S. Humanlike immune response of human leukocyte antigen-DR3 transgenic mice to staphylococcal enterotoxins: A novel model for superantigen vaccines. *J. Infect. Dis.* **2002**, *185*, 1754–1760.

115. Roy, C.J.; Warfield, K.L.; Welcher, B.C.; Gonzales, R.F.; Larsen, T.; Hanson, J.; David, C.S.; Krakauer, T.; Bavari, S. Human leukocyte antigen-DQ8 transgenic mice: A model to examine the toxicity of aerosolized staphylococcal enterotoxin B. *Infect. Immun.* **2005**, *73*, 2452–2460.

116. Tilahun, A.Y.; Marietta, E.V.; Wu, T.T.; Patel, R.; David, C.S.; Rajagopalan, G. Human leukocyte antigen class II transgenic mouse model unmasks the significant extrahepatic pathology in toxic shock syndrome. *Am. J. Pathol.* **2011**, *178*, 2760–2772.

117. Darenberg, J.; Soderquist, B.; Normark, B.H.; Norrby-Teglund, A. Differences in potency of intravenous polyspecific immunoglobulin G against streptococcal and staphylococcal superantigens: Implications for therapy of toxic shock syndrome. *Clin. Infect. Dis.* **2004**, *38*, 836–842.

118. Bavari, S.; Ulrich, R.G.; LeClaire, R.D. Cross-reactive antibodies prevent the lethal effects of *Staphylococcus aureus* superantigens. *J. Infect. Dis.* **1999**, *180*, 1365–1369.

119. Grumann, D.; Ruotsalainen, E.; Kolata, J.; Kuusela, P.; Jarvinen, A.; Kontinen, V.P.; Broker, B.M.; Holtfreter, S. Characterization of infecting strains and superantigen-neutralizing antibodies in Staphylococcus aureus bacteremia. *Clin. Vaccine Immunol.* **2001**, *18*, 487–493.

120. Parsonnet, J.; Hansmann, M.A.; Seymour, J.L.; Delaney, M.L.; Dubois, A.M.; Modern, P.A.; Jones, M.B.; Wild, J.E.; Onderdonk, A.B. Persistence survey of toxic shock syndrome toxin-1 producing *Staphylococcus aureus* and serum antibodies to this superantigen in five groups of menstruating women. *BMC Infect. Dis.* **2010**, *10*, 249–256.

121. Kansal, R.; Davis, C.; Hansmann, M.; Seymour, J.; Parsonnet, J.; Modern, P.; Gilbert, S.; Kotb, M. Structural and functional properties of antibodies to the superantigen TSST-1 and their relationship to menstrual toxic shock syndrome. *J. Clin. Immunol.* **2007**, *27*, 327–338.

122. Tilahun, M.E.; Rajagopalan, G.; Shah-Mahoney, N.; Lawlor, R.G.; Tilahun, A.Y.; Xie, C.; Natarajan, K.; Margulies, D.H.; Ratner, D.I.; Osborne, B.A.; *et al.* Potent neutralization of staphylococcal enterotoxin B by synergistic action of chimeric antibodies. *Infect. Immun.* **2010**, *78*, 2801–2811.

123. Larkin, E.A.; Stiles, B.G.; Ulrich, R.G. Inhibition of toxic shock by human monoclonal antibodies against staphylococcal enterotoxin B. *PLoS One* **2010**, *5*, e13253.

124. Varshney, A.K.; Wang, X.; Cook, E.; Dutta, K.; Scharff, M.D.; Goger, M.J.; Fries, B.C. Generation, characterization, and epitope mapping of neutralizing and protective monoclonal antibodies against staphylococcal enterotoxin B-induced lethal shock. *J. Biol. Chem.* **2011**, *286*, 9737–9747.

125. Bavari, S.; Dyas, B.; Ulrich, R.G. Superantigen vaccines: A comparative study of genetically attenuated receptor-binding mutants of staphylococcal enterotoxin A. *J. Infect. Dis.* **1996**, *174*, 338–345.

126. Inskeep, T.K.; Stahl, C.; Odle, J.; Oakes, J.; Hudson, L.; Bost, K.L.; Piller, K.J. Oral vaccine formulations stimulate mucosal and systemic antibody responses against staphylococcal enterotoxin B in a piglet model. *Clin. Vaccine. Immunol.* **2010**, *17*, 1163–1169.

127. Arad, G.; Levy, R.; Hillman, D.; Kaempfer, R. Superantigen antagonist protects against lethal shock and defines a new domain for T-cell activation. *Nat. Med.* **2000**, *6*, 414–421.

128. Hamad, A.R.; Marrack, P.; Kappler, J.W. Transcytosis of staphylococcal superantigen toxins. *J. Exp. Med.* **1997**, *185*, 1447–1454.

129. Saha, B.; Jaklic, B.; Harlan, D.M.; Gray, G.S.; June, C.H.; Abe, R. Toxic shock syndrome toxin-1 induced death is prevented by CTLA4Ig. *J. Immunol.* **1996**, *157*, 3869–3875.

130. Ramachandran, G.; Tulapurkar, M.E.; Harris, K.M.; Arad, G.; Shirvan, A.; Shemesh, R.; Detolla, L.J.; Benazzi, C.; Opal, S.M.; Kaempfer, R.; Cross, A.S. A peptide antagonist of CD28 signaling attenuates toxic shock and necrotizing soft-tissue infection induced by Streptococcus pyogenes. *J. Infect. Dis.* **2013**, *207*, 1869–1877.

131. Sun, J.; Law, G.P.; Bridges, C.C.; McKallip, R.J. CD44 as a novel target for treatment of staphylococcal enterotoxin B-induced acute inflammatory lung injury. *Clin. Immunol.* **2012**, *144*, 41–52.

132. Geller-Hong, E.; Möllhoff, M.; Shiflett, P.R.; Gupta, G. Design of chimeric receptor mimics with different TcRVβ isoforms: Type-specific inhibition of superantigen pathogenesis. *J. Biol. Chem.* **2004**, *279*, 5676–5684.

133. Wang, N.; Mattis, D.M.; Sundberg, E.J.; Schlievert, P.M.; Kranz, D.M. A single, engineered protein therapeutic agent neutralizes exotoxins from both *Staphylococcus aureus* and *Streptococcus pyogenes*. *Clin. Vaccine. Immunol.* **2010**, *17*, 1781–1789.

134. Liu, D.; Liu, X.Y.; Robinson, D.; Burnett, C.; Jackson, C.; Seele, L.; Veach, R.A.; Downs, S.; Collins, R.D.; Ballard, R.W.; *et al.* Suppression of staphylococcal enterotoxin B-induced toxicity by a nuclear import inhibitor. *J. Biol. Chem.* **2004**, *279*, 19239–19246.

135. Krakauer, T.; Buckley, M. Dexamethasone attenuates staphylococcal enterotoxin B-induced hypothermic response and protects mice from superantigen-induced toxic shock. *Antimicrob. Agents Chemother.* **2006**, *50*, 391–395.

136. Tilahun, A.Y.; Theuer, J.E.; Patel, R.; David, C.S.; Rajagopalan, G. Detrimental effect of the proteasome inhibitor, bortezomib in bacterial superantigen- and lipopolysaccharide-induced systemic inflammation. *Mol. Ther.* **2010**, *18*, 1143–1154.

137. Kissner, T.L.; Ruthel, G.; Alam, S.; Mann, E.; Ajami, D.; Rebek, M.; Larkin, E.; Fernandez, S.; Ulrich, R.G.; Ping, S.; *et al.* Therapeutic inhibition of pro-inflammatory signaling and toxicity to staphylococcal enterotoxin B by a synthetic dimeric BB-loop mimetic of MyD88. *PLos One* **2012**, *7*, e40773.

138. Kissner, T.L.; Moisan, L.; Mann, E.; Alam, S.; Ruthel, G.; Ulrich, R.G.; Rebek, M.; Rebek, J., Jr.; Saikh, K.U. A small molecule that mimics the BB-loop in the Toll interleukin-1 (IL-1) receptor domain of MyD88 attenuates staphylococcal enterotoxin B-induced pro-inflammatory cytokine production and toxicity in mice. *J. Biol. Chem.* **2011**, *286*, 31385–31396.

139. Krakauer, T. Suppression of endotoxin- and staphylococcal exotoxin-induced cytokines and chemokines by a phospholipase C inhibitor in human peripheral blood mononuclear cells. *Clin. Diagn. Lab. Immunol.* **2001**, *8*, 449–453.

140. Tschaikowsky, K.J.; Schmidt, J.; Meisner, M. Modulation of mouse endotoxin shock by inhibition of phosphatidylcholine-specific phospholipase C. *J. Pharmacol. Exp. Ther.* **1999**, *285*, 800–804.

141. Jo, D.; Liu, D.; Yao, S.; Collins, R.D.; Hawiger, J. Intracellular protein therapy with SOCS3 inhibits inflammation and apoptosis. *Nat. Med.* **2005**, *11*, 892–898.

142. Krakauer, T.; Buckley, B. Intranasal rapamycin rescues mice from staphylococcal enterotoxin B-induced shock. *Toxins* **2012**, *4*, 718–728.

143. Tilahun, A.Y.; Karau, M.J.; Clark, C.R.; Patel, R.; Rajagopalan, G. The impact of tacrolimus on the immunopathogenesis of with staphylococcal enterotoxin-induced systemic inflammatory response syndrome and pneumonia. *Microbes Infect.* **2012**, *14*, 528–536.

144. Krakauer, T.; Buckley, M. The potency of anti-oxidants in attenuating superantigen-induced proinflammatory cytokines correlates with inactivation of NFκB. *Immunopharmacol. Immunotoxicol.* **2008**, *30*, 163–179.

145. Krakauer, T.; Buckley, M. Efficacy of two FDA-approved drug combination in a mouse model of staphylococcal enterotoxin B-induced shock. *Mil. Med.* **2013**, *178*, 1024–1028.

146. LeClaire, R.D.; Kell, W.; Bavari, S.; Smith, T.; Hunt, R.E. Protective effects of niacinamide in staphylococcal enterotoxin B induced toxicity. *Toxicology* **1996**, *107*, 69–81.

147. Krakauer, T.; Stiles, B.G. Pentoxifylline inhibits staphylococcal superantigen induced toxic shock and cytokine release. *Clin. Diagn. Lab. Immunol.* **1999**, *6*, 594–598.

148. Sprung, C.L.; Goodman, S.; Weiss, Y.G. Steriod therapy of septic shock. *Crit. Care Clin.* **2009**, *25*, 825–834.

149. Mohty, M.; Brissot, E.; Savani, B.N.; Gaugler, B. Effects of bortezomib on the immune system: A focus on immune regulation. *Biol. Blood Marrow Transplant.* **2013**, doi:10.1016/j.bbmt.2013.05.011.

150. Krakauer, T. Differential inhibitory effects of interleukin-10, interleukin-4, and dexamethasone on staphylococcal enterotoxin-induced cytokine production and T cell activation. *J. Leuko. Biol.* **1995**, *57*, 450–454.

151. Krakauer, T. Comparative potency of green tea and red wine polyphenols in attenuating staphylococcal superantigen-induced immune responses. *Am. J. Biomed. Sci.* **2012**, doi:10.5099/aj120200157.

152. Watson, J.L.; Vicario, M.; Wang, A.; Moreto, M.; McKay, D.M. Immune cell activation and subsequent epithelial dysfunction by staphylococcal enterotoxin B is attenuated by the green tea polyphenol (−)-epigallocatechin gallate. *Cell. Immunol.* **2005**, *237*, 7–16.

153. Rieder, S.A.; Nagarkatti, P.; Nagarkatti, M. Identification of multiple anti-inflammatory pathways triggered by resveratrol leading to amelioration of staphylococcal enterotoxin B-induced lung inflammation. *Br. J. Pharmacol.* **2012**, *167*, 1244–1258.

154. Krakauer, T. Nuclear factor-κB: Fine-tuning a central integrator of diverse biologic stimuli. *Int. Rev. Immunol.* **2008**, *27*, 286–292.

155. See, R.H.; Chow, A.W. Staphylococcal toxic shock syndrome toxin 1-induced tumor necrosis factor alpha and interleukin-1ß secretion by human peripheral blood monocytes and T lymphocytes is differentially suppressed by protein kinase inhibitors. *Infect. Immun.* **1992**, *60*, 3456–3459.

156. Schafer, P.H.; Wang, L.; Wadsworth, S.A.; Davis, J.E.; Siekierka, J.J. T cell activation signals up-regulate p38 mitogen-activated protein kinase activity and induce TNF-alpha production in a manner distinct from LPS activation of monocytes. *J. Immunol.* **1999**, *162*, 659–668.

157. Komisar, J.L.; Weng, C.F.; Oyejide, A.; Hunt, R.E.; Briscoe, C.; Tseng, J. Cellular and cytokine responses in the circulation and tissue reactionsin the lung of rhesus monkeys (*Macaca mulatta*) pretreated with cyclosporine A and challenged with staphylococcal enterotoxin B. *Toxicol. Pathol.* **2001**, *29*, 369–378.

158. Tilahun, M.E.; Kwan, A.; Natarajan, K.; Quinn, M.; Tilahun, A.Y.; Xie, C.; Margulies, D.H.; Osborne, B.A.; Goldsby, R.A.; Rajagopalan, G. Chimeric anti- staphylococcal enterotoxin B antibodies and lovastatin act synergistically to provide *in vivo* protection against lethal doses of SEB. *PLoS One* **2011**, *6*, e27203.

159. Bean, A.G.; Freiberg, R.A.; Andrade, S.; Menon, S.; Zlotnik, A. Interleukin 10 protects mice against staphylococcal enterotoxin B-induced lethal shock. *Infect. Immun.* **1993**, *61*, 4937–4939.

160. Florquin, S.; Amraoui, Z.; Abramowicz, D.; Goldman, M. Systemic release and protective role of IL-10 in staphylococcal enterotoxin B-induced shock in mice. *J. Immunol.* **1994**, *153*, 2618–2623.

161. Krakauer, T.; Buckley, M. Doxycycline is anti-inflammatory and inhibits staphylococcal exotoxin-induced cytokines and chemokines. *Antimicrob. Agents Chemother.* **2003**, *47*, 3630–3633.

162. Pichereau, S.; Moran, J.J.; Hayney, M.S.; Shukla, S.K.; Sakoulas, G.; Rose, W.E. Concentration-dependent effects of antimicrobials on Staphylococcus aureus toxin-mediated cytokine production from peripheral blood mononuclear cells. *J. Antimicrob. Chemohter.* **2012**, *67*, 123–129.

163. Hiwatashi, Y.; Maeda, M.; Fukushima, H.; Onda, K.; Tanaka, S.; Utsumi, H.; Hirano, T. Azithromycin suppresses proliferation, interleukin production and mitogen-activated protein kinases in human peripheral-blood mononuclear cells stimulated with bacterial superantigen. *J. Pharm. Pharmacol.* **2011**, *63*, 1320–1326.

164. Krakauer, T. Caspase inhibitors attenuate superantigen-induced inflammatory cytokines, chemokines and T-cell proliferation. *Clin. Diagn. Lab. Immunol.* **2004**, *11*, 621–624.

165. Paul, S.M.; Lewis-Hall, F. Drugs in search of diseases. *Sci. Trans. Med.* **2013**, *186*, 1–3.

166. Arrowsmith, J. Trial watch: Phase III and submission failures; 2007–2010. *Nat. Rev. Drug Discov.* **2011**, *10*, 87–92.

Soluble T Cell Receptor Vβ Domains Engineered for High-Affinity Binding to Staphylococcal or Streptococcal Superantigens

Preeti Sharma, Ningyan Wang and David M. Kranz

Abstract: *Staphylococcus aureus* and group A *Streptococcus* secrete a collection of toxins called superantigens (SAgs), so-called because they stimulate a large fraction of an individual's T cells. One consequence of this hyperactivity is massive cytokine release leading to severe tissue inflammation and, in some cases, systemic organ failure and death. The molecular basis of action involves the binding of the SAg to both a T cell receptor (TCR) on a T cell and a class II product of the major histocompatibility complex (MHC) on an antigen presenting cell. This cross-linking leads to aggregation of the TCR complex and signaling. A common feature of SAgs is that they bind with relatively low affinity to the variable region (V) of the beta chain of the TCR. Despite this low affinity binding, SAgs are very potent, as each T cell requires only a small fraction of their receptors to be bound in order to trigger cytokine release. To develop high-affinity agents that could neutralize the activity of SAgs, and facilitate the development of detection assays, soluble forms of the Vβ regions have been engineered to affinities that are up to 3 million-fold higher for the SAg. Over the past decade, six different Vβ regions against SAgs from *S. aureus* (SEA, SEB, SEC3, TSST-1) or *S. pyogenes* (SpeA and SpeC) have been engineered for high-affinity using yeast display and directed evolution. Here we review the engineering of these high-affinity Vβ proteins, structural features of the six different SAgs and the Vβ proteins, and the specific properties of the engineered Vβ regions that confer high-affinity and specificity for their SAg ligands.

Reprinted from *Toxins*. Cite as: Sharma, P.; Wang, N.; Kranz, D.M. Soluble T Cell Receptor Vβ Domains Engineered for High-Affinity Binding to Staphylococcal or Streptococcal Superantigens. *Toxins* **2014**, *6*, 556-574.

1. Overview

Over the past 30 years, the family of exotoxins expressed by *S. aureus* and group A *Streptococcus* known as "superantigens" (SAgs) [1] has been studied extensively at the molecular and structural levels. There are 24 SAgs known to be expressed by *S. aureus*, and 11 SAgs known to be expressed by group A *Streptococcus* [2–5]. Despite their sequence diversity, these toxins exhibit a canonical structural motif that consists of two domains, a smaller N-terminal domain with two β-sheets and a larger C-terminal domain with a central α-helix and a five-stranded β-sheet [5–8]. This canonical structure has presumably allowed SAgs to maintain their ability to interact with a T cell receptor Vβ domain on one side of the molecule and a class II product of the MHC on another side [6]. This dual binding is required for activation of T cells and subsequent cytokine release, as monovalent binding of a ligand to the TCR is not sufficient for signaling. SAg-mediated crosslinking with MHC allows multiple MHC-bound SAg molecules to form a multivalent TCR complex, thereby initiating signaling [6,9,10].

The pathogenic function of SAgs is not clear, although it is likely related to their ability to dysregulate an immune response, or perhaps to generate a cytokine milieu that is favorable for survival of the organism. While the precise functional or evolutionary advantage of expressing a large family of SAgs with extensive sequence diversity is unclear, one clinical consequence has been that antibodies generated against one of the SAgs are not likely to cross-react with most of the other SAgs, thereby limiting the ability of an individual to neutralize multiple toxins [11]. Understanding the clinical correlates of SAg expression are further complicated because of the varied prevalence of individual SAg genes among different bacterial isolates, especially of *S. aureus* [12,13]. Most of the SAgs, including staphyloccocal enterotoxin A (SEA), SEB, SEC, and toxic shock syndrome toxin-1 (TSST-1), are encoded on variable genetic elements [14–17]. Thus, some strains express one or more SAgs while other strains can express a different pattern of SAgs. Finally, there is additional complexity because there is variation in SAg protein expression levels, with some evidence that SAgs SEB, SEC and TSST-1 may be expressed at higher levels than the other SAgs, due to transcriptional regulation [18].

Despite this variability in prevalence and expression levels, it is clear that the potency of SAgs is a direct cause of disease or at the least exacerbates a host of diseases. These include toxic shock syndrome (TSS), pneumonia, purpura fulminans, severe atopic dermatitis, and endocarditis [19–25]. While TSS is the disease most often associated with SAgs, especially TSST-1, the frequency of staphylococcal or streptococcal infections in specific tissues (e.g., lung, skin, soft tissue) results in SAg-mediated, hyper-inflammatory reactions at these sites [26,27]. Specific inhibition of such severe tissue inflammation could be a useful adjunct to treatment of these diseases.

Given the considerable structural information about SAgs and their interaction with Vβ domains, we embarked over ten years ago on an effort to engineer soluble versions of the Vβ domains against various SAgs for the purpose of developing potent neutralizing agents that could suppress the hyper-inflammatory properties of SAgs [28]. A similar receptor-based strategy has worked for neutralizing the effects of TNF-α with the soluble TNF-α receptor/immunoglobulin fusion Etanercept (trade name Enbrel) [29,30]. Because of the low affinity of SAgs for their Vβ receptors, we reasoned that effective neutralization would require the generation of higher affinity variants of the Vβ, which would outcompete toxin engagement by TCRs bearing any Vβ region since the same binding epitope on the SAg is used regardless of the Vβ region expressed by the T cell. This affinity maturation has been accomplished using a directed evolution process and yeast display [31,32], an approach that has yielded, to date, high-affinity Vβ proteins against the six SAgs SEA, SEB, SEC3, TSST-1, SpeA, and SpeC [28,33–37]. Several of these have been used successfully in animal models of *S. aureus* infections involving (TSS), pneumonia, skin disease, and endocarditis [20,33,35,36,38]. Here we review features of the entire collection of high-affinity Vβ domains.

2. Structural Features of the Superantigens

SAgs are structurally homologous, even though the primary sequences of the proteins are diverse [1,5]. These proteins are globular and range between 20 and 30 kilodaltons [5]. Although known staphylococcal and streptococcal SAgs have been classified into five groups based on differences in their amino acid sequences [5], here we focus on the six SAgs that have been the targets for high-affinity Vβ regions engineered by yeast display (see below). Sequence alignments of specific members of these groups (TSST-1 from Group I; SEB, SEC3 and SpeA from Group II; SEA from Group III and SpeC from Group IV) are shown in Figure 1. These six SAgs have 10% to 65% sequence identity among each other. SEB, SEC3 and SpeA are 50% to 65% identical, and perhaps it is not surprising that all three stimulate T cells with the same Vβ, mouse Vβ8.2 [5]. Although SpeC belongs to a different group than SEB, SEC3 and SpeA, it exhibits significant sequence homology with these proteins in specific regions, despite having overall low sequence identity (21% identity with SEB and SEC3, and 24% identity with SpeA). Overall, TSST-1 is the most distant and contains the lowest level of sequence homology (and 7% to 20% identity) to these SAgs.

Examination of the aligned sequences shows that there are several linear stretches of amino acids that are more similar among the six SAgs. Examination of their co-crystal structures with the variable regions of beta chain (Vβ) of TCRs or with class II MHC ligands suggest that these regions however are among the least homologous (Figure 1). These include residues near the N-terminus centered around position 20 which is part of the epitope for binding the Vβ region of the TCR [39–42]. Other Vβ binding regions are found between residues 90 to 95, and near the C-terminus of the protein sequences (positions 215–220), which also appear to lack the same degree of homology as flanking regions which are involved the structural framework for the SAgs. Thus, the residues in these regions, and also their atomic interactions with the cognate Vβ, most often differ and thereby account for the specificity of the interactions between SAg and the TCR. However, it is important to note that SAgs that employ a zinc-dependent binding site for interaction with class II MHC ligands use conserved residues for coordinating zinc ion (Figure 1).

SEA, SEB, SEC3 and SpeA also contain a homologous region (residues 45 to 55 in Figure 1) that serves as a binding site for class II MHC [43–47]. These same four SAgs, but not SpeC and TSST-1, possess a characteristic cystine loop of 9 to 19 residues [48,49], which has been implicated in emetic activity of SAgs. Toxins such as SpeC and TSST-1 that lack the cystine loop have been shown to exhibit reduced to no emetic activity [50].

Although many of the staphylococcal and streptococcal SAgs possess overall low sequence identity, their structures possess striking similarity. The canonical structure consists of a N-terminal, β-barrel containing domain and a C-terminal domain containing a β-grasp motif and an α-helix which spans the center of the structure, connecting the two domains [5] (Figure 2). In the past two decades, a number of crystal structures of SAg have been solved (Table 1, only those that are relevant to this review are shown). Co-crystal structures with the Vβ region of TCR or class II MHC, have provided the basis for understanding SAg interactions with receptors on T cells and antigen presenting cells. The different modes of interaction of each SAg with these receptors reveal the diversity in mechanisms of binding to Vβ and MHC-II, which is particularly intriguing considering they possess highly conserved three-dimensional folds.

Figure 1. Sequence alignment of various superantigen sequences. The sequence of each superantigen (SAg) was obtained from the PDB file corresponding to its crystal structure. Multiple sequence alignment [CLUSTAL W (1.81)] was performed using "Biology WorkBench" online tool. Positions with homologous amino acids in three or more SAg sequences are highlighted in yellow or green. Residues in red are involved in forming the characteristic disulfide loop in certain SAg. Residues underlined in red and in blue are involved in binding to Vβ and class II MHC, respectively. Residues in bold are involved in binding zinc.

```
                    10        20        30        40        50        60
                    |         |         |         |         |         |
SEA_1SXT_A   SEKSEEINEKDLRKKSELQGTALGNLKQIYYYNEKAKTENKESHDQFLQHTILFKGFFTD
SEB_1SBB_B   ESQPDPKPDELHKSSKFTGLMEN--MKVLYDDNHVSAINVKSIDQFLYFDLIYSIKDTK
SEC3_1JCK_B  ESQPDPMPDDLHKSSEFTGTMGN--MKYLYDDHYVSATKVKSVDKFLAHDLIYNINDKK
SpeA_1L0X_B  QQDPDPSQLHRSS-LVKNLQN--IYFLYEGDPVTHENVKSVDQLLSHDLIYNVSGP-
SpeC_1KTK_A  DSKKDISNVKSDLLYAYTITP------YDYKDCRVNFSTTHTLNIDTQKYRGKD--
TSST1_3MFG_A STNDNIKDLLDWYSSGSDTFTN----SEVLDNSLGSMRIKNTDGSISLIIFPSP---
                .   :   .              . ... .

                    70        80        90       100       110       120
                    |         |         |         |         |         |
SEA_1SXT_A   HSWYNDLLVDFDSKDIVDKYKGKKVDLYGAYYGYQCAGGTPN----------KTACMYGG
SEB_1SBB_B   LGNYDNVRVEFKNKDLADKYKDKYVDVFGANYYYQCYFSKKTNDINSHQTDKRKT-CMYGG
SEC3_1JCK_B  LNNYDKVKTELLNEDLANKYKDEVVDVYGSNYYVNCYFSSKD---NVGKVTSGKT-CMYGG
SpeA_1L0X_B  --NYDKLKTELKNQEMATLFKDKNVDIYGVEYYHLCYLSEN--------AERSA-CIYGG
SpeC_1KTK_A  ----YYISSEMSYEASQKFKRDDHVDVFGLFYILNSHTG----------------EYIYGG
TSST1_3MFG_A --YYSPAFTKGEKVDLNTKRTKKSQHTSEGTYIHFQIS--------------------G
                .             .   .   *                              *

                   130       140       150       160       170       180
                    |         |         |         |         |         |
SEA_1SXT_A   VTLHDNNRLTEEKKVPINLWLDGKQNTVPLETVKTNKKNVTVQELDLQARRYLQEKYNLY
SEB_1SBB_B   VTEHNGNQLD--KYRSITVRVFEDGKNLLSFDVQTNKKKVTAQELDYLTRHYLVKNKKLY
SEC3_1JCK_B  ITKHEGNHFDNGNLQNVLIRVYENKRNTISFEVQTDKKSVTAQELDIKARNFLINKKNLY
SpeA_1L0X_B  VTNHEGNHLE--IPKKIVVKVSIDGIQSLSFDINTNKKMVTAQELDYKVRKYLTDNKQLY
SpeC_1KTK_A  ITPAQNNKVN--HKLLGNLFISGESQQNLNNKIILEKDIVTFQEIDFKIRKYLMDNYKIY
TSST1_3MFG_A VTNTEKLPTP--IELPLKVKVHGKD-SPLKYWPKFDKKQLAISTLDFEIRHQLTQIHGLY
                :*   :            :  :     . :*. :: . :*   *. *. .   :*

                   190       200       210       220       230       240
                    |         |         |         |         |         |
SEA_1SXT_A   NSDVFDGKVQRGLIVFHTSTEPSVNYDLFGAQG--QYSNTLLRIYRDNKTINSE-NMHID
SEB_1SBB_B   EFNNSP--YETGYIKFIE-NENSFWYDMMPAPGDKFDQSKYLMMYNDNKMVDSK-DVKIE
SEC3_1JCK_B  EFNSSP--YETGYIKFIESNGNTFWYDMMPAPGDKFDQSKYLMIYKDNKMVDSK-SVKIE
SpeA_1L0X_B  TNGPSK--YETGYIKFIPKNKESFWFDFFPEP--EFTQSKYLMIYKDNETLDSN-TSQIE
SpeC_1KTK_A  DATSP---YVSGRIEIGTKDGKHEQIDLFDSPN-EGTRSDIFAKYKDNRIINMKNFSHFD
TSST1_3MFG_A RSSDKT----GGYWKITMNDGSTYQSDLS---------KKFEYNTEKPPINIDEIKTIE
                *   :       *:           :   :: :: .   ::
```

```
                   246
                    |
SEA_1SXT_A   IYLYTS
SEB_1SBB_B   VYLTTKKK
SEC3_1JCK_B  VHLTTKNG
SpeA_1L0X_B  VYLTTK
SpeC_1KTK_A  IYL-EK
```

Consensus key:
* - single, fully conserved residue
: - conservation of strong groups
. - conservation of weak groups
- - no consensus

Four different modes of interaction of SAgs with class II MHC have been described: (1) SEB, SEC, SpeA bind to class II MHC alpha chain with a single, low affinity binding site that is located in the N-terminal domain of the protein. This binding is independent of the peptide located in the groove of the MHC-II molecule; (2) TSST-1 on the other hand uses a peptide-dependent binding mechanism to interact with low affinity to the class II MHC alpha chain through the TSST-1 N-terminal binding domain; (3) SpeC binds to the beta chain of class II MHC with high affinity, in a zinc-dependent manner through the C-terminal domain of SpeC; (4) SEA contains both a low affinity binding site and a high affinity, zinc-dependent site which could possible involve cross-linking of MHC molecules [43–45,47,51,52]. Structural features of SAg binding to the Vβ domains are described below.

Figure 2. Two-domain architecture of superantigens. The canonical structure of SAg consists of two domains. The N-terminal domain (red) consists of a β barrel motif and C-terminal domain (blue) consists of a β-grasp motif and an α-helix which spans the center of the structure.

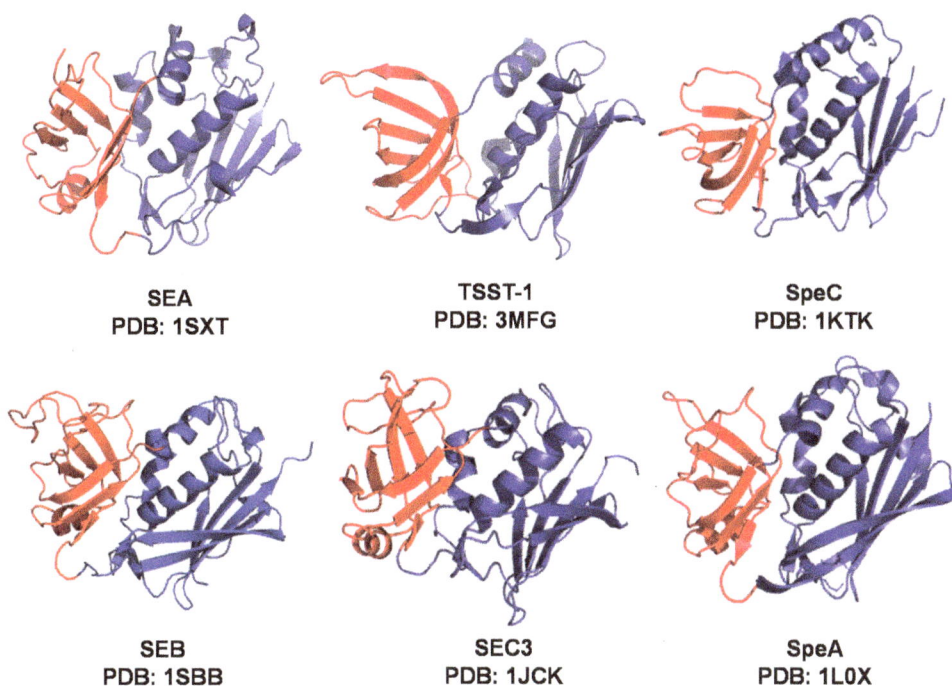

SEA
PDB: 1SXT

TSST-1
PDB: 3MFG

SpeC
PDB: 1KTK

SEB
PDB: 1SBB

SEC3
PDB: 1JCK

SpeA
PDB: 1L0X

Table 1. Crystal structures of staphylococcal and streptococcal superantigens and their complexes with Vβ domains.

Organism	SAg	Crystal Structure (PDB code, ligand)	Year	Reference
S. aureus	SEA	1ESF (co-crystallized with Cd^{2+})	1995	[53]
S. aureus		1SXT (co-crystallized with Zn^{2+})	1996	[54]
S. aureus	SEB	3SEB	1998	[55]
S. aureus		1SBB (co-crystallized with mVβ8.2)	1998	[39]
S. aureus		3R8B (co-crystallized with affinity matured mVβ8.2 mutant G5-8)	2011	[56]
S. aureus	SEC3	1CK1 (co-crystallized with Zn^{2+})	2002	[57]
S. aureus		1JCK (co-crystallized with mVβ8.2)	1996	[40]
S. aureus		2AQ3 (co-crystallized with affinity matured mVβ8.2 mutant L2CM)	2005	[58]
S. aureus	TSST1	2QIL	1996	[59]
S. aureus		2IJ0 (co-crystallized with affinity matured hVβ2.1 mutant D10)	2007	[42]
S. aureus		3MFG (co-crystallized with hVβ2.1 stabilized wild-type EP-8)	2011	[56]
S. pyogenes	SpeA	1FNU (co-crystallized with Cd^{2+}) 1FNV (co-crystallized with Cd^{2+}) 1FNW (co-crystallized with Cd^{2+})	2000	[60]
S. pyogenes		1L0X (co-crystallized with mVβ8.2) 1L0Y (co-crystallized with mVβ8.2 and Zn^{2+})	2002	[41]
S. pyogenes	SpeC	1AN8	1997	[61]
S. pyogenes		1KTK (co-crystallized with hVβ2.1)	2002	[41]

3. Engineering High-Affinity T Cell Receptor Vβ Domains against Superantigens SEA, SEB, SEC3, TSST-1, SpeA, and SpeC

Except for staphylococcal enterotoxin H (SEH), which has been shown to interact primarily with the variable region of alpha chain of the TCR [62,63], most SAgs are known to specifically interact with variable regions of TCR beta chain. The hallmark feature of SAgs is that they stimulate T-cells that bear a specific subset of variable regions in their beta chains (Vβ) [64,65]. Interestingly, despite their potent activity, SAgs are known to bind to their cognate Vβ receptors with low affinity (K_D values in the micromolar range) (Table 2 and [66–68]). In order to develop an antagonist that can effectively neutralize their toxic effects *in vivo*, a panel of six soluble, high-affinity TCR Vβ mutants have been engineered [28,33–37]. These Vβ mutants bind to one of six key staphylococcal and streptococcal SAgs (SEA, SEB, SEC3, TSST1, SpeA, and SpeC), at the same epitope as the wild type receptors but with much higher affinity, in the picomolar to nanomolar range. These represent 1000 to 3,000,000-fold increases in affinity, compared to wild-type (Table 2). Unlike antibodies, which could bind to any epitope of the SAg, engineering of the Vβ ensures that the neutralizing agent binds the identical SAg epitope as the wild-type receptor, thereby ensuring that direct competition and corresponding neutralization occurs.

Table 2. High-affinity Vβ domains that bind to various superantigens.

Organism	SAg	WT Vβ		High affinity Vβ		Improvement in affinity (fold)	References
		Name	Affinity (μM)	Name	Affinity (pM)		
S. aureus	SEA	Human Vβ22	100	FL	4,000	25,000	[37]
S. aureus	SEB	Mouse Vβ8.2	144	G5-8	50	2,880,000	[33,66]
S. aureus	SEC3	Mouse Vβ8.2	3	L3	3,000	1,000	[36,66]
S. aureus	TSST1	Human Vβ2.1	2.3	D10	180	13,000	[34]
S. pyogenes	SpeA	Mouse Vβ8.2	6	KKR	270	22,000	[35,66]
S. pyogenes	SpeC	Human Vβ2.1	20	HG_FI	500	40,000	[41,69]

All high-affinity Vβ mutants were engineered using yeast display technology (Figure 3) [31,32] and directed evolution. The process involved use of a wild type Vβ from TCRs known to be stimulated by the SAg of interest. This Vβ gene was cloned into the yeast display vector in frame with the yeast mating protein, Aga2 to be displayed on the yeast surface. The Vβ gene was flanked by hemagglutinin (HA) and c-myc tags, which served as probes of the protein expression. Unlike many antibody variable domains, wild type Vβ domains typically require that one or more key mutations be engineered into the protein in order to be expressed on the yeast cell surface [70,71]. To accomplish this, the Vβ gene was subjected to error prone PCR to introduce random mutations, and the library was selected by fluorescence-activated cell sorting (FACS) with a conformation-specific anti-Vβ antibody (these are typically available commercially against most of the human Vβ regions). Anti-Vβ antibody is used, rather than fluorophore-labeled SAg at this stage, as the affinity of the SAgs with wild type Vβ are so low that detection by flow cytometry is not possible. The mutated Vβ region that allows it to be expressed on the surface of yeast is often called a stabilized Vβ as it has been shown that such mutations yield stabilized, soluble domains [72] (Figure 4).

Stabilized Vβ region genes served as templates for either additional random mutagenesis or for site-directed mutagenesis to generate libraries (Figure 4) with mutations in the putative SAg-binding sites. Typically, selection of the sites for mutagenesis was guided by the crystal structure of Vβ:SAg complexes. If structural information was not available, such as with the recent engineering of Vβ against SEA [37], the residues in CDR2 were chosen to generate the first generation site-directed mutagenesis libraries because of its central role in the interaction of other Vβ regions with SAgs, as observed in Vβ:SAg crystal structures [39–42]. To construct the site-directed mutagenesis libraries, amino acid positions were encoded by randomized codons (NNS) in primers, and cloned by PCR using overlapping primers. PCR products were transformed into yeast cells by homologous recombination, yielding library sizes of 10^6–10^8 transformants.

Pre-selected degenerate libraries typically exhibit no detectable binding with SAg by flow cytometry due to their low affinity, or loss of binding, by the great majority of mutants. To select for the rare Vβ mutants that exhibit higher affinity binding, libraries were subjected to several rounds of selection with a decreasing concentration of biotinylated-SAg, followed by staining with fluorescently-labeled streptavidin, and fluorescence-activated cell sorting (FACS). In each round of selection, a small fraction of cells that exhibited the top 1% fluorescence upon selection with SAg of interest, were collected and expanded for subsequent rounds of screening. When a distinct yeast population with positive SAg-staining emerged after 3~4 rounds of selection, yeast cells were plated and higher affinity clones were isolated, characterized and sequenced. In the recent engineering of Vβ22 against SEA, mutations isolated from the CDR2 library alone were capable of increasing the SEA affinity by 25,000-fold [37]. Even more strikingly, with the engineering of mouse Vβ8.2 against SEB, mutations from one CDR2 library accounted for a 220,000-fold increase of Vβ affinity with SEB (from 144 μM wide type affinity to 650 pM for G2-5) [33]. These results further validated the essential role of CDR2 loop in Vβ:SAg interactions (see below).

Figure 3. Schematic of yeast display system for engineering high-affinity Vβ domains against Superantigens. The Vβ libraries with various mutations are fused to the C terminus of the yeast mating protein Aga-2 to be displayed on the yeast cell surface. HA and c-myc tags are included in the fusion gene to probe and quantify the Vβ protein expression level. Fluorescent ligands include either a monoclonal antibody to the Vβ region or the SAg.

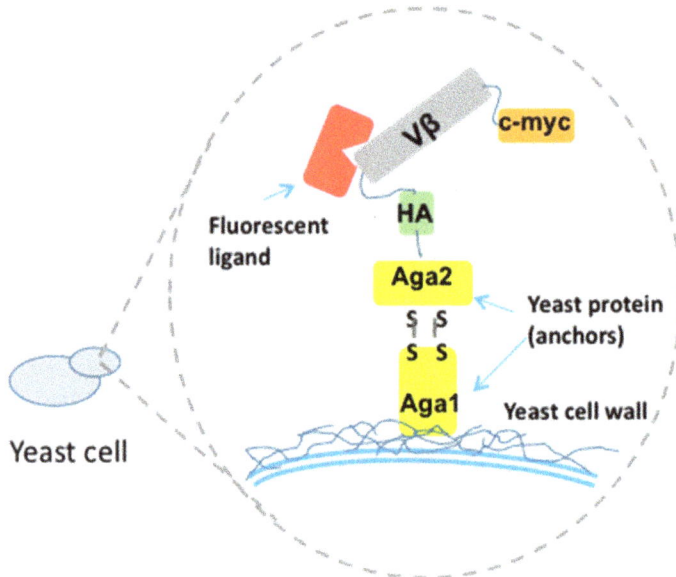

Following the initial selection, additional mutagenized libraries were often constructed in regions (CDR1, HV4 and FR3) that flank the CDR2 tertiary structure, using one or a combination of the first generation lead mutants as templates. After further affinity- or off-rate-based selections, mutants exhibited a more modest 10 to ~100 fold further increase in affinity. The specific region(s) where higher affinity mutations were successfully isolated for each pair of Vβ/SAg reflected, in part, the diverse binding modes of the SAgs with their cognate Vβs. Ultimately, mutations isolated from multiple libraries could be combined to generate the highest affinity mutants that yielded 1000 to 3,000,000-fold increases in affinity with targeted SAgs compared to the wild type Vβ.

Figure 4. General flow chart for the cloning, display and engineering of high-affinity Vβ neutralizing agents by yeast display. A Vβ clone that is specific for the SAg of interest is cloned into the yeast display vector (Figure 3) and used to generate libraries of mutants. The libraries are selected with fluorescently-labeled ligands, (e.g., conformation-specific anti-Vβ antibodies or the SAg of interest). Multiple rounds of selections are conducted to enrich the Vβ mutants with desired properties, which serve as templates for subsequent library design and screening to achieve desirable stability or affinity of Vβ with SAg.

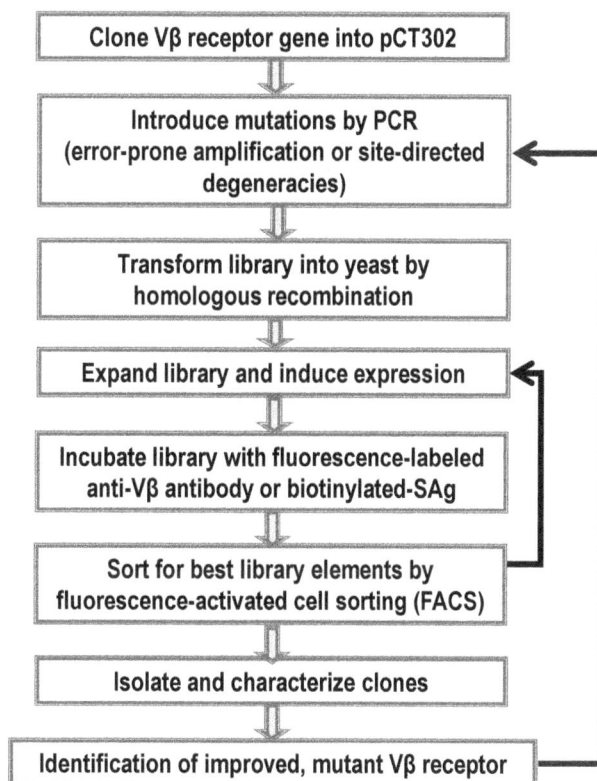

4. Topology of Vβ:Superantigen Interactions

Crystal structures of five out of six SAgs discussed in this review have been solved in complex with their cognate Vβ receptor ligand (Figure 5B–F) [39–42]. In general, the Vβ receptor docks in the cleft between the two domains of the SAg and uses its hypervariable loops (CDRs), or specific framework regions for engagement (Figure 5). Co-crystal structures of different SAg with their cognate Vβ ligands indicate that Vβ domains interact with the SAgs with considerable diversity in positioning and in interaction chemistries. However, the CDR2 loop of Vβ appears to be central to each SAg-TCR interaction [73,74] (Figure 5). Other regions of the Vβ appear to play important, but supporting roles, in the binding energy and specificity for the SAg [34,73,75].

SEB, SEC3 and SpeA (Group II SAg) are more structurally similar and each has been co-crystallized with murine Vβ8.2 (mVβ8.2) [39–41]. As indicated, these three SAgs possess 50%–65% sequence identity and they engage with the Vβ8.2 region of the TCR using similar residues (Figure 1), thereby determining their specificity for mVβ8.2. Accordingly, SEB, SEC3 and SpeA interact with mVβ8.2 with similar topologies (Figure 5), and they engage in intermolecular contacts primarily with CDR2 (accounting for 50%, 63% and 33% of total contacts respectively), HV4 and to some extent framework (FR) regions.

The mechanisms by which these three SAgs interact with mVβ8.2 are largely dependent on the common conformation of CDR2 and HV4, although SpeA forms a distinct contact via its E94 residue, by forming hydrogen bonds with N28 of CDR1 loop of mVβ8.2 [41]. Since SEB and SEC3 depend primarily on interactions with main chain atoms of Vβ-CDR2, their Vβ binding specificity is considerably reduced. However, SpeA:Vβ8.2 interaction specificity appears to be enhanced because the interface involves H-bonds between side chain atoms from both SpeA and the Vβ molecule.

In contrast, SpeC interacts with human Vβ2.1 (hVβ2.1) with more extensive use of the Vβ region, engaging all of the hypervariable loops. The specificity of the SpeC:hVβ2.1 interaction is increased by numerous H-bonds and van der Waals interactions with both main chain and side chain atoms of hVβ2.1. Non-canonical amino acid insertions in CDR1 and CDR2, and the presence of an extended CDR3 loop (at least in some β-chains), also increase the specificity of hVβ2.1 for SpeC [41].

Human Vβ2.1 is also the highly restricted target of TSST-1. Thus, both SpeC and TSST-1 interact with hVβ2.1 and both engage residues in CDR2 to make contacts with Vβ. Although the two toxins engage a few common residues in CDR2, each also uses other distinct, non-overlapping regions for binding and for achieving specificity. TSST-1 uniquely binds FR3 while SpeC engages with Vβ in a distinct mode by making extensive contacts involving residues from CDR1, HV4, FR2, FR3 and CDR3 in Vβ [41,42]. The specificity of TSST-1 for hVβ2.1 has been attributed to the involvement of hVβ2.1 FR3 residues E61 and K62. It has been speculated that TSST-1 does not activate T-cells bearing other Vβ domains because 75% of all other human TCR Vβ regions possess a proline at position 61, resulting in reduced conformational flexibility; this reduced flexibility could prevent the specific conformation required for interaction with TSST-1. In addition, the absence of a residue at position 62 in 50% of human TCR Vβ domains also contributes to the high specificity of TSST-1 for hVβ2.1 [42]. Finally, the residues that TSST-1 uses to interact with Vβ2.1 share little homology with residues that other SAgs (including SpeC) interact with their cognate Vβ ligand (Figure 1), which further enhances TSST-1 specificity towards hVβ2.1. Recently, the molecular basis of the extreme Vβ specificity of TSST-1 was determined to be the combination of both the non-canonical conformation adopted by CDR2 region of the Vβ along with residues Y56 and K62 on FR3 region [74].

Figure 5. Co-crystal structures of six superantigens with cognate Vβ domain of the T cell receptor. Except for SEA, the co-crystal structures of SAg (blue) with their cognate Vβ ligand (gray) are available in PDB (Table 1). Residues of SAg interacting with Vβ are indicated in teal. Various regions of Vβ are colored as follows: CDR1 (green); CDR2 (Red); CDR3 (orange); HV4 (purple); FR2 (olive) and FR3 (yellow). Interacting residues of Vβ and SAg are displayed in stick configurations. (**A**) The SEA crystal structure was manually docked with mouse Vβ16 crystal structure (PDB: 4ELK), that is 66% identical to human Vβ22 protein sequence; (**B**) Co-crystal structure of TSST-1 with human Vβ2.1 mutant, EP-8; (**C**) Co-crystal structure of SpeC with human Vβ2.1; (**D**) Co-crystal structure of SEB with mouse Vβ8.2; (**E**) Co-crystal structure of SEC3 with mouse Vβ8.2; (**F**) Co-crystal structure of SpeA with mouse Vβ8.2.

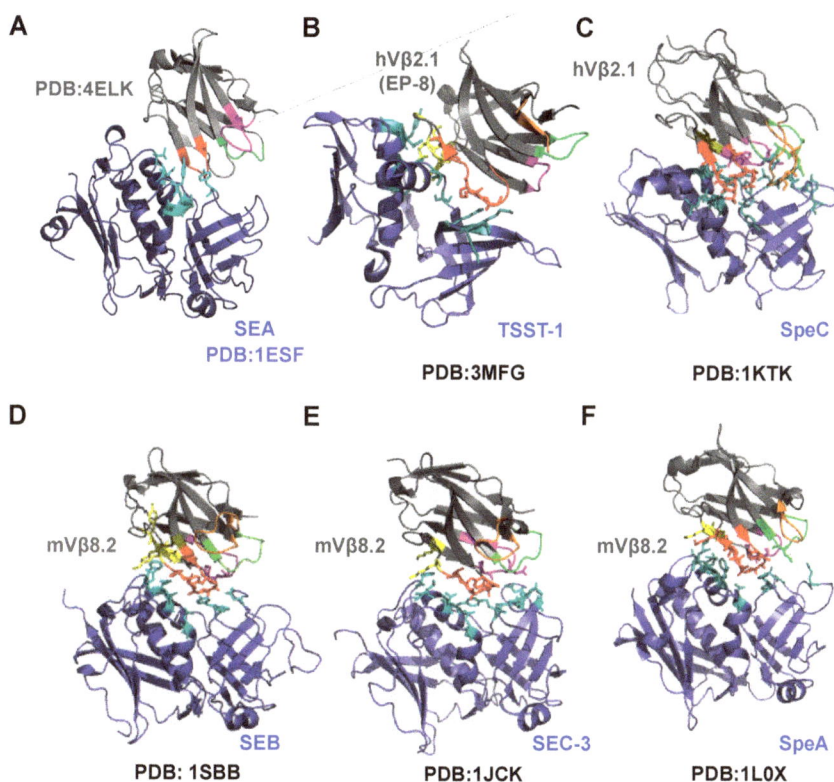

5. Structural Basis of High-Affinity and Specificity of the Vβ:SAg Interactions

Although CDR2 regions have served as the predominant site for improving the affinities of Vβ domains for binding to their SAgs, it is clear that other regions can also serve to enhance affinity through structural changes in each Vβ:SAg interface. The involvement of regions other than CDR2, also can contribute to the high level of specificity in Vβ:SAg interactions exhibited by the high-affinity Vβ mutants. Several structures have been solved of SAgs in complex with engineered, high affinity Vβ domains, thereby providing an understanding of the interactions that confer both higher affinity and specificity [42,56,58]. Not surprisingly, multiple factors, including increases in van der Waals interactions, hydrogen bonds, hydrophobic interactions, cooperativity, and conformational flexibility have all been shown to be involved. Here we discuss the generation and structural basis of high affinity for select Vβ domains, which have been engineered over the past decade.

In order to generate a high affinity Vβ mutant for binding SEC3, mVβ8.2 was displayed on the surface of yeast and mutations were introduced by error-prone PCR, followed by site directed mutagenesis to combine mutations. One resulting mutant, called L2CM (K_D = 7 nM) (also called mL2.1/A52V, a first generation variant of L3, Table 1) exhibited ~450 fold increase in affinity compared to the wild type [28]. The structural basis of the SEC3:Vβ interaction has been studied extensively by alanine scan mutagenesis [76], and the high-affinity interaction with L2CM has been examined for binding energetics [77] and crystallization of various L2CM variants [58]. Although L2CM contained nine mutations, only four (A52V, S54N, K66E, Q72H) were energetically significant [77]. Structural analysis [58] indicated that the A52V mutation in CDR2, allowed an increase in hydrophobic contact area and also induced conformational changes in Q72 of the Vβ. The S54N mutation in CDR2 participated in affinity maturation by allowing recruitment of water molecules to SEC3:Vβ interface hence mediating contacts between $N26^{Vβ}$ and $N24^{Vβ}$ with D204, K205 and F206 in SEC3. Residue K66 appeared to be conformationally restrained in the SEC3:wtVβ8.2 structure, but it adopted a more extended conformation when mutated to glutamate. This also resulted in loss of van der Waals interactions with SEC3 and unfavorable change in enthalpy of binding but a highly favorable entropic change, resulting in a higher affinity complex [77]. Mutations Q72H and A52V were shown to be involved in inducing subtle conformational changes in hypervariable loops, thereby affecting how CDR1 residues, and CDR2 residue 52, interacted with SEC3. Although the A52V mutation had a dominant effect in affinity maturation by mediating restructuring events of the hypervariable loops, Q72H had a minor but significant contribution to affinity maturation [58]. Recently, L2CM was further engineered by incorporating additional mutations in CDR1, HV4 and framework regions to obtain mutant L3 (K_D = 3 nM) (Table 1 and [36]).

Similarly, mVβ8.2 was engineered for binding to SEB with a remarkable 3-million fold increase in affinity relative to wild-type mVβ8.2 (Table 1 and [33]). The engineered protein (G5-8) was crystallized in complex with SEB [56]. The structural details of this complex indicated that lengthening CDR1 loop by incorporation of a serine residue at CDR1 residue 27a, and incorporation of additional mutations N28Y and H29F, resulted in a distinct conformation of CDR1 loop. These mutations resulted in an increase in intermolecular contacts with SEB. Y28 in G5-8 was involved in pi stacking interaction with R110 in SEB and in H-bond interaction with N60. Additionally, two mutations (A52I and G53R) acquired in CDR2, resulted in replacement of residues with smaller side chains to relatively larger side chains which resulted in an increase in van der Waals contacts and H-bond formation with N31, N60 and N88 in SEB. Overall, it was concluded that an increase in the number of intermolecular contacts between G5-8 and SEB resulted in the significant increase in binding affinity [56].

Subsequently, mVβ8.2 was engineered for high affinity (K_D = 270 pM) for SpeA, using yeast display [35]. Key mutations, which were responsible for affinity maturation, were acquired in CDR2 (G53K, S54H), CDR1 (N30K) and in HV4 (Q72R). The resulting mutant, KKR showed a

22,000 fold affinity improvement compared to wt Vβ8.2 (K_D = 6 μM) [41]. Although there is no crystal structure of SpeA with the high affinity Vβ, the authors used revertant mutants and energy minimized computer modeling of the mutated Vβ -SpeA complex to propose the basis of this affinity maturation. The analysis revealed that the side chain of arginine acquired at position 53 in mutant KKR could be accommodated in a binding pocket in SpeA and promoted favorable interactions with side chain oxygen of Y90 and E88 of SpeA. Not only could this account for the affinity maturation, it could contribute to a lack of cross reactivity with SEC3, and the ability of mutant KKR to cross-react with SEB with high-affinity.

The mutant of hVβ2.1 called D10 was engineered for high affinity (K_D = 180 pM) against TSST-1, using yeast display (Table 1 and [34]). D10 contained 14 mutations relative to stabilized wt Vβ2.1 (EP-8). Of these 14 mutations, four were found to be energetically significant: three mutations in CDR2 (at residues 51, 52a and 53) and one mutation in FR3 (at residue 61). Surprisingly, positive cooperativity was observed between the distant mutations in CDR2 and FR3 [75]. Crystal structure analysis indicated that changes in intermolecular contacts, buried surface and/or shape complementarity were not the primary driving factor in affinity maturation of hVβ2.1 to TSST-1. Instead, altered conformational flexibility of D10 was proposed to have resulted in affinity increase by linking CDR2 and FR3 at the Vβ:SAg interface [42]. Using a similar approach, the hVβ2.1 region gene has also been engineered for high-affinity binding to SpeC (manuscript in preparation).

Finally, the hVβ22 region was engineered recently for high affinity binding (K_D = 4 nM) to SEA using yeast display [37]. The engineered mutant called FL contained ten mutations, of which five were located in CDR2. In the absence of a crystal structure of SEA with cognate Vβ, the authors suggested that the structural basis of high affinity may be improved electrostatic interactions (due to mutations N52E and E53D in CDR2), and pi stacking interactions involving N51Y in CDR2 with Y94 and Y205 in SEA's putative TCR binding site.

6. High-Affinity Vβ Domains as Neutralizing Agents

The first study to validate the use of soluble, high-affinity Vβ domains in the neutralization of SAg activity was performed with the Vβ L2CM against the SAg SEC3. This work showed that soluble L2CM, but not soluble wild-type Vβ8, was able to completely inhibit SEC3-mediated T cell cytolysis at nanomolar concentrations [28]. This same high-affinity Vβ was fused to a class II MHC molecule in an attempt to increase the avidity of the interaction with SEC3 [78]. Although the fusion was shown to inhibit SEC3 activity *in vitro*, this study did not show whether the fusion had greater activity than the high-affinity Vβ alone.

Subsequent studies showed that the high-affinity Vβ against SEB, G5-8, but not the wild-type Vβ8.2, was able to inhibit both *in vitro* and *in vivo* activities of SEB [33]. In this study, the *in vitro* activity was shown to be progressively improved in comparing different generations of mutants, with K_D values from 100 μM to 50 pM. For example, the 50 pM G5-8 protein was more effective at inhibition (*i.e.*, had a lower IC_{50}) than the 650 pM G2-5 protein. Furthermore, the G5-8 protein was given to rabbits intravenously at the same time or after SEB administration, in an LPS-enhancement model of lethality, and the protein was able to prevent death even at concentrations close to stoichiometric with the SEB. In the same study, G5-8 administered daily to rabbits implanted with pumps containing SEB was able to prevent temperature increases and lethality due to the SAg. In a rabbit model of skin disease, G5-8 was able to inhibit the hypersensitivity reactions caused by SEB [38].

An in-frame fusion of the high-affinity G5-8 against SEB and the high-affinity D10 against TSST-1 yielded a single 30kDa protein, expressed in *E. coli*, was able to completely inhibit the *in vitro* activity of both SEB and TSST-1 [79], raising the possibility that multiple SAgs might be neutralized with a single therapeutic. An alternative approach is to identify a high-affinity Vβ domain that cross-reacts with multiple SAgs. While this has not been possible for structurally distinct SAgs (e.g., SEB and TSST-1), it has been shown that Vβ domains against SEB (e.g., G5-8) cross-reacted with high-affinity against SpeA, and that Vβ domains were capable of inhibiting both SEB and SpeA in the LPS enhancement models [35]. More recent findings showed that it is possible to engineer a cross-reactive neutralizing Vβ (L3) against both SEC3 and SEB [36].

The greatest clinical potential of soluble, high-affinity Vβ domains, aside from possible applications in biodefense, would be in serious diseases caused by *S. aureus*. The first study to show that high-affinity Vβ proteins were effective in diseases caused by *S. aureus* (*i.e.*, rather than the purified toxins), involved a rabbit model of pneumonia [20]. Rabbits receiving an intrabronchial inoculation (2×10^9 cells) of *S. aureus* USA400 strain CA-MRSA c99-529 (SEB$^+$) were protected from death when treated with 100 μg of G5-8, administered intravenously on a daily basis. Subsequent studies have shown that the high-affinity Vβ L3 against SEC3 also protected rabbits exposed to an SEC-positive strain of MRSA (USA400 MW2) in the pneumonia model [36]. Interestingly, the same L3 protein was capable of significantly reducing the bacterial burden of the MRSA (USA400 MW2) strain in an infective endocarditis model [36]. These pre-clinical studies suggest that these small Vβ proteins could be used intravenously, with antibiotics, to manage staphylococcal diseases that involve SAgs. The diversity of SAgs among different strains of *S. aureus* will likely require that diagnostics be developed for detection of the specific SAgs in patients, or that a multi-targeted therapeutic that can neutralize many of the SAgs be developed.

Acknowledgements

We thank past and present members of the Kranz lab for helpful discussions. This work was supported by various National Institutes of Health grants over the past decade, including R43 AI102432 (to David M. Kranz) and a grant from the National Institutes of Health-supported Great Lakes Regional Center for Excellence in Biodefense and Emerging Diseases (U54 AI57153 to David M. Kranz).

Conflicts of Interest

David M. Kranz co-founded a company called ImmuVen that has acquired rights from the University of Illinois for some of the T cell receptors engineered in his lab.

References

1. Marrack, P.; Kappler, J. The staphylococcal enterotoxins and their relatives. *Science* **1990**, *248*, 705–711.

2. Spaulding, A.R.; Salgado-Pabón, W; Kohler, P.L.; Horswill, A.R.; Leung, D.Y.; Schlievert, P.M. Staphylococcal and streptococcal superantigen exotoxins. *Clin. Microbiol. Rev.* **2013**, *26*, 422–447.

3. Krakauer, T.; Stiles, B.G. The staphylococcal enterotoxin (SE) family: SEB and siblings. *Virulence* **2013**, *4*, 759–773.

4. Dinges, M.M.; Orwin, P.M.; Schlievert, P.M. Exotoxins of *Staphylococcus aureus*. *Clin. Microbiol. Rev.* **2000**, *13*, 16–34.

5. McCormick, J.K.; Yarwood, J.M.; Schlievert, P.M. Toxic shock syndrome and bacterial superantigens: An update. *Ann. Rev. Microbiol.* **2001**, *55*, 77–104.

6. Li, H.; Llera, A.; Malchiodi, E. L.; Mariuzza, R. A. The structural basis of T cell activation by superantigens. *Ann. Rev. Immunol.* **1999**, *17*, 435–466.

7. Baker, M.D.; Acharya, K.R. Superantigens: Structure-function relationships. *Int. J. Med. Microbiol.* **2004**, *293*, 529–537.

8. Papageorgiou, A.C.; Acharya, K.R. Microbial superantigens: From structure to function. *Trends Microbiol.* **2000**, *8*, 369–375.

9. Fraser, J.D.; Proft, T. The bacterial superantigen and superantigen-like proteins. *Immunol. Rev.* **2008**, *225*, 226–243.

10. Pless, D.D.; Ruthel, G.; Reinke, E.K.; Ulrich, R.G.; Bavari, S. Persistence of zinc-binding bacterial superantigens at the surface of antigen-presenting cells contributes to the extreme potency of these superantigens as T-cell activators. *Infect. Immun.* **2005**, *73*, 5358–5366.

11. Bavari, S.; Ulrich, R.G.; LeClaire, R.D. Cross-reactive antibodies prevent the lethal effects of *Staphylococcus aureus* superantigens. *J. Infect. Dis.* **1999**, *180*, 1365–1369.

12. Varshney, A.K.; Mediavilla, J.R.; Robiou, N.; Guh, A.; Wang, X.; Gialanella, P.; Levi, M.H.; Kreiswirth, B.N.; Fries, B.C. Diverse enterotoxin gene profiles among clonal complexes of *Staphylococcus aureus* isolates from the Bronx, New York. *Appl. Environ. Microbiol.* **2009**, *75*, 6839–6849.

13. Hu, D.L.; Omoe, K.; Inoue, F.; Kasai, T.; Yasujima, M.; Shinagawa, K.; Nakane, A. Comparative prevalence of superantigenic toxin genes in meticillin-resistant and meticillin-susceptible *Staphylococcus aureus* isolates. *J. Med. Microbiol.* **2008**, *57*, 1106–1112.

14. Lindsay, J.A.; Ruzin, A.; Ross, H.F.; Kurepina, N.; Novick, R.P. The gene for toxic shock toxin is carried by a family of mobile pathogenicity islands in *Staphylococcus aureus*. *Mol. Microbiol.* **1998** *29*, 527–543.

15. Novick, R.P.; Schlievert, P.; Ruzin, A. Pathogenicity and resistance islands of staphylococci. *Microbes Infect.* **2001**, *3*, 585–594.

16. Fitzgerald, J.R.; Monday, S.R.; Foster, T.J.; Bohach, G.A.; Hartigan, P.J.; Meaney, W.J.; Smyth, C.J. Characterization of a putative pathogenicity island from bovine *Staphylococcus aureus* encoding multiple superantigens. *J. Bacteriol.* **2001**, *183*, 63–70.

17. Betley, M.J.; Mekalanos, J.J. Staphylococcal enterotoxin A is encoded by phage. *Science* **1985**, *229*, 185–187.

18. Derzelle, S.; Dilasser, F.; Duquenne, M.; Deperrois, V. Differential temporal expression of the staphylococcal enterotoxins genes during cell growth. *Food Microbiol.* **2009**, *26*, 896–904.

19. Bachert, C.; Gevaert, P.; Zhang, N.; van Zele, T.; Perez-Novo, C. Role of staphylococcal superantigens in airway disease. *Chem. Immunol. Allergy* **2007**, *93*, 214–236.

20. Strandberg, K.L.; Rotschafer, J.H.; Vetter, S.M.; Buonpane, R.A.; Kranz, D.M.; Schlievert, P.M. Staphylococcal superantigens cause lethal pulmonary disease in rabbits. *J. Infect. Dis.* **2010**, *202*, 1690–1697.

21. Schlievert, P.M.; Case, L.C.; Strandberg, K.L.; Abrams, B.B.; Leung, D.Y. Superantigen profile of *Staphylococcus aureus* isolates from patients with steroid-resistant atopic dermatitis. *Clin. Infect. Dis.* **2008**, *46*, 1562–1567.

22. Macias, E.S.; Pereira, F.A.; Rietkerk, W.; Safai, B. Superantigens in dermatology. *J. Am. Acad. Dermatol.* **2011**, *64*, 455–472; quiz 473–454.

23. Bohach, G.A.; Fast, D.J.; Nelson, R.D.; Schlievert, P.M. Staphylococcal and streptococcal pyrogenic toxins involved in toxic shock syndrome and related illnesses. *Crit. Rev. Microbiol.* **1990**, *17*, 251–272.

24. Lappin, E.; Ferguson, A.J. Gram-positive toxic shock syndromes. *Lancet Infect. Dis.* **2009**, *9*, 281–290.

25. Dhodapkar, K.; Corbacioglu, S.; Chang, M.W.; Karpatkin, M.; DiMichele, D. Purpura fulminans caused by group A beta-hemolytic *Streptococcus* sepsis. *J. Pediatr.* **2000**, *137*, 562–567.

26. Tilahun, A.Y.; Holz, M.; Wu, T.T.; David, C.S.; Rajagopalan, G. Interferon gamma-dependent intestinal pathology contributes to the lethality in bacterial superantigen-induced toxic shock syndrome. *PLoS ONE* **2011**, *6*, doi: 10.1371/journal.pone.0016764.

27. Tilahun, A.Y.; Karau, M.J.; Clark, C.R.; Patel, R.; Rajagopalan, G. The impact of tacrolimus on the immunopathogenesis of staphylococcal enterotoxin-induced systemic inflammatory response syndrome and pneumonia. *Microbes Infect.* **2012**, *14*, 528–536.

28. Kieke, M.C.; Sundberg, E.; Shusta, E.V.; Mariuzza, R.A.; Wittrup, K.D.; Kranz, D.M. High affinity T cell receptors from yeast display libraries block T cell activation by superantigens. *J. Mol. Biol.* **2001**, *307*, 1305–1315.

29. Garrison, L.; McDonnell, N. Etanercept: therapeutic use in patients with rheumatoid arthritis. *Ann. Rheum. Dis.* **1999**, *58*, 165–169.

30. Combe, B. Update on the use of etanercept across a spectrum of rheumatoid disorders. *Biologics* **2008**, *2*, 165–173.

31. Boder, E.T.; Wittrup, K.D. Yeast surface display for directed evolution of protein expression, affinity, and stability. *Meth. Enzymol.* **2000**, *328*, 430–444.

32. Boder, E.T.; Wittrup, K.D. Yeast surface display for screening combinatorial polypeptide libraries. *Nat. Biotechnol.* **1997**, *15*, 553–557.

33. Buonpane, R.A.; Churchill, H.R.; Moza, B.; Sundberg, E.J.; Peterson, M.L.; Schlievert, P.M.; Kranz, D.M. Neutralization of staphylococcal enterotoxin B by soluble, high-affinity receptor antagonists. *Nat. Med.* **2007**, *13*, 725–729.

34. Buonpane, R.A.; Moza, B.; Sundberg, E.J.; Kranz, D.M. Characterization of T cell receptors engineered for high affinity against toxic shock syndrome toxin-1. *J. Mol. Biol.* **2005**, *353*, 308–321.

35. Wang, N.; Mattis, D.M.; Sundberg, E.J.; Schlievert, P.M.; Kranz, D.M. A single, engineered protein therapeutic agent neutralizes exotoxins from both *Staphylococcus aureus* and *Streptococcus pyogenes*. *Clin. Vaccine Immunol.* **2010**, *17*, 1781–1789.

36. Mattis, D.M.; Spaulding, A.R.; Chuang-Smith, O.N.; Sundberg, E.J.; Schlievert, P.M.; Kranz, D.M. Engineering a soluble high-affinity receptor domain that neutralizes staphylococcal enterotoxin C in rabbit models of disease. *Protein Eng. Des. Sel.* **2013**, *26*, 133–142.

37. Sharma, P.; Postel, S.; Sundberg, E.J.; Kranz, D.M. Characterization of the Staphylococcal enterotoxin A: Vβ receptor interaction using human receptor fragments engineered for high affinity. *Protein Eng. Des. Sel.* **2013**, *26*, 781–789.

38. John, C.C.; Niermann, M.; Sharon, B.; Peterson, M.L.; Kranz, D.M.; Schlievert, P.M. Staphylococcal toxic shock syndrome erythroderma is associated with superantigenicity and hypersensitivity. *Clin. Infect. Dis.* **2009**, *49*, 1893–1896.

39. Li, H.; Llera, A.; Tsuchiya, D.; Leder, L.; Ysern, X.; Schlievert, P.M.; Karjalainen, K.; Mariuzza, R.A. Three-dimensional structure of the complex between a T cell receptor beta chain and the superantigen staphylococcal enterotoxin B. *Immunity* **1998**, *9*, 807–816.

40. Fields, B.A.; Malchiodi, E.L.; Li, H.; Ysern, X.; Stauffacher, C.V.; Schlievert, P.M.; Karjalainen, K.; Mariuzza, R.A. Crystal structure of a T-cell receptor b-chain complexed with a superantigen. *Nature* **1996**, *384*, 188–192.

41. Sundberg, E.J.; Li, H.; Llera, A.S.; McCormick, J.K.; Tormo, J.; Schlievert, P.M.; Karjalainen, K.; Mariuzza, R.A. Structures of two streptococcal superantigens bound to TCR beta chains reveal diversity in the architecture of T cell signaling complexes. *Structure* **2002**, *10*, 687–699.

42. Moza, B.; Varma, A.K.; Buonpane, R.A.; Zhu, P.; Herfst, C.A.; Nicholson, M.J.; Wilbuer, A.K.; Seth, N.P.; Wucherpfennig, K.W.; McCormick, J.K.; *et al.* Structural basis of T-cell specificity and activation by the bacterial superantigen TSST-1. *EMBO J.* **2007**, *26*, 1187–1197.

43. Petersson, K.; Thunnissen, M.; Forsberg, G.; Walse, B. Crystal structure of a sea variant in complex with MHC class II reveals the ability of Sea to crosslink MHC molecules. *Structure* **2002**, *10*, 1619–1626.

44. Jardetzky, T.S.; Brown, J.H.; Gorga, J.C.; Stern, L.J.; Urban, R.G.; Chi, Y.; Stauffacher, C.; Strominger, J.; Wiley, D.C. Three-dimensional structure of a human class II histocompatibility molecule complexed with superantigen. *Nature* **1994**, *368*, 711–718.

45. Sundberg, E.J.; Andersen, P.S.; Schlievert, P.M.; Karjalainen, K.; Mariuzza, R.A. Structural, energetic, and functional analysis of a protein-protein interface at distinct stages of affinity maturation. *Structure* **2003**, *11*, 1151–1161.

46. Papageorgiou, A.C.; Collins, C.M.; Gutman, D.M.; Kline, J.B.; O'Brien, S.M.; Tranter, H.S.; Acharya, K.R. Structural basis for the recognition of superantigen streptococcal pyrogenic exotoxin A (SpeA1) by MHC class II molecules and T-cell receptors. *EMBO J.* **1999**, *18*, 9–21.

47. Li, Y.; Li, H.; Dimasi, N.; McCormick, J.K.; Martin, R.; Schuck, P.; Schlievert, P.M.; Mariuzza, R.A. Crystal structure of a superantigen bound to the high-affinity, zinc- dependent site on MHC class II. *Immunity* **2001**, *14*, 93–104.

48. Wang, X.; Xu, M.; Cai, Y.; Yang, H.; Zhang, H.; Zhang, C. Functional analysis of the disulphide loop mutant of staphylococcal enterotoxin C2. *Appl. Microbiol. Biotechnol.* **2009**, *82*, 861–871.

49. Hovde, C.J.; Marr, J.C.; Hoffmann, M.L.; Hackett, S.P.; Chi, Y.I.; Crum, K.K.; Stevens, D.L.; Stauffacher, C.V.; Bohach, G.A. Investigation of the role of the disulphide bond in the activity and structure of staphylococcal enterotoxin C1. *Mol. Microbiol.* **1994** *13*, 897–909.

50. Schlievert, P.M.; Jablonski, L.M.; Roggiani, M.; Sadler, I.; Callantine, S.; Mitchell, D.T.; Ohlendorf, D.H.; Bohach, G.A. Pyrogenic toxin superantigen site specificity in toxic shock syndrome and food poisoning in animals. *Infect. Immun.* **2000**, *68*, 3630–3634.

51. Kim, J.; Urban, R.G.; Strominger, J.L.; Wiley, D.C. Toxic shock syndrome toxin-1 complexed with a class II major histocompatibility molecule HLA-DR1. *Science* **1994**, *266*, 1870–1874.

52. Tiedemann, R.E.; Urban, R.J.; Strominger, J.L.; Fraser, J.D. Isolation of HLA-DR1.(staphylococcal enterotoxin A)2 trimers in solution. *Proc. Natl. Acad. Sci. USA* **1995**, *92*, 12156–12159.

53. Schad, E.M.; Zaitseva, I.; Zaitsev, V.N.; Dohlsten, M.; Kalland, T.; Schlievert, P.M.; Ohlendorf, D.H.; Svensson, L.A. Crystal structure of the superantigen staphylococcal enterotoxin type A. *EMBO J.* **1995**, *14*, 3292–3301.

54. Sundstrom, M.; Hallen, D.; Svensson, A.; Schad, E.; Dohlsten, M.; Abrahmsen, L. The Co-crystal structure of staphylococcal enterotoxin type A with Zn^{2+} at 2.7 A resolution. Implications for major histocompatibility complex class II binding. *J. Biol. Chem.* **1996**, *271*, 32212–32216.

55. Papageorgiou, A.C.; Tranter, H.S.; Acharya, K.R. Crystal structure of microbial superantigen staphylococcal enterotoxin B at 1.5 A resolution: implications for superantigen recognition by MHC class II molecules and T-cell receptors. *J. Mol. Biol.* **1998**, *277*, 61–79.

56. Bonsor, D.A.; Postel, S.; Pierce, B.G.; Wang, N.; Zhu, P.; Buonpane, R.A.; Weng, Z.; Kranz, D.M.; Sundberg, E.J. Molecular basis of a million-fold affinity maturation process in a protein-protein interaction. *J. Mol. Biol.* **2011**, *411*, 321–328.

57. Chi, Y.I.; Sadler I.; Jablonski, L.M.; Callantine, S.D.; Deobald, C.F.; Stauffacher, C.V.; Bohach, G.A. Zinc-mediated dimerization and its effect on activity and conformation of staphylococcal enterotoxin type C. *J. Biol. Chem.* **2002**, *277*, 22839–22846.

58. Cho, S.; Swaminathan, C.P.; Yang, J.; Kerzic, M.C.; Guan, R.; Kieke, M.C.; Kranz, D.M.; Mariuzza, R.A.; Sundberg, E.J. Structural basis of affinity maturation and intramolecular cooperativity in a protein-protein interaction. *Structure* **2005**, *13*, 1775–1787.

59. Papageorgiou, A.C.; Brehm, R.D.; Leonidas, D.D.; Tranter, H.S.; Acharya, K.R. The refined crystal structure of toxic shock syndrome toxin-1 at 2.07 A resolution. *J. Mol. Biol.* **1996**, *260*, 553–569.

60. Earhart, C.A.; Vath, G.M.; Roggiani, M.; Schlievert, P.M.; Ohlendorf, D.H. Structure of streptococcal pyrogenic exotoxin A reveals a novel metal cluster. *Protein Sci.* **2000**, *9*, 1847–1851.

61. Roussel, A.; Anderson, B.F.; Baker, H.M.; Fraser, J.D.; Baker, E.N. Crystal structure of the streptococcal superantigen SPE-C: Dimerization and zinc binding suggest a novel mode of interaction with MHC class II molecules. *Nat. Struct. Biol.* **1997**, *4*, 635–643.

62. Saline, M.; Rödström, K.E.; Fischer, G.; Orekhov, V.Y.; Karlsson, B.G.; Lindkvist-Petersson, K. The structure of superantigen complexed with TCR and MHC reveals novel insights into superantigenic T cell activation. *Nat. Commun.* **2010**, *1*, doi:10.1038/ncomms1117.

63. Petersson, K.; Pettersson, H.; Skartved, N.J.; Walse, B.; Forsberg, G. Staphylococcal enterotoxin H induces V alpha-specific expansion of T cells. *J. Immunol.* **2003**, *170*, 4148–4154.

64. Kappler, J.; Kotzin, B.; Herron, L.; Gelfand, E.W.; Bigler, R.D.; Boylston, A.; Carrel, S.; Posnett, D.N.; Choi, Y.; Marrack, P. V beta-specific stimulation of human T cells by staphylococcal toxins. *Science* **1989**, *244*, 811–813.

65. Thomas, D.; Dauwalder, O.; Brun, V.; Badiou, C.; Ferry, T.; Etienne, J.; Vandenesch, F.; Lina, G. *Staphylococcus aureus* superantigens elicit redundant and extensive human Vbeta patterns. *Infect. Immun.* **2009**, *77*, 2043–2050.

66. Malchiodi, E.L.; Eisenstein, E.; Fields, B.A.; Ohlendorf, D.H.; Schlievert, P.M.; Karjalainen, K.; Mariuzza, R.A. Superantigen binding to a T cell receptor b chain of known three-dimensional structure. *J. Exp. Med.* **1995**, *182*, 1833–1845.

67. Khandekar, S.S.; Bettencourt, B.M.; Wyss, D.F.; Naylor, J.W.; Brauer, P.P.; Huestis, K.; Dwyer, D.S.; Profy, A.T.; Osburne, M.S.; Banerji, J.; *et al.* Conformational integrity and ligand binding properties of a single chain T-cell receptor expressed in Escherichia coli. *J. Biol. Chem.* **1997**, *272*, 32190–32197.

68. Andersen, P.S.; Geisler, C.; Buus, S.; Mariuzza, R.A.; Karjalainen, K. Role of TCR-ligand affinity in T cell activation by bacterial superantigens. *J. Biol. Chem.* **2001**, *276*, 33452–33457.

69. Wang, N.; Kranz, D.M. University of Illinois, Urbana, IL, USA, Unpublished work, 2013.

70. Kieke, M.C.; Shusta, E.V.; Boder, E. T.; Teyton, L.; Wittrup, K.D.; Kranz, D.M. Selection of functional T cell receptor mutants from a yeast surface- display library. *Proc. Natl. Acad. Sci. USA* **1999**, *96*, 5651–5656.

71. Richman, S.A.; Aggen, D.H.; Dossett, M.L.; Donermeyer, D.L.; Allen, P.M.; Greenberg, P.D.; Kranz, D.M. Structural features of T cell receptor variable regions that enhance domain stability and enable expression as single-chain ValphaVbeta fragments. *Mol. Immunol.* **2009**, *46*, 902–916.

72. Shusta, E.V.; Kieke, M.C.; Parke, E.; Kranz, D.M.; Wittrup, K.D. Yeast polypeptide fusion surface display levels predict thermal stability and soluble secretion efficiency. *J. Mol. Biol.* **1999**, *292*, 949–956.

73. Sundberg, E.J.; Deng, L.; Mariuzza, R.A. TCR recognition of peptide/MHC class II complexes and superantigens. *Semin. Immunol.* **2007**, *19*, 262–271.

74. Nur-ur Rahman, A.K.; Bonsor, D.A.; Herfst, C.A.; Pollard, F.; Peirce, M.; Wyatt, A.W.; Kasper, K.J.; Madrenas, J.; Sundberg, E.J.; McCormick, J.K. The T cell receptor beta-chain second complementarity determining region loop (CDR2beta) governs T cell activation and Vbeta specificity by bacterial superantigens. *J. Biol. Chem.* **2011**, *286*, 4871–4881.

75. Moza, B.; Buonpane, R.A.; Zhu, P.; Herfst, C.A.; Rahman, A.K.; McCormick, J.K.; Kranz, D.M.; Sundberg, E.J. Long-range cooperative binding effects in a T cell receptor variable domain. *Proc. Natl. Acad. Sci. USA* **2006**, *103*, 9867–9872.

76. Churchill, H.R.; Andersen, P.S.; Parke, E.A.; Mariuzza, R.A.; Kranz, D.M. Mapping the energy of superantigen Staphylococcus enterotoxin C3 recognition of an alpha/beta T cell receptor using alanine scanning mutagenesis. *J. Exp. Med.* **2000**, *191*, 835–846.

77. Yang, J.; Swaminathan, C.P.; Huang, Y.; Guan, R.; Cho, S.; Kieke, M.C.; Kranz, D.M.; Mariuzza, R.A.; Sundberg, E.J. Dissecting cooperative and additive binding energetics in the affinity maturation pathway of a protein-protein interface. *J. Biol. Chem.* **2003**, *278*, 50412–50421.

78. Hong-Geller, E.; Mollhoff, M.; Shiflett, P.R.; Gupta, G. Design of chimeric receptor mimics with different TcRVbeta isoforms. Type-specific inhibition of superantigen pathogenesis. *J. Biol. Chem.* **2004**, *279*, 5676–5684.

79. Yang, X.; Buonpane, R.A.; Moza, B.; Rahman, A.K.; Wang, N.; Schlievert, P.M.; McCormick, J.K.; Sundberg, E.J.; Kranz, D.M. Neutralization of multiple staphylococcal superantigens by a single-chain protein consisting of affinity-matured, variable domain repeats. *J. Infect. Dis.* **2008**, *198*, 344–348.

Staphylococcal enterotoxins in the Etiopathogenesis of Mucosal Autoimmunity within the Gastrointestinal Tract

MaryAnn Principato and Bi-Feng Qian

Abstract: The staphylococcal enterotoxins (SEs) are the products of *Staphylococcus aureus* and are recognized as the causative agents of classical food poisoning in humans following the consumption of contaminated food. While illness evoked by ingestion of the SE or its producer organism in tainted food are often self-limited, our current understanding regarding the evolution of *S. aureus* provokes the utmost concern. The organism and its associated toxins, has been implicated in a wide variety of disease states including infections of the skin, heart, sinuses, inflammatory gastrointestinal disease, toxic shock, and Sudden Infant Death Syndrome. The intricate relationship between the various subsets of immunocompetent T cells and accessory cells and the ingested material found within the gastrointestinal tract present daunting challenges to the maintenance of immunologic homeostasis. Dysregulation of the intricate balances within this environment has the potential for extreme consequences within the host, some of which are long-lived. The focus of this review is to evaluate the relevance of staphylococcal enterotoxin in the context of mucosal immunity, and the underlying mechanisms that contribute to the pathogenesis of gastrointestinal autoimmune disease.

Reprinted from *Toxins*. Cite as: Principato, M.A.; Qian, B.-F. Staphylococcal enterotoxins in the Etiopathogenesis of Mucosal Autoimmunity within the Gastrointestinal Tract. *Toxins* **2014**, *6*, 1471-1489.

1. The Staphylococcal enterotoxins and Disease

The staphylococcal enterotoxins (SEs), produced by *Staphylococcus aureus*, are recognized as etiologic agents of food poisoning in man, and as potent immunologic superantigens, that produce clinically important syndromes based on their capacity to induce T cell activation and proliferation that lead to the production of significant quantities of proinflammatory mediators. Early recognition of their characteristic as superantigens promoted theoretical considerations that these microbial contaminants might contribute to the underlying causes of self-reactive autoimmune disease due to the nature of the interaction between the microbial superantigen and host cells [1,2]. The SEs have since been implicated in the induction of several human autoimmune exacerbations that include allergic airway disease [3,4], atopic dermatitis [5,6], inflammatory arthritic conditions [3], and colitis [7]. Indeed, patients exhibiting dust mite-induced allergic rhinitis or asthma demonstrate significant levels of IgE antibodies to SEA, SEB, and SEC [8] as well as elevated levels of serum eosinophil cationic protein, an indicator of asthma and rhinitis. More recently, a murine model demonstrated that primary dermal sensitization with both peanut extract and SEB, supported a Th2-mediated IL-4 dependent secondary response to peanut extract, presenting a possible mechanism for the induction of some food allergies [9]. Taken together, the superantigenic staphylococcal enterotoxins provide multiple mechanisms for the induction of organ-specific and systemic autoimmune disease, in addition to classic pathogenic states that includes food poisoning.

The staphylococcal enterotoxins (SEs) have long been implicated in food-borne diseases, and are produced by enterotoxigenic strains of the pathogen *Staphylococcus aureus* [10,11]. Several enterotoxigenic toxins are attributed to *S. aureus* which include the classic serotypes SEA, SEB, SEC, SED, and SEE [12,13]. Molecular techniques have revealed additional closely related enterotoxins including SEG, SEH [14,15], SEI, SEJ, SEK, SEL, SEM [16,17], SEN, SEO, SEP, SER [18,19] and SEU [18,20]. These products have been identified following isolation from food samples implicated in food poisoning scenarios and clinical samples derived from affected individuals [16,18,19,21–25]. So far, more than 20 different types of SEs and enterotoxin –like molecules (SEl) have been described (SEA-SEIV) that share phylogenetic relationship, structure, and sequence similarities [26,27]. It should be noted that while several of the newly identified enterotoxins were identified in various contamination scenarios, it is not yet known whether all of the identified enterotoxins directly induce emetic disease in humans. Further, their possible role in the development or induction of autoimmune disease has not been described. These toxic proteins are composed of approximately 220–240 amino acids and demonstrate an averaged molecular size of 25 kD. They are encoded on plasmids, bacteriophages, pathogenicity islands, and mobile genetic elements [28,29], and demonstrate a remarkable conservation of structural architecture [30,31].

Surveillance of food-borne illness by the Centers for Disease Control and Prevention (CDC) has revealed that more than 200 known diseases are transmitted through food, which cause approximately 76 million cases, 325,000 hospitalizations, and 5000 deaths in the United States annually [32]. However, because many cases of food-borne diseases may not seek medical attention or come to the notice of public health authorities, it is likely that the true incidence is significantly greater than that estimated from investigations of suspected food-borne disease outbreaks. Among the bacteria that can cause food-borne illness, *Staphylococcus aureus* is particularly noteworthy due to the ability to produce toxins that are etiologic agents of gastroenteritis, implicated in autoimmune dysregulation, and lack any FDA-approved therapeutic measures or vaccines to counter the organism or its products. *Staphylococcus aureus* has been most often implicated in staphylococcal food poisoning which is usually attributed to the improper handling of food. It is an encapsulated Gram-positive facultative anaerobic bacteria whose pathogenicity is supported by its multiple virulence factors that include leukocidins, hemolysins, antibiotic resistance(s), exfoliative toxins, toxic shock syndrome toxin-1 (TSST1), and multiple homologous enterotoxins. The bacterium is linked to food poisoning and multiple human illnesses such as endocarditis and pneumonia, and skin infections including toxic shock boils, cellulitis, scalded skin syndrome, and has been implicated in Sudden Infant Death Syndrome [33,34]. Up to 30%–50% of healthy people can harbor the bacteria in the nostrils, on skin, and on hair [22,35]. Interestingly, it is only in a small proportion of the food poisoning outbreaks that the same strain of disease-causing bacteria has been identified from the food handler, food and/or the victim [36,37]. The bacteria may maintain steady growth in a wide range of temperatures, pHs, and sodium chloride concentrations [38]. These characteristics enable the bacteria to contaminate and spread in a great variety of food, including foods that require extensive manipulation during processing. Indeed, the SEB toxin in particular has been characterized as being particularly resistant to destruction by heating [25,39] and is able to withstand storage in acidic foods [40]. The ingested dose of SE required to induce illness in humans has not been determined. Early volunteer studies reported that administration of 50 ug [41] into volunteers weighing 145 lbs and less induced illness. However nanogram quantities have been estimated to induce illness [24,42], and the degree of severity can vary with individual sensitivity. In a large US outbreak caused by ingestion of SEA-contaminated chocolate milk [24], the mean concentration of the toxin in a 400-mL container was approximately 0.5 ng/mL. The usual incubation period of staphylococcal food poisoning is between 2 and 6 h [41], depending on the amount of toxin ingested. Typical disease is characterized by cramping abdominal pain, nausea, vomiting, sometimes followed by diarrhea. Physical examination may reveal dehydration and hypotension if fluid loss has been significant. Routine laboratory testing may detect an electrolyte disturbance. Approximately 10% of the individuals with the food-borne disease need to be hospitalized for further assessment and management. Mortality is rare as demonstrated by a study on 7126 patients where the case fatality rate was as low as 0.03% and the death was exclusively seen in the elderly group [36]. Several mechanisms have been proposed to explain how SEs cause enteric illness: (1) the release of proinflammatory cytokines as a result of SE-induced superantigenic T cell proliferation; (2) the binding of SEs to intestinal mast cells that leads to degranulation [43,44]; and (3) a direct effect upon the intestinal epithelium affecting gut transit [38]. Elegant studies using cultured porcine jejunal sections have demonstrated the binding of SEA and SEB toxins onto the surface of the enterocyte microvillus, possibly mediated by binding onto digalactosylceramide residues. This research further demonstrated that the toxins appear within sub-apical punctae, indicative of entry into the enterocytes via apical endocytosis within the endosomes [45]. It should be noted however, that the pathological events following SE introduction into the intestinal area have been shown to be due to the combined effect of SE toxins and toxins that disrupt the epithelial barrier by inducing enterocyte-cytopathic toxins produced by *Staphylococcus aureus* [46]. In addition, a 5-HT or serotonin-mediated pathway has been reported recently [47].

2. Immunity within the Gut Associated Lymphoid Tissue

The gastrointestinal tract is typically characterized by a large surface area that is responsible for the digestion and absorption of ingested nutrients. This function is aided by the intestine's mucosal lining, whose absorptive surface is increased by the presence of villi which are composed of a single layer of epithelial cells containing a rich network of capillaries and lymphatics and project into the lumen. The intestinal lumen normally contains ingested material consisting of degraded dietary products, commensal microbial flora, and any ingested contaminants including pathogenic bacteria and their products, viruses, fungi, or parasites. Thus, absorption of essential nutrients and host immune defense, two apparently divergent processes, must occur in the intestinal mucosa. In order to provide a healthy microenvironment for the normal physiological activities of the gut, the immune system within the gut-associated lymphoid tissue (GALT) must: (1) generate immunologic tolerance towards nutrients and the commensal microflora; and (2) recognize and abolish infectious agents and potentially injurious toxins [48–50].

Beneath the epithelial layer of the mammalian gastrointestinal tract lies a rich source of immunocompetent cells that comprise a significant portion of the body's T cells. While the peripheral immune system contains effector T lineage cells bearing the $\alpha\beta$ T cell receptor (TCR) which are composed of either class II-restricted CD4$^+$ T cells or class I-restricted CD8$^+$T cells, the intraepithelial lymphocytes are distinguished by the predominant presence of homodimeric CD8$\alpha\alpha^+$ T cells and T lineage cells containing the $\gamma\delta$TCR [51]. The $\gamma\delta$-T cell enriched Intraepithelial lymphocytes (IEL) function as a surveillance system for damaged or infected epithelial cells, and may modulate local immune responses by controlling cellular traffic and limiting mucosal access of inflammatory cells. Substantial numbers of $\alpha\beta$TCR T cells are present in lamina propria of the gastrointestinal tract that display an activated/memory phenotype consisting mainly of MHC class II-restricted CD4$^+$ T helper (Th) cells. However, the intestinal mucosa harbors all of the major Th-cell subsets (Th1, Th2, Th17, and Tfh) that are defined by their lineage-specific transcription factor expression, cytokine production, and subsequent immune function. Additionally, redundant regulatory strategies include naturally occurring and adaptive CD4$^+$CD25$^+$Foxp3$^+$ regulatory T cells. A functionally specialized population of CD103$^+$ dendritic cells that are enriched in the lamina propria of intestine and mesenteric lymph nodes (MLN) are highly effective in promoting the conversion of naive CD4$^+$ T cells into Foxp3$^+$ T cells in an antigen-specific manner and in maintaining the stability of preexisting Foxp3$^+$ cell population [52,53]. Foxp3$^+$ Treg cells are pivotal in the control of intestinal homeostasis and their action is achieved via a dual mechanism involving direct cell-to-cell interactions or the release of regulatory cytokines, TGF-β and IL-10 [54–56]. It is worthy of note that mice orally tolerized to a specific antigen may produce a great number of converted Foxp3$^+$ T cells in the Peyer's Patches (PP), lamina propria, and mesenteric lymph nodes [53].

The innate immune cells of the intestinal mucosa includes macrophages, dendritic cells, the M cells found on the dome epithelium of Peyer's Patches and recognized for their endocytic activity, that are involved in critical activities pertaining to the initiation of pathogen defense, maintenance of tissue homeostasis [57–59]. Gastrointestinal macrophages are distinguished by possessing cell surface Fc receptors that bind the Fc portion of IgG immunoglobulin, complement C3b and C3d receptors, MHC Class I and Class II, Toll like receptors (TLR) [60,61], the F4/80 marker [62], cytokine receptors and the CX$_3$CR1 receptor for fractalkine (CX3CL1). The chemokine receptor CX$_3$CR1 is widely expressed by macrophages, dendritic cells, T cells, and intestinal epithelial cells. Importantly, dendritic cells are involved in the control of Treg development and function in the gut. Research by Neiss and coworkers has demonstrated the critical importance of CX$_3$CR1$^+$ dendritic cells (DC) in the surveillance and defense against pathogens since these cells can directly sample the intestinal lumen by use of their transepithelial dendrites [63], representing a pathway that is distinct from previously established pathways of antigen transit from the gastrointestinal lumen involving the specialized M cells of the epithelial layer.

During a gastrointestinal immune response, ingested antigens in the lumen enter the Peyer's Patches via the specialized epithelial cells known as M cells present in the epithelial layer overlaying the PP. The M cells endocytose antigen, effectively transporting the antigen into the interior of the PP where dendritic cells and phagocytic cells process the antigen via intracellular metabolic and proteolytic degradation and modification. The resulting fragment [58,64], is transported to the surface of the cell where it is presented in conjunction with the major histocompatibility (MHC) gene molecule. The presentation of antigen is critical for the activation of the appropriate responding T cell, which contains a great variability of gene sequences, which must be rearranged to configure a mature, functional, TCR. This permits the specific recognition of the presented peptide sequence by the TCR of the responding T cell, and provides for the development of the adaptive immune response, initiating T cell activation and differentiation into antigen-specific effector cells. Antigen-specific FoxP3$^+$CD4$^+$ are critical for the control of effector responses in an antigen specific manner [65]. Following activation, the activated T cells migrate to the gut lamina propria effector site via an appropriate homing mechanism mediated by addressins-integrins [66]. The route of antigen sampling, the form and concentration of antigen, and the host's immunologic status (age, previous sensitizations) will determine the course of immune responsiveness [67].

Superantigenic Response in the Gastrointestinal Mucosa

In contrast, an immune response driven by a superantigen presents the scenario in which the requirements for antigen internalization, processing, and expression are circumvented due to the characteristic binding of this class of molecule. The defining characteristic of a superantigen is its unconventional binding to the MHC class II molecules, and to non-antigen-specific sequences found on the β chain of the T cell receptors (TCR) [68] of humans and mice. The staphylococcal enterotoxins are recognized as superantigens [69], as is the staphylococcal toxic shock toxin [70], the Mls murine self-antigens [71], and Mycoplasma arthritidis mitogen [72]. They are distinguished not only by their characteristic binding, but also by the ensuing induction of a potent T cell proliferation. SEB, as a superantigen, possesses the ability to bind directly to the α chain of major histocompatibility complex (MHC) class II glycoprotein [73], outside the peptide-binding groove of antigen presenting cells, effectively bypassing normal antigen processing and presenting mechanisms. However, the binding of some SAgs to MHC proteins can be distinguished by an additional zinc-mediated, higher-affinity binding site on the HLA DR β chain mediated by the conserved H81 histidine residue [74,75]. This interaction has been reported with SEA, SED, SEE, SEH [74–76], and can result in a crosslinking of the MHC on the surface of antigen-presenting cells [77]. The molecular interface of SEB with the αβT cell receptor occurs by SEB's attachment to the external Vβ domain in the TCR, resulting in the stimulation of a large proportion of CD4$^+$ and CD8$^+$ T cells (5%–30%) without regard to antigenic specificity or repertoire [27,31,78]. Thus, bound SEB interfaces with 5 residues (Tyr50, Ala52, Gly53, Thr55, and Ser54) within the complementarity determining region 2 (CDR2) of the TCR Vβ, 5 residues within FR3 (Ala67, Lys66, Tyr65, Lys57, Glu56), and to a lesser extent, two residues (Pro70, Ser71) within hypervariable region 4 (HV4), and a single His47 residue within framework region 2 (FR2). In contrast, the binding sites for SEC3 include 6 residues within CDR2 (Tyr50, Gly51, Ala52, Gly53, Ser54, Thr55), 3 residues within FR3 (Glu56, Lys57, Lys66), and 2 residues (Pro70, Ser71) within HV4 [79]. Significantly, these differences demonstrate that the specificity of the SAg binding to the TCR is primarily determined by complementarity determining region 2 (CDR2) of the TCR Vβ chain, and framework region 3 (FR3). The activated T cells proliferate and release massive proinflammatory cytokines that can lead to the development of clinical symptoms such as fever, swollen lymph nodes. Critically, the immunotoxic effects of the SEs are ultimately dependent upon the presence of reactive T lymphocytes. Thus, T cell-deficient nude mice have been shown to be protected from SE-induced shock, weight loss, and death [80]. Severe combined immunodeficiency (SCID) mice, which are deficient in T and B cells, appear to be resistant to the enteropathic effects of SEB. In this instance, intraperitoneal administration of SEB into either immunologically intact normal Balb/c mice or reconstituted SCID mice induces a jejunal histopathology characterized by reduced villus height, increased crypt depth, and upregulated MHC class II expression. However, the SE-induced enteropathy can only be achieved when the SCID mice are reconstituted with either mixed lymphocytes or with CD4$^+$ cells [81,82].

3. Immunopathology of the Staphylococcal enterotoxins within the Intestinal Mucosa

When ingested, SEs of the classic serotypes function both as potent gastrointestinal toxins and superantigens; however, they also possess the ability to traverse the intact intestinal epithelium. This was demonstrated using a mouse model using B10.BR mice that were inoculated orally with either SEA or SEB. Blood serum obtained from these animals induced IL2 production in T hybridomas bearing the appropriate surface TCR, providing strong evidence that the toxin(s) readily crossed the intact intestinal epithelium. Using an *in vitro* culture system, Hamad and coworkers further showed that MHC class II-negative human intestinal epithelial cells (Caco-2) can transcytose SEB and other bacterial superantigenic toxins in a dose-dependent manner [83]. Interestingly, mutated SEB containing either a mutation in the TCR biding site (N23) or a mutation in the area that binds the MHC class II molecule (F44) were weak stimulators of T cell proliferation [84]. In their study, Hamad and coworkers demonstrated a marked reduction in transcytosis, indicating the importance of these residues in transcytosis [83]. SEB exposure has also been shown to reduce the expression of mucosal tight-junction and adherent-junction proteins, leading to increased permeability and intestinal secretion [85,86]. The disturbance of the intestinal epithelial tight junction is probably associated with the enhanced secretion of IFN-γ and tumor necrosis factor (TNF) from lymphocytes. SEB treatment of rabbit intestinal segments instigates a rapid and large amount of structural destruction, and the damaging effect of SEB appeared to be more prominent in the small, rather than in the large, intestine. The primary target of the toxin was the epithelial cells along the length of the villi, followed by the lamina propria [87]. When ingested, SEs may cause emesis, exaggerated intestinal peristalsis, and pronounced alteration in intestinal mucosa structure when administered enterally in monkeys, dogs, or pigs [27]. The extent of the response reflects the toxin's superantigenic capacity, by which SEs activate gastrointestinal T cells and provoke a "cytokine storm". The numerous cytokines released include Th1 (IL-2 and IFN-γ), Th2 (IL-4), and Th17 (IL-17) cytokines from CD4$^+$ T lymphocytes as well as IL-1β, IL-6, IL-8, and TNF from activated macrophages and other cell types. These cytokines may act as chemoattractants and induce expression of adhesion molecules, favoring localization of diverse immune cells responding to SE challenge. Notably, distinct cytokines stimulate functional maturation of immune cells and boost their response to SEs. In addition, many cytokines can affect intestinal epithelial cell functions, particularly ion and water transport. Early research in which rats were fed, portrayed the striking events induced by the toxin within the gastrointestinal tract; specifically, a rapid inflammatory response, confined primarily to stomach and duodenum [88]. The upper gastrointestinal tract of the challenged rats displayed a marked infiltration of different types of inflammatory cells in the epithelium and lamina propria, with some evidence of subepithelial edema and watery exudate containing cells and mucus. In a mouse model, intragastric administration of SEB has been shown to induce an early activation and expansion of responsive Vβ8$^+$ T cells in PP and MLN [89]. RT-PCR analysis of cytokine mRNA in purified Vβ8$^+$ T cells showed that SEB significantly upregulated the mRNA expression of IL-2 and IFN-γ. In the MLN, mRNA specific for IL-2 was increased by 100-fold and IFN-γ mRNA was increased 20-fold. In the spleen, mRNA for IL-2 was increased 10-fold, while IFN-γ was increased 5-fold. Data derived from this laboratory has demonstrated that ingestion of SEB by C57Bl/10J mice can result in a dose-dependent induction of hyperplastic proliferation in peripheral lymphoid organs such as the spleen (Figure 1), resulting in a hyperplastic proliferation that alters the normal follicular architecture in spleen and Peyer's Patches (Figure 2) by 6 days after ingestion of the toxin [90] and a marked increase in apoptotic events as evidenced in both the spleen and Peyer's Patches of treated mice when compared to normals. Further, ingestion of the toxin will also induce dramatic changes with respect to the localization of B220$^+$ cells in the PP of the gastrointestinal tract [91]. In other research, intraperitoneal administration of SEB has triggered enterocolitis in mice [81,82] and rats [85,92,93], marked by intestinal recruitment of innate immune cells, activation of T helper lymphocytes, and enhanced release of TNF and IFN-γ. Intraepithelial lymphocytes are thought to play an important role in the pathophysiologic response to Staphylococcal infection. SEB challenge has also been shown to increase the γδ T lymphocyte population in PP, lamina propria, and epithelium [92]. Further, in the presence of SEB, human jejunal IEL have exhibited an enhanced *in vitro* cytotoxic effect on human C1R B-lymphoblastoid cells and IFN-γ pretreated colonic adenocarcinoma cell clone (HT-29). This response was greater than what is induced by using IEL that were activated with either phytohaemagglutinin (PHA)-, IL-2-, or anti-TCR antibody. In this report, the enhanced cytotoxicity demonstrated by SEB-treated IEL, did not engage MHC class II, TCR, or CD1d [94].

4. Staphylococcal enterotoxins and Immune Tolerance

Immune tolerance demonstrates a specific suppression of cellular or humoral responses that can be induced by repeated administration of an antigen, either at very large doses or at small doses that are below the stimulation threshold [67]. The two primary mechanisms that account for tolerance are clonal deletion (anergy) of antigen-specific effector cells and active suppression by regulatory T cells [95–97]. Oral tolerance is one of the most important forms of the induced tolerance, which is driven by prior administration of antigen by the oral route and presumably evolved to prevent destructive reactivity to normally occurring food proteins and commensal bacterial antigens in the intestinal mucosa. Multiple factors may contribute to the development of tolerance, including antigen properties, route of exposure, and host susceptibility and age. However, the antigen's dose is critical to the development of tolerance [98]. Larger doses of antigens lead to anergy [99] or deletion [97] of antigen-specific T cell clones, while induction of tolerance by small doses is mediated by regulatory cells (T reg). *In vivo* exposure of mice to SEs is has been implicated in inducing peripheral T cell tolerance. Mucosal introduction (by feeding) of SEA to neonatal mice has been shown to support the development of oral tolerance to OVA protein [100]. In these experiments, neonatal mice were exposed to SEA and fed OVA as adults, and intranasally challenged. Significantly, the animals demonstrated reduced production of serum IgE antibodies to the ingested challenge antigen and reduced inflammatory processes within the lungs. Other researchers have shown that oral administration of SEA and myelin basic protein results in an increase in Treg populations and IL-10 production [101].

Figure 1. Staphylococcal enterotoxin B (SEB) induces apoptosis in mouse Spleen. (**A**) Normal splenic tissue from a 12 week-old C57BL/10J mouse as seen in 10× magnification and 20× (**B**); Representative Peyer's patch tissue excised from a 12 week-old C57BL/10J mouse gavaged with SEB is shown in 20× magnifcation (**C**) and 40× (**D**). Tissues were excised 6 days following oral treatment. Immunohistochemical TUNEL staining of the tissues was performed where darkly stained cells are indicative of apoptosis. Note the darkly-stained apoptotic cells in the expanded follicular area of the SEB-treated spleen section. Staining was performed using a Trevigen TACS^R TdT In Situ Apoptosis Detection Kit.

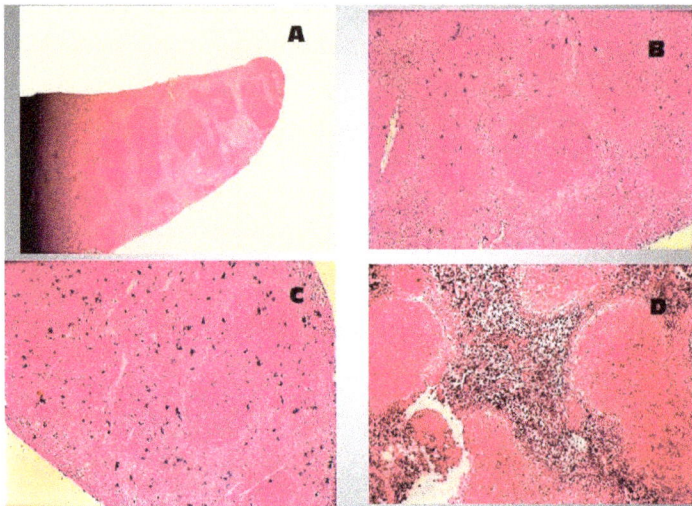

Figure 2. Staphylococcal enterotoxin B (SEB) induces apoptosis in mouse Peyer's patches. (**A**) Normal Peyer's patch tissue from a 12 week-old C57BL/10J mouse demonstrating germinal center containing lymphocytes; (**B**) Representative Peyer's patch tissue excised from a 12 week-old C57BL/10J mouse gavaged with SEB. Tissues were excised 6 days following oral treatment. Immunohistochemical TUNEL staining of the tissues was performed where darkly stained cells are indicative of apoptosis. Note the darkly-stained apoptotic cells in the expanded follicular area of the SEB-treated Peyer's patch section. Staining was performed using a Trevigen TACS^R TdT In Situ Apoptosis Detection Kit.

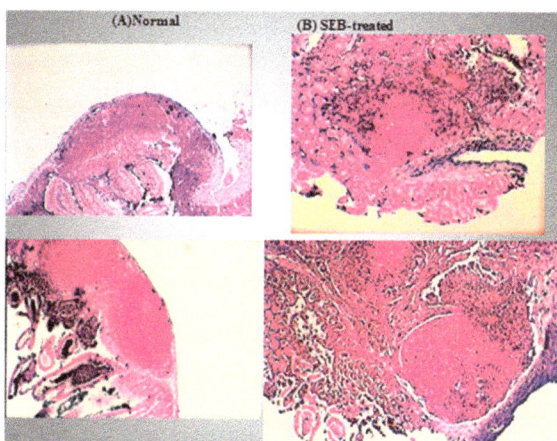

Mice injected with single dose of SEA demonstrate a rapid production of the Th1 proinflammatory cytokines IL-2, TNF, and IFN-γ, and serum IL-10 is not detectable. Repeated SEA challenges generated high levels of IL-10, but the production of IL-2, TNF, and IFN-γ was impaired and gradually eliminated. Interestingly, pretreatment of the mice with neutralizing anti-IL-10 mAb before the SEA challenges resulted in an increased IFN-γ and TNF response [102]. When stimulated with peptides *in vitro*, spleen cells from SEA-treated mice exhibited profound blocks in proliferation and IL-2 responses as compared to naïve spleen cells. Both SEA-treated CD4$^+$ cells and supernatants from cultures of stimulated SEA-treated spleen cells had the capacity to inhibit IL-2 production by stimulated naïve spleen cells [103]. The supernatant-mediated suppression could be blocked by addition of the antibodies against IL-10 and TGF-β. However, only anti-TGF-γ antibody could partially restore the IL-2 production in co-cultures with the SEA-treated spleen cells, suggesting that efficient suppression of the co-cultured primary-stimulated spleen cells might be largely mediated by other mechanisms not involving IL-10 and TGF-β, such as the direct contact of primary effector cells with the regulatory cells within the SEA-treated cell population.

Repeated SEA or SEB exposure may induce anergy and regulatory T cell function in mice. The *in vivo*-tolerized CD4$^+$ T cells are functionally similar to the natural Tregs, expressing elevated levels of CTLA and suppressing T cell proliferation [104,105]. Continuous stimulation of human CD4$^+$CD25$^-$ cells or CD8$^+$CD25$^-$ cells with SEB *in vitro* also leads to generation of adaptive Foxp3-expressing T cells with potent immuno-suppressive properties [106,107]. Spleen cells from mice made tolerant to SEA and/or SEB by repeated injection of the toxins were able to transfer their state of unresponsiveness to naive syngeneic recipient mice *in vivo* and to primary-stimulated T cells *in vitro* [108]. The production of IL-2 and IFN-γ following SE stimulation was greatly impaired in the animals adoptively transferred with the unresponsive spleen cells. Similarly, in the presence of CD4$^+$ T cells from the SE-tolerant animals, the primary-stimulated T cells showed significantly reduced cytokine responses to SE *in vitro* [108]. The mechanism of the regulatory function of the SE-unresponsive cells likely involves the expression of CTLA and the secretion of IL-10 [109]. Further study demonstrated that in mice made unresponsive to SEB after chronic exposure to low doses of the toxin, the reactive CD4$^+$TCR-Vβ8$^+$ T cells with regulatory function contained two subpopulations, *i.e.*, CD4$^+$CD25$^+$ Treg and CD4$^+$CD25$^-$ T cells [110]. The Treg cells control the induction phase of the tolerant state and control primary SEB-induced T cell proliferation. The tolerant state could not be reached in thymectomized CD25$^+$ cell-depleted mice and repeated injection of SEB was unable to protect these animals from lethal toxic shock. In contrast, CD4$^+$CD25$^-$ T regulatory cells which did not express Foxp3 and CTLA, have been shown to exert a regulatory effect upon the SEB response. Namely, addition of purified CD4$^+$CD25$^-$ from SEB-tolerant mice to primary cultures of normal spleen cells cultured with SEB resulted in a reduction of SEB-induced proliferation. These results indicate the regulatory role of the CD4$^+$CD25$^-$ cells in the maintenance of the SEB tolerant state. However, a recent study from Eroukhmanoff *et al.* argued that repeated immunization with SEB did not really upregulate Foxp3 expression. The increased frequency of Foxp3$^+$ cells in spleen and MLN of immune tolerant mice was actually due to a reduced number of antigen-reactive conventional CD4$^+$ T cells rather than the conversion of these cells to Foxp3$^+$ Treg [111]. Breakdown of oral tolerance may lead to the development of some autoimmune diseases, some of which may include inflammatory bowel disease processes, such as ulcerative colitis and Crohn's disease, and celiac disease.

5. Staphylococcal enterotoxins in Inflammatory Bowel Diseases

The etiologies of inflammatory bowel diseases such as ulcerative colitis and Crohn's disease remain unclear but an uncontrolled T cell-mediated immune response to non-pathogenic commensal luminal bacteria in a genetically susceptible host has been proposed [112]. A growing body of evidence suggests that the SEs are implicated in the pathogenesis of inflammatory bowel diseases. A subgroup of Crohn's disease patients display an increased number of Vβ8-expressing T cells, both CD4$^+$ and CD8$^+$, in peripheral blood and MLN [113]. A significant over-expression of Vβ5.1 and Vβ8 gene segments was also determined after long-term cultivation of colonic biopsies from Crohn's disease patients *versus* controls [114]. In that study, the cultures were set up under conditions that favor T cell growth. Since neither feeder cells nor any exogenous antigens that may modify Vβ gene expression were applied, T cell activation was solely dependent on biopsy-derived endogenous stimulants and the results indicated a prior *in vivo* exposure to Vβ8-selective superantigens. Indeed, *in vitro* stimulation of colonic explants from patients with Crohn's disease or ulcerative colitis with SEB or SEE led to the expansion of T cells expressing Vβ8 and a concomitant inflammatory cytokine release; the response was greater in the inflamed than non-inflamed tissues [114–116]. Interestingly, the cytotoxic function of Vβ8$^+$ cells seemed to be compromised in Crohn's disease patients [115]. While Vβ8$^+$ T cells seem to play a role in Crohn's disease, selective expansion of T cells bearing TCR-Vβ4 has been demonstrated in patients with ulcerative colitis. Patients with skewed TCR-Vβ4-expressing cells are reported to have longer disease duration as compared with patients with low level of TCR-Vβ4 [117]. In this instance, Streptococcal mitogenic exotoxin Z-2 (SMEZ-2), produced by Streptococcus pyogene, is known to preferentially activate the TCR-Vβ4-bearing T cells. Patients with skewed TCR-Vβ4demonstrate significantly higher level of anti-SMEZ-2 antibodies, suggesting an intimate relationship between the bacterial toxin and the bowel disease status. Another pathway for the induction of inflammatory bowel disease has been associated with pre-existing chronic rhinosinusitis by a mechanism of swallowing sinusitis-derived SEB. Early studies documented an increased frequency of enterotoxigenic *Staphylococcus aureus* in the nares of individuals diagnosed with allergic rhinitis [118]. Patients with both ulcerative colitis and chronic rhinosinusitis present significant sinus infection with *Staphylococcus aureus*, accompanied by high levels of SEB in the sinus wash fluids, anti-SEB antibody in the sera, and anti-SEB positive-cells in the colonic mucosa [7]. After functional endoscopic sinus surgery for rhinosinusitis, these patients showed ameliorated intestinal inflammation. Bacteriologic analysis of removed sinus mucosal specimens showed that the colony number of cultured *Staphylococcus aureus* correlated with the decrease in disease severity of ulcerative colitis. When cultured *in vitro* with SEB, colonic biopsy-derived mast cells from the patients with both ulcerative colitis and chronic rhinosinusitis demonstrated massive degranulation and release of histamine and tryptase, as compared to the patients with ulcerative colitis only, or the healthy controls. Similarly, introducing sinusitis-derived SEB to murine gastrointestinal tract also caused increased colonic epithelial permeability and impaired mucosal barrier function [119]. Mice sensitized by intra-gastric gavage of ovalbumin (OVA) in the presence of SEB-containing sinus wash fluid from patients with chronic rhinosinusitis and then challenged with OVA, developed mucosal immunopathology in the colon, evidenced by marked degranulation of mast cells and eosinophils, infiltration of inflammatory cells, mucosal ulceration, and abscess formation in the lamina propria. In contrast, the inflammatory state was not induced in the mice sensitized to OVA only or when anti-SEB antibody was added to the sensitizing mixture. Likewise, intra-rectal administration of SEA and SEB resulted in colonic inflammation in a time- and dose-dependent manner. Moreover this treatment aggravated the illness in mice recovering from dextran-sodium sulfate-induced colitis [120]. In a SCID mouse model of colitis, feeding SEB to mice reconstituted with CD4$^+$CD45RB^{high} T cells resulted in an earlier onset of intestinal inflammation and more severe symptoms, which was accompanied by activation and expansion of SEB-reactive CD4$^+$Vβ8$^+$ T cells and marked impairment in CD4$^+$CD25$^+$Foxp3$^+$ Treg cell development .

6. Conclusions

The SEs exert remarkable responsiveness in the gastrointestinal tract where they provoke diverse innate and adaptive immune events involving both exaggerated inflammatory and tolerant states. SEs are not only incriminated in food poisoning episodes but are also implicated in various inflammatory and autoimmune conditions such as those reviewed in this paper. That the toxins are able to exert such a diversity of activities in the digestive tract is extraordinary, but is evidently associated with the unique local microenvironment. Despite intensive efforts, the nature of the intertwined interaction between SEs and the different varieties of cells, cytokines, signaling molecules, and transcription factors in intestinal mucosa is yet to be fully understood. The precise mechanisms that trigger the inflammatory or tolerance processes are fundamentally unknown. Although it remains to be established whether targeting SEs, either to block their superantigenicity or to use as a tool to generate immune tolerance to disease-inducing proteins, is appropriate or adequate, this approach has the potential to provide a therapeutic benefit in human food-borne and immune-based gastrointestinal diseases.

Acknowledgments

The authors wish to thank Jeffrey Yourick and Robert Sprando for the review of the manuscript. This work was supported in part by a fund from the US Food and Drug Administration Commissioner's Fellowship program.

Conflicts of Interest

The authors report no declarations of interest and do not have any financial or personal association with any individuals or organizations that could inappropriately influence the submitted work.

Declare

This article reflects the views of the authors and should not be construed to represent views or policies of the U.S. Food and Drug Administration.

References

1. Friedman, S.M.; Tumang, J.R.; Crow, M.K. Microbial superantigens as etiopathogenic agents in autoimmunity. *Rheum. Dis. Clin. N. Am.* **1993**, *19*, 207–222.
2. Friedman, S.M.; Posnett, D.N.; Tumang, J.R.; Cole, B.C.; Crow, M.K. A potential role for microbial superantigens in the pathogenesis of systemic autoimmune disease. *Arthr. Rheum.* **1991**, *34*, 468–480.
3. Lee, J.H.; Lin, Y.T.; Yang, Y.H.; Wang, L.C.; Chiang, B.L. Increased levels of serum-specific immunoglobulin E to staphylococcal enterotoxin A and B in patients with allergic rhinitis and bronchial asthma. *Int. Arch. Allergy Immunol.* **2005**, *138*, 305–311.
4. Bachert, C.; Gevaert, P.; van Cauwenberge, P. Staphylococcus aureus enterotoxins: A key in airway disease? *Allergy* **2002**, *57*, 480–487.
5. Orfali, R.L.; Sato, M.N.; Takaoka, R.; Azor, M.H.; Rivitti, E.A.; Hanifin, J.M.; Aoki, V. Atopic dermatitis in adults: Evaluation of peripheral blood mononuclear cells proliferation response to Staphylococcus aureus enterotoxins A and B and analysis of interleukin-18 secretion. *Exp. Dermatol.* **2009**, *18*, 628–633.

6. Yudate, T.; Yamada, H.; Tezuka, T. Role of staphylococcal enterotoxins in pathogenesis of atopic dermatitis: Growth and expression of T cell receptro Vbeta of peripheral blood mononuclear cells stimulated by enterotoxins A and B. *J. Dermatol. Sci.* **1996**, *13*, 63–70.

7. Yang, P.-C.; Liu, T.; Wang, B.-Q.; Zhang, T.-Y.; An, Z.-Y.; Zheng, P.-Y.; Tian, D.-F. Rhinosinusitis derived Staphylococcal enterotoxin B possibly associates with pathogenesis of ulcerative colitis. *BMC Gastroenterol.* **2005**, *5*, doi:10.1186/1471-230X-5-28.

8. Rossi, R.E.; Monasterolo, G. Prevalence of serum IgE antibodies to the Staphylococcus aureus enterotoxins (SAE, SEB, SEC, SED, TSST-1) in patients with persistent allergic rhinitis. *Int. Arch. Allergy Immunol.* **2004**, *133*, 261–266.

9. Forbes-Blom, E.; Camberis, M.; Prout, M.; Tang, S.C.; le Gros, G. Staphylococcal-derived superantigen enhances peanut induced Th2 responses in the skin. *J. Br. Soc. Allergy Clin. Immunol.* **2012**, *42*, 305–314.

10. Bergdoll, M.S. Importance of Staphylococci that produce nanogram quantities of enterotoxin. *Zbl. Bakt.* **1995**, *282*, 1–6.

11. Breckinridge, J.C.; Bergdoll, M.S. Outbreak of food-borne gastroenteritis due to a coagulase-negative enterotoxin-producing staphylococcus. *N. Engl. J. Med.* **1971**, *284*, 541–543.

12. Marrack, P.; Kappler, J. The staphylococcal enterotoxins and their relatives. *Science* **1990**, *248*, 705–711.

13. Krakauer, T. Update on staphylococcal superantigen-induced signaling pathways and therapeutic interventions. *Toxins* **2013**, *5*, 1629–1654.

14. Jarraud, S.; Peyrat, M.A.; Lim, A.; Tristan, A.; Bes, M.; Mougel, C.; Etienne, J.; Vandenesch, F.; Bonneville, M.; Lina, G. EGC, a highly prevalent operon of enterotoxin gene, forms a putative nursery of superantigens in Staphylococcus aureus. *J. Immunol.* **2001**, *166*, 669–677.

15. Su, Y.C.; Wong, A.C. Identification and purification of a new staphylococcal enterotoxin, H. *Appl. Environ. Microbiol.* **1995**, *61*, 1438–1443.

16. Chiang, Y.C.; Chang, L.T.; Lin, C.W.; Yang, C.Y.; Tsen, H.Y. PCR primers for the detection of staphylococcal enterotoxins K, L, and M and survey of staphylococcal enterotoxin types in Staphylococcus aureus isolates from food poisoning cases in Taiwan. *J. Food Prot.* **2006**, *69*, 1072–1079.

17. Orwin, P.M.; Leung, D.Y.; Donahue, H.L.; Novick, R.P.; Schlievert, P.M. Biochemical and biological properties of Staphylococcal enterotoxin K. *Infect. Immun.* **2001**, *69*, 360–366.

18. Chiang, Y.-C.; Liao, W.-W.; Fan, C.-M.; Pai, W.-Y.; Chiou, C.-S.; Tsen, H.-Y. PCR detection of Staphylococcal enterotoxins (SEs) N, O, P, Q, R, U, and survey of SE types in Staphylococcus aureus isolates from food-poisoning cases in Taiwan. *Int. J. Food Microbiol.* **2008**, *121*, 66–73.

19. Omoe, K.; Imanishi, K.; Hu, D.-L.; Kato, H.; Takahashi-Omoe, H.; Nakane, A.; Uchiyama, T.; Shinagawa, K. Biological properties of staphylococcal enterotoxin-like toxin type R. *Infect. Immun.* **2004**, *72*, 3664–3667.

20. Letertre, C.; Perelle, S.; Dilasser, F.; Fach, P. Identification of a new putative enterotoxin SEU encoded by the egc cluster of Staphylococcus aureus. *J. Appl. Microbiol.* **2003**, *95*, 38–43.

21. Asao, T.; Kumeda, Y.; Kawai, T.; Shibata, T.; Oda, H.; Haruki, K.; Nakazawa, H.; Kozaki, S. An extensive outbreak of staphylococcal food poisoning due to low-fat milk in Japan: Estimation of enterotoxin A in the incriminated milk and powdered skim milk. *Epidemiol. Infect.* **2003**, *130*, 33–40.

22. Becker, K.; Friedrich, A.W.; Lubritz, G.; Weilert, M.; Peters, G.; von Eiff, C. Prevalence of genes encoding pyrogenic toxin superantigens and exfoliative toxins among strains of Staphylococcus aureus isolated from blood and nasal specimens. *J. Clin. Microbiol.* **2003**, *41*, 1434–1439.

23. Do Carmo, L.S.; Cummings, C.; Linardi, V.R.; Dias, R.S.; De Souza, J.M.; De Sena, M.J.; Dos Santos, D.A.; Shupp, J.W.; Pereira, R.K.P.; Jett, M. A case study of a massive staphylococcal food poisoning incident. *Foodborne Pathog. Dis.* **2004**, *1*, 241–246.

24. Evenson, M.L.; Hinds, M.W.; Bernstein, R.S.; Bergdoll, M.S. Estimation of human dose of staphylococcal enterotoxin A from a large outbreak of staphylococcal food poisoning involving chocolate milk. *Int. J. Food Microbiol.* **1988**, *7*, 311–316.

25. Levine, W.C.; Bennett, R.W.; Choi, Y.; Henning, K.J.; Rager, J.R.; Hendricks, K.A.; Hopkins, D.P.; Gunn, R.A.; Griffin, P.M. Staphylococcal food poisoning caused by imported canned mushrooms. *J. Infect. Dis.* **1996**, *173*, 1263–1267.

26. Hennekinne, J.-A.; Ostyn, A.; Guillier, F.; Herbin, S.; Prufer, A.-L.; Dragacci, S. How should staphylococcal food poisoning outbreaks be characterized? *Toxins* **2010**, *2*, 2106–2116.

27. Pinchuk, I.V.; Beswick, E.J.; Reyes, V.E. Staphylococcal enterotoxins. *Toxins* **2010**, *2*, 2177–2197.

28. Omoe, K.; Hu, D.-L.; Takahashi-Omoe, H.; Nakane, A.; Shinagawa, K. Comprehensive analysis of classical and newly described staphlyococcal superantigenic toxin genes in Staphylococcus aureus isolates. *FEMS Microbiol. Lett.* **2005**, *246*, 191–198.

29. Yarwood, J.M.; McCormick, J.K.; Paustian, M.L.; Orwin, P.M.; Kapur, V.; Schlievert, P.M. Characterization and expression analysis of Staphylococcus aureus pathogenicity island 3. Implications for the evolution of staphylococcal pathogenicity islands. *Biol. Chem.* **2002**, *277*, 13138–13147.

30. Fleischer, B.; Schrezenmeier, H. T cell stimulation by staphylococcal enterotoxins. Clonally variable response and requirement for major histocompatibility complex class II molecules on accessory or target cells. *J. Exp. Med.* **1988**, *167*, 1697–1707.

31. Kappler, J.; Kotzin, B.; Herron, L.; Gelfand, E.W.; Bigler, R.D.; Boylston, A.; Carrel, S.; Posnett, D.N.; Choi, Y.; Marrack, P. V beta-specific stimulation of human T cells by staphylococcal toxins. *Science* **1989**, *244*, 811–813.

32. Mead, P.S.; Slutsker, L.; Dietz, V.; McCaig, L.F.; Bresee, J.S.; Shapiro, C.; Griffin, P.M.; Tauxe, R.V. Food-related illness and death in the United States. *Emerg. Infect. Dis.* **1999**, *5*, 607–625.

33. Zorgani, A.; Essery, S.D.; Madani, O.A.; Bentley, A.J.; James, V.S.; MacKenzie, D.A.C.; Keeling, J.W.; Rambaud, C.; Hilton, J.; Blackwell, C.C. Detection of pyrogenic toxins of *Staphylococcus aureus* in sudden infant death syndrome. *FEMS Immunol. Med. Microbiol.* **1999**, *25*, 103–108.

34. Principato, M. Infant Formulas and Feeding:Risks Associated with Staphylococcus aureus and Its Enterotoxins. In *Dietary and Nutritional Aspects of Bottle Feeding*; Preedy, V., Ed.; Wageningen Academic Publishers: Wageningen, The Netherlands, 2014.

35. Lawrynowicz-Paciorek, M.; Kochman, M.; Piekarska, K.; Grochowska, A.; Windyga, B. The distribution of enterotoxin and enterotoxin-like genes in *Staphylococcus aureus* strains isolated from nasal carriers and food samples. *Int. J. Food Microbiol.* **2007**, *117*, 319–323.

36. Holmberg, S.D.; Blake, P.A. Staphylococcal food poisoning in the United States. New facts and old misconceptions. *JAMA* **1984**, *251*, 487–489.

37. Udo, E.E.; Al-Bustan, M.A.; Jacob, L.E.; Chugh, T.D. Enterotoxin production by coagulase-negative staphylococci in restaurant workers from Kuwait City may be a potential cause of food poisoning. *J. Med. Microbiol.* **1999**, *48*, 819–823.

38. Le Loir, Y.; Baron, F.; Gautier, M. Staphylococcus aureus and food poisoning. *Genet. Mol. Res.* **2003**, *2*, 63–76.

39. Read, R., Jr.; Bradshaw, J. Staphylococcal enterotoxin B thermal inactivation in milk. *J. Dairy Sci.* **1966**, *49*, 202–203.

40. Principato, M.; Boyle, T.; Njoroge, J.; Jones, R.L.; O'Donnell, M. Effect of thermal processing during yogurt production upon the detection of staphylococcal enterotoxin B. *J. Food Protect.* **2009**, *72*, 2212–2216.

41. Raj, H.D.; Bergdoll, M.S. Effect of enterotoxin B on human volunteers. *J. Bacteriol.* **1969**, *98*, 833–834.

42. Ikeda, T.; Tamate, N.; Yamaguchi, K.; Makino, S. Mass outbreak of food poisoning disease caused by small amounts of staphylococcal enterotoxins A and H. *Appl. Environ. Microbiol.* **2005**, *71*, 2793–2795.

43. Balaban, N.; Rasooly, A. Staphylococcal enterotoxins. *Int. J. Food Microbiol.* **2000**, *61*, 1–10.

44. Dinges, M.M.; Orwin, P.M.; Schlievert, P.M. Exotoxins of *Staphylococcus aureus*. *Clin. Microbiol. Rev.* **2000**, *13*, 16–34.

45. Danielsen, E.M.; Hansen, G.H.; Karlsdottir, E. Staphylococcus aureus enterotoxins A and B: Binding to the enterocyte brush border and uptake by perturbation of the apical endocytic membrane traffic. *Histochem. Cell Biol.* **2013**, *139*, 513–524.

46. Edwards, L.A.; O'Neill, C.; Furman, M.A.; Hicks, S.; Torrente, F.; Pérez-Machado, M.; Wellington, E.M.; Phillips, A.D.; Murch, S.H. Enterotoxin-producing staphylococci cause intestinal inflammation by a combination of direct epithelial cytopathy and superantigen-mediated T cell activation. *Inflamm. Bowel Dis.* **2012**, *18*, 624–640.

47. Hu, D.-L.; Zhu, G.; Mori, F.; Omoe, K.; Okada, M.; Wakabayashi, K.; Kaneko, S.; Shinagawa, K.; Nakane, A. Staphylococcal enterotoxin induces emesis through increasing serotonin release in intestine and it is downregulated by cannabinoid receptor 1. *Cell Microbiol.* **2007**, *9*, 2267–2277.

48. Luongo, D.; D'Arienzo, R.; Bergamo, P.; Maurano, F.; Rossi, M. Immunomodulation of gut-associated lymphoid tissue: Current perspectives. *Int. Rev. Immunol.* **2009**, *28*, 446–464.

49. Van Wijk, F.; Cheroutre, H. Intestinal T cells: Facing the mucosal immune dilemma with synergy and diversity. *Semin. Immunol.* **2009**, *21*, 130–138.

50. Strober, W.; Kelsall, B.; Marth, T. Oral tolerance. *J. Clin. Immunol.* **1998**, *18*, 1–30.

51. LeFrancois, L.; Lynn, P. *Basic Aspects of Intraepithelial Lymphocytes Immunobiology*; Academic Press: San Diego, CA, USA, 1999.

52. Coombes, J.L.; Siddiqui, K.R.; Arancibia-Carcamo, C.V.; Hall, J.; Sun, C.M.; Belkaid, Y.; Powrie, F. A functionally specialized population of mucosal CD103+ DCs induces Foxp3+ regulatory T cells via a TGF-beta and retinoic acid-dependent mechanism. *J. Exp. Med.* **2007**, *204*, 1757–1764.

53. Sun, C.M.; Hall, J.A.; Blank, R.B.; Bouladoux, N.; Oukka, M.; Mora, J.R.; Belkaid, Y. Small intestine lamina propria dendritic cells promote de novo generation of Foxp3 T reg cells via retinoic acid. *J. Exp. Med.* **2007**, *204*, 1775–1785.

54. Chen, W.; Jin, W.; Hardegen, N.; Lei, K.J.; Li, L.; Marinos, N.; McGrady, G.; Wahl, S.M. Conversion of peripheral CD4+CD25− naive T cells to CD4+CD25+ regulatory T cells by TGF-beta induction of transcription factor Foxp3. *J. Exp. Med.* **2003**, *198*, 1875–1886.

55. Li, M.O.; Flavell, R.A. TGF-beta: A master of all T cell trades. *Cell* **2008**, *134*, 392–404.

56. Marie, J.C.; Liggitt, D.; Rudensky, A.Y. Cellular mechanisms of fatal early-onset autoimmunity in mice with the T cell-specific targeting of transforming growth factor-beta receptor. *Immunity* **2006**, *25*, 441–454.

57. Aggeler, J.; and Werb, Z. Initial events during phagocytosis by macrophages viewed from the outside and inside the cell: Membrane-particle interactions and clathrin. *J. Cell Biol.* **1982**, *94*, 613–623.

58. Unanue, E.R.; Allen, P.M. The basis for the immunoregulatory role of macrophages and other accessory cells. *Science* **1987**, *236*, 551–557.

59. Unanue, E.R.; Ungewickell, E.; Branton, D. The binding of clathrin triskelions to membranes from coated vesicles. *Cell* **1981**, *26*, 439–446.

60. Kawai, T.; Akira, S. The role of pattern-recogntion receptors in innate immunity: Update on toll-like receptors. *Nat. Immunol.* **2010**, *11*, 373–384.

61. Janeway, C.A., Jr.; Medzhitov, R. Innate immune recognition. *Annu. Rev. Immunol.* **2002**, *20*, 197–216.

62. Gordon, S.; Hamann, J.; Lin, H.-H.; Stacey, M. F4/80 and the related adhesion-GPCRs. *Eur. J. Immunol.* **2011**, *41*, 2472–2476.

63. Neiss, J.H.; Brand, S.; Giu, X.; Landsman, L.; Jung, S.; McCormick, B.A.; Vyas, J.; Boes, M.; Plooegh, H.; Fox, J. CX3CR1-mediated dendritic cell access to the intestinal lumen and bacterial clearance. *Science* **2005**, *307*, 254–258.

64. Zeigler, K.; Unanue, E.R. Identification of a macrophage antigen-processing event required for I-region-restricted antigen presentation to T lymphocytes. *J. Immunol.* **1981**, *127*, 1869–1875.

65. Hadis, U.; Wahl, B.; Schulz, O.; Hardtke-Wolenski, M.; Schippers, A.; Wagner, N.; Muller, W.; Sparwasser, T.; Forster, R.; Pabst, O. Intestinal tolerance requires gut homing and expansion of FoxP3+ regulatory T cells in the lamina propria. *Immunity* **2011**, *34*, 237–246.

66. McGhee, J.R.; Kunisawa, J.; Kiyono, H. Gut lymphocyte migration: We are halfway 'home'. *Trends Immunol.* **2007**, *28*, 150–153.

67. Mayer, L.; Sperber, K.; Chan, L.; Child, J.; Toy, L. Oral tolerance to protein antigens. *Allergy* **2001**, *56*, 12–15.

68. Dellabona, P.; Peccoud, J.; Kappler, J.; Marrack, P.; Benoist, C.; Mathis, D. Superantigens interact with MHC class II molecules outside of the antigen groove. *Cell* **1990**, *62*, 1115–1121.

69. White, J.; Herman, A.; Pullen, A.M.; Kubo, R.; Kappler, J.W.; Marrack, P. The V beta specific superantigen staphylococcal enterotoxin B: Stimulation of mature T cells and clonaldeletion in neonatal mice. *Cell* **1989**, *56*, 27.

70. Acharya, K.R.; Passalacqua, E.F.; Jones, E.Y.; Harlos, K.; Stuart, D.I.; Brehm, R.D.; Tranter, H.S. Structural basis of superantigen action inferred from crystal structure of toxic-shock syndrome toxin-1. *Nature* **1994**, *367*, 94–97.

71. MacDonald, H.R.; Schnieder, R.; Lees, R.K.; Howe, R.C.; Acha-Orbea, H.; Festenstein, F.; Zinkernagel, R.M. T cell receptor V beta use predicts reactivity andtolerance to Mls-encoded antigens. *Nature* **1988**, *332*, 40–45.

72. Atkins, C.L.; Cole, B.C.; Sullivan, G.J.; Washburn, L.R.; Wiley, B.B. Stimulation of mouse lymphocytes by a mitogen derived from M. arthriditis. *J. Immunol.* **1986**, *137*, 1581.

73. Jardetzky, T.S.; Brown, J.H.; Gorga, J.C.; Stern, L.J.; Urban, R.G.; Chi, Y.I.; Stauffacher, C.; Strominger, J.L.; Wiley, D.C. Three-dimensional structure of a human class II histocompatibility molecule complexed with superantigen. *Nature* **1994**, *368*, 711–718.

74. Hudson, K.R.; Tiedemann, R.E.; Urban, R.G.; Lowe, S.C.; Strominger, J.L.; Fraser, J.D. Staphylococcal enterotoxin A has two cooperative binding sites on major histocompatibility complex class II. *J. Exp. Med.* **1995**, *182*, 711–720.

75. Sundström, M.; Abrahmsén, L.; Antonsson, P.; Mehindate, K.; Mourad, W.; Dohlsten, M. The crystal structure of staphylococcal enterotoxin type D reveals Zn^{2+}-mediated homodimerization. *EMBO J.* **1996**, *15*, 6832–6840.

76. Proft, T.; Fraser, J.D. Bacterial superantigens. *Clin. Exp. Immunol.* **2003**, *133*, 299–306.

77. Tiedemann, R.E.; Fraser, J.D. Cross-linking of MHC class II molecules by staphylococcal enterotoxin A is essential for antigen-presenting cell and T cell activation. *J. Immunol.* **1996**, *157*, 3958–3966.

78. Krakauer, T. Chemotherapeutics targeting immune activation by staphylococcal superantigens. *Med. Sci. Monit.* **2005**, *11*, 290–295.

79. Li, H.; Llera, A.; Tsuchiya, D.; Leder, L.; Ysern, X.; Schlievert, P.M.; Karjalainen, K.; Mariuzza, R.A. Three-dimensional structure of the complex between a T cell receptor β chain and the superantigen staphylococcal enterotoxin B. *Immunity* **1998**, *9*, 807–816.

80. Marrack, P.; Blackman, M.; Kushnir, E.; Kappler, J. The toxicity of staphylococcal enterotoxin B in mice is mediated by T cells. *J. Exp. Med.* **1990**, *171*, 455–464.

81. Benjamin, M.A.; Lu, J.; Donnelly, G.; Dureja, P.; McKay, D.M. Changes in murine jejunal morphology evoked by the bacterial superantigen Staphylococcus aureus enterotoxin B are mediated by $CD4^+$ T cells. *Infect. Immun.* **1998**, *66*, 2193–2199.

82. McKay, D.M.; Benjamin, M.A.; Lu, J. $CD4^+$ T cells mediate superantigen-induced abnormalities in murine jejunal ion transport. *Am. J. Physiol.* **1998**, *275*, G29–G38.

83. Hamad, A.R.; Marrack, P.; Kappler, J.W. Transcytosis of staphylococcal superantigen toxins. *J. Exp. Med.* **1997**, *185*, 1447–1454.

84. Kappler, J.; Herman, A.; Clements, J.; Marrack, P. Mutations defining functional regions of hte superantigen staphylcoccal enterotoxin B. *J. Exp. Med.* **1992**, *175*, 387–396.

85. Moreto, M.; Perez-Bosque, A. Dietary plasma proteins, the intestinal immune system, and the barrier functions of the intestinal mucosa. *J. Anim. Sci.* **2009**, *87*, E92–E100.

86. Perez-Bosque, A.; Moreto, M. A rat model of mild intestinal inflammation induced by Staphylococcus aureus enterotoxin B. *Proc. Nutr. Soc.* **2010**, *69*, 447–453.

87. Kamaras, J.; Murrell, W.G. The effect of bacterial enterotoxins implicated in SIDS on the rabbit intestine. *Pathology* **2001**, *33*, 187–196.

88. Beery, J.T.; Taylor, S.L.; Schlunz, L.R.; Freed, R.C.; Bergdoll, M.S. Effects of staphylococcal enterotoxin A on the rat gastrointestinal tract. *Infect. Immun.* **1984**, *44*, 234–240.

89. Spiekermann, G.M.; Nagler-Anderson, C. Oral administration of the bacterial superantigen staphylococcal enterotoxin B induces activation and cytokine production by T cells in murine gut-associated lymphoid tissue. *J. Immunol.* **1998**, *161*, 5825–5831.

90. Principato, M.A.. Tissue-specific immune responsiveness to Staphylococcal enterotoxin in the aged gut lymphatics. *FASEB J.* **1999**, *13*, A291.

91. Principato, M. Gastrointestinal Immunoregulation and the Challenges of Nanotechnology in Foods. In *Food Industry*; Muzzalupo, I., Ed.; InTech: Rijeka , Croatia, 2013.

92. Perez-Bosque, A.; Miro, L.; Polo, J.; Russell, L.; Campbell, J.; Weaver, E.; Crenshaw, J.; Moreto, M. Dietary plasma proteins modulate the immune response of diffuse gut-associated lymphoid tissue in rats challenged with Staphylococcus aureus enterotoxin B. *J. Nutr.* **2008**, *138*, 533–537.

93. Perez-Bosque, A.; Pelegri, C.; Vicario, M.; Castell, M.; Russell, L.; Campbell, J.M.; Quigley, J.D., 3rd, Polo, J.; Amat, C.; Moreto, M. Dietary plasma protein affects the immune response of weaned rats challenged with S. aureus Superantigen B. *J. Nutr.* **2004**, *134*, 2667–2672.

94. Roberts, A.I.; Blumberg, R.S.; Christ, A.D.; Brolin, R.E.; Ebert, E.C. Staphylococcal enterotoxin B induces potent cytotoxic activity by intraepithelial lymphocytes. *Immunology* **2000**, *101*, 185–190.

95. Faria, A.M.; Weiner, H.L. Oral tolerance. *Immunol. Rev.* **2005**, *206*, 232–259.

96. Friedman, A.; Weiner, H.L. Induction of anergy or active suppression following oral tolerance is determined by antigen dosage. *Proc. Natl. Acad. Sci. USA* **1994**, *91*, 6688–6692.

97. Chen, Y.; Inobe, J.; Marks, R.; Gonnella, P.; Kuchroo, V.K.; Weiner, H.L. Peripheral deletion of antigen-reactive T cells in oral tolerance. *Nature* **1995**, *376*, 177–180.

98. Weiner, H.L.; da Cunha, A.P.; Quintana, F.; Wu, H. Oral tolerance. *Immunol. Rev.* **2011**, *241*, 241–259.

99. Whitacre, C.C.; Gienapp, I.E.; Orosz, C.G.; Bitar, D.M. Oral tolerance in experimental autoimmune encephalomyelitis. III. Evidence for clonal anergy. *J. Immunol.* **1991**, *147*, 2155–2163.

100. Lonnqvist, A.; Ostman, S.; Almqvist, N.; Hultkrantz, S.; Telemo, E.; Wold, A.E.; Rask, C. Neonatal exposure to staphylococcal superantigen improves induction of oral tolerance in a mouse model of airway allergy. *Eur. J. Immunol.* **2009**, *39*, 447–456.

101. Miron, N.; Feldrihan, V.; Berindan-Neagoe, I.; Cristea, V. The role of staphylococcal enterotoxin A in acheiving oral tolerance to myelin basic protein in adult mice. *Immunol. Invest.* **2014**, *43*, 267–277.

102. Sundstedt, A.; Hoiden, I.; Rosendahl, A.; Kalland, T.; van Rooijen, N.; Dohlsten, M. Immunoregulatory role of IL-10 during superantigen-induced hyporesponsiveness *in vivo*. *J. Immunol.* **1997**, *158*, 180–186.

103. Miller, C.; Ragheb, J.A.; Schwartz, R.H. Anergy and cytokine-mediated suppression as distinct superantigen-induced tolerance mechanisms *in vivo*. *J. Exp. Med.* **1999**, *190*, 53–64.

104. Grundstrom, S.; Cederbom, L.; Sundstedt, A.; Scheipers, P.; Ivars, F. Superantigen-induced regulatory T cells display different suppressive functions in the presence or absence of natural CD4$^+$CD25$^+$ regulatory T cells *in vivo*. *J. Immunol.* **2003**, *170*, 5008–5017.

105. Schartner, J.M.; Singh, A.M.; Dahlberg, P.E.; Nettenstrom, L.; Seroogy, C.M. Recurrent superantigen exposure *in vivo* leads to highly suppressive CD4$^+$CD25$^+$ and CD4$^+$CD25$^-$ T cells with anergic and suppressive genetic signatures. *Clin. Exp. Immunol.* **2009**, *155*, 348–356.

106. Mahic, M.; Henjum, K.; Yaqub, S.; Bjornbeth, B.A.; Torgersen, K.M.; Tasken, K.; Aandahl, E.M. Generation of highly suppressive adaptive CD8(+)CD25(+)FOXP3(+) regulatory T cells by continuous antigen stimulation. *Eur. J. Immunol.* **2008**, *38*, 640–646.

107. Mahic, M.; Yaqub, S.; Bryn, T.; Henjum, K.; Eide, D.M.; Torgersen, K.M.; Aandahl, E.M.; Tasken, K. Differentiation of naive CD4$^+$ T cells into CD4$^+$CD25$^+$FOXP3$^+$ regulatory T cells by continuous antigen stimulation. *J. Leukoc. Biol.* **2008**, *38*, 1111–1117.

108. Noel, C.; Florquin, S.; Goldman, M.; Braun, M.Y. Chronic exposure to superantigen induces regulatory CD4(+) T cells with IL-10-mediated suppressive activity. *Int. Immunol.* **2001**, *13*, 431–439.

109. Feunou, P.; Vanwetswinkel, S.; Gaudray, F.; Goldman, M.; Matthys, P.; Braun, M.Y. Foxp3$^+$CD25$^+$ T regulatory cells stimulate IFN-gamma-independent CD152-mediated activation of tryptophan catabolism that provides dendritic cells with immune regulatory activity in mice unresponsive to staphylococcal enterotoxin B. *J. Immunol.* **2007**, *179*, 910–917.

110. Feunou, P.; Poulin, L.; Habran, C.; Le Moine, A.; Goldman, M.; Braun, M.Y. CD4$^+$CD25$^+$ and CD4$^+$CD25$^-$ T cells act respectively as inducer and effector T suppressor cells in superantigen-induced tolerance. *J. Immunol.* **2003**, *171*, 3475–3484.

111. Eroukhmanoff, L.; Oderup, C.; Ivars, F. T-cell tolerance induced by repeated antigen stimulation: Selective loss of Foxp3$^-$ conventional CD4 T cells and induction of CD4 T-cell anergy. *Eur. J. Immunol.* **2009**, *39*, 1078–1087.

112. Sartor, R.B. Microbial influences in inflammatory bowel diseases. *Gastroenterology* **2008**, *134*, 577–594.

113. Posnett, D.N.; Schmelkin, I.; Burton, D.A.; August, A.; McGrath, H.; Mayer, L.F. T cell antigen receptor V gene usage. Increases in V beta 8⁺ T cells in Crohn's disease. *J. Clin. Invest.* **1990**, *85*, 1770–1776.

114. Kelsen, J.; Agnholt, J.; Hoffmann, H.J.; Kaltoft, K.; Dahlerup, J.F. Increased expression of TCR vbeta5.1 and 8 in mucosal T-cell lines cultured from patients with Crohn disease. *Scand. J. Gastroenterol.* **2004**, *39*, 238–245.

115. Baca-Estrada, M.E.; Wong, D.K.; Croitoru, K. Cytotoxic activity of V beta 8+ T cells in Crohn's disease: The role of bacterial superantigens. *Clin. Exp. Immunol.* **1995**, *99*, 398–403.

116. Dionne, S.; Laberge, S.; Deslandres, C.; Seidman, E.G. Modulation of cytokine release from colonic explants by bacterial antigens in inflammatory bowel disease. *Clin. Exp. Immunol.* **2003**, *133*, 108–114.

117. Shiobara, N.; Suzuki, Y.; Aoki, H.; Gotoh, A.; Fujii, Y.; Hamada, Y.; Suzuki, S.; Fukui, N.; Kurane, I.; Itoh, T.; *et al.* Bacterial superantigens and T cell receptor beta-chain-bearing T cells in the immunopathogenesis of ulcerative colitis. *Clin. Exp. Immunol.* **2007**, *150*, 13–21.

118. Gittelman, P.D.; Jacobs, J.B.; Lebowitz, A.S.; Tierno, P.M., Jr. Staphylococcus aureus nasal carriage in patients with rhinosinusitis. *Laryngoscope* **1991**, *101*, 733–737.

119. Yang, P.C.; Wang, C.S.; An, Z.Y. A murine model of ulcerative colitis: Induced with sinusitis-derived superantigen and food allergen. *BMC Gastroenterol.* **2005**, *5*, doi:10.1186/1471-230X-5-6.

120. Lu, J.; Wang, A.; Ansari, S.; Hershberg, R.M.; McKay, D.M. Colonic bacterial superantigens evoke an inflammatory response and exaggerate disease in mice recovering from colitis. *Gastroenterology* **2003**, *125*, 1785–1795.

Treatment with the Hyaluronic Acid Synthesis Inhibitor 4-Methylumbelliferone Suppresses SEB-Induced Lung Inflammation

Robert J. McKallip, Harriet F. Hagele and Olga N. Uchakina

Abstract: Exposure to bacterial superantigens, such as staphylococcal enterotoxin B (SEB), can lead to the induction of acute lung injury/acute respiratory distress syndrome (ALI/ARDS). To date, there are no known effective treatments for SEB-induced inflammation. In the current study we investigated the potential use of the hyaluronic acid synthase inhibitor 4-methylumbelliferone (4-MU) on staphylococcal enterotoxin B (SEB) induced acute lung inflammation. Culturing SEB-activated immune cells with 4-MU led to reduced proliferation, reduced cytokine production as well as an increase in apoptosis when compared to untreated cells. Treatment of mice with 4-MU led to protection from SEB-induced lung injury. Specifically, 4-MU treatment led to a reduction in SEB-induced HA levels, reduction in lung permeability, and reduced pro-inflammatory cytokine production. Taken together, these results suggest that use of 4-MU to target hyaluronic acid production may be an effective treatment for the inflammatory response following exposure to SEB.

Reprinted from *Toxins*. Cite as: McKallip, R.J.; Hagele, H.F.; Uchakina, O.N. Treatment with the Hyaluronic Acid Synthesis Inhibitor 4-Methylumbelliferone Suppresses SEB-Induced Lung Inflammation. *Toxins* **2013**, *5*, 1814-1826.

1. Introduction

Exposure to staphylococcal enterotoxin B (SEB) resulting from infection with *Staphylococcal aureus* can result in life threatening complications due to activation of up to 40% of naïve T and possibly NKT cells [1–4]. This exaggerated response leads to the production of a number of pro-inflammatory cytokines including IL-1β, IL-2, IL-6, IFN-γ, and TNF-α [5,6], which can lead to endothelial cell injury, acute lung injury (ALI), acute respiratory distress syndrome (ARDS), and vascular collapse (shock) [7]. Due to its potential to cause widespread disease, its universal availability and ease of production, and dissemination SEB is currently listed by the Centers for Disease Control and Prevention (CDC) as a category B select agent. Currently, there are no known effective treatments for these conditions [8].

Modulation of the extracellular matrix can play an important role in the regulation of the inflammatory response. For example, a number of reports demonstrate that increased production hyaluronic acid is associated with various inflammatory conditions [9–11]. Under normal non-inflammatory conditions, hyaluronic acid exists primarily in its high molecular weight form (HMW-HA). However, under inflammatory condition low molecular weight hyaluronic acid (LMW-HA) accumulates [12]. Additional evidence suggests that LWM-HA has pro-inflammatory activity while HWM-HA has anti-inflammatory properties.

In recent work examining a possible role of HA in SEB-induced vascular damage we revealed that following SEB exposure, there was an increase in the level of HA in the lungs and that treatment with a HA blocking peptide led to a significant reduction in SEB-induced lung injury [13]. In the current study, we tested the hypothesis that inhibition of hyaluronic acid production will lead to a reduction in lung inflammation following exposure to SEB. To test this hypothesis we used the hyaluronic acid synthesis inhibitor 4-MU, and examined its effect on SEB-induced acute lung inflammation (ALI). Knowledge gained from this study will advance our understanding of the role of HA in SEB-mediated vascular damage and may ultimately lead to significantly improved treatment of symptoms associated with SEB exposure.

2. Results and Discussion

2.1. 4-MU Inhibits SEB-Induced Leukocyte Proliferation and Cytokine Production in Vitro

Exposure to SEB leads to activation of lymphocytes, which is characterized by elevated proliferation as well as increased production of inflammatory cytokines. Studies were conducted to determine whether culturing lymphocytes with 4-MU had an effect on SEB-induced proliferation or cytokine production. To this end, spleen cells were cultured with various concentrations of 4-MU (0.1, 0.5, or 1.0 mM) or vehicle control and stimulated with SEB. The effects of 4-MU on SEB-induced proliferation were determined 48 h later by MTT assay and revealed that treatment with 4-MU at concentrations as low as 0.1 mM significantly reduced the proliferative response (Figure 1A). Next, the effects of 4-MU on cytokine production were examined by measuring the levels of cytokines in the culture supernatants following 48 h of culture using a cytometric bead assay. The results demonstrated that treatment with 4-MU significantly inhibited SEB-induced cytokine production (Figure 1B). Taken together, these results demonstrate that culturing spleen cells with 4-MU significantly inhibits the inflammatory response to SEB.

2.2. 4-MU Treatment Leads to Increased Apoptosis in SEB-Exposed T lymphocytes in Vitro

Mechanistically, 4-MU treatment has been shown to induce cell death in various cancers as well as in through the induction of apoptosis through and effect on HA production [14–16]. Furthermore, previous results from our laboratory demonstrated that SEB exposure leads to increased production of HA in the lungs and in separate studies we demonstrated an important role of CD44 in lymphocyte activation-induced cell death (AICD), suggesting the possibility that SEB-induced HA production may protect immune cells from apoptosis through binding CD44 [2,13]. This possibility is further supported by other studies that demonstrated a protective role of CD44 in lymphocyte survival [17,18]. Therefore, we examined whether treatment of SEB-exposed spleen cells with 4-MU had an effect on the induction of apoptosis. To this end, spleen cells were exposed to SEB and then treated with 4-MU (0.1 and 0.5 mM) or vehicle control. The levels of apoptosis were determined 48 h later by Annexin V/PI (Figure 2A) and TUNEL (Figure 2B) assays. The results demonstrated that treatment with 4-MU at concentrations as low as 0.1 mM led to an increase in the levels of apoptosis suggesting that 4-MU may act to suppress the immune response to SEB through the induction of apoptosis. This effect was specific for SEB-exposed spleen cells as no significant increase in apoptosis was seen in niave spleen cells following 4-MU treatment (data not shown). In addition, we examined whether treatment with 4-MU had an effect on apoptosis in specific leukocytes subsets. Spleen cells were exposed to SEB and then treated with 4-MU (0.5 mM) or vehicle control. The effect on the specific leukocyte subsets was determined 48 h later by staining the cells with fluorescently-labeled phenotype-specific mAbs followed by TUNEL staining. The results demonstrated that 4-MU treatment had little effect on B cell, dendritic cells, or NK cells, but specifically induced apoptosis in the T cell populations (Figure 2C).

Figure 1. 4-MU inhibits SEB-induced leukocyte proliferation and cytokine production *in vitro*. Spleen cells from C57BL/6 mice were treated with 4-MU (0.1, 0.5, and 1.0 mM) or vehicle control (DMSO) and then stimulated with SEB (2 µg/mL). The effect of 4-MU on the proliferative response and cytokine production was determined 48 h later by MTT assay (**A**) and cytometric bead array (**B**), respectively. Asterisks indicate statistically significant difference when compared with vehicle controls, $p \leq 0.05$.

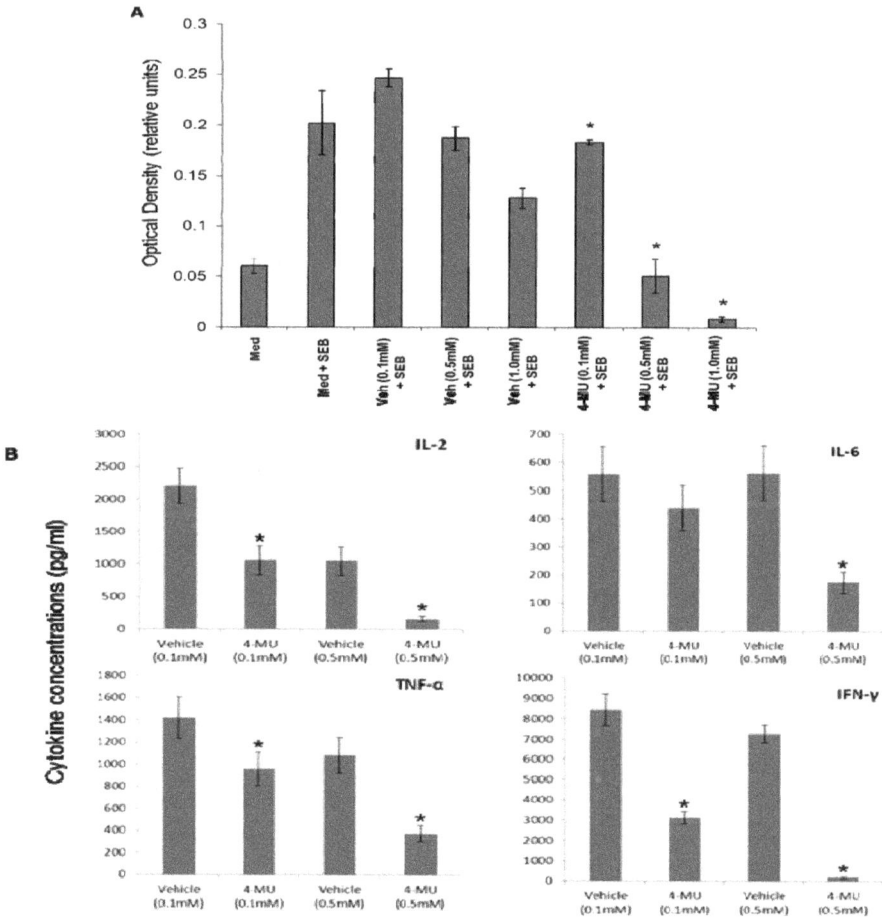

2.3. 4-MU Treatment Suppresses SEB-Induced Hyaluronic Acid Synthase Expression and Accumulation of Soluble HA in the Lungs

Previously, we reported that exposure to SEB led to an increase in the levels of soluble HA in the lungs of mice and that treatment with a HA blocking peptide could protect mice from SEB-induced lung injury [13]. It has been reported that 4-MU can reduce hyaluronic acid production through inhibition of the mRNA expression of hyaluronic acid synthases or through the depletion of UDP-GlcUA which are essential for HA synthesis [13,19–21]. In the current study we examined whether treatment with 4-MU had any effect on the hyaluronic acid synthase expression (HAS) or soluble HA levels in the lungs of SEB-exposed mice. Initially, experiments were conducted to examine the effects of 4-MU treatment on the expression of HAS in the lungs of SEB-exposed mice. To date three isoforms of hyaluronic acid synthase (HAS-1, HAS-2, and HAS-3) have been identified [22]. Exposure to SEB led to increased expression of all three isoforms, which was significantly inhibited by 4-MU treatment (Figure 3A). Next, we examined the effects of 4-MU on soluble levels of HA in the lungs of SEB-exposed mice. The results demonstrated that exposure to SEB led to significantly elevated levels of soluble HA in the lungs and that this increase was significantly reduced by treatment with 4-MU (Figure 3B). Taken together, these results demonstrated that SEB exposure alters lung levels of soluble HA possibly through increased expression of HAS mRNA and that treatment with 4-MU can significantly inhibit the accumulation of HA by reducing HAS mRNA expression.

Figure 2. 4-MU treatment leads to increased apoptosis in SEB-exposed leukocytes *in vitro*. Spleen cells from C57BL/6 mice were treated with 4-MU (0.1, and 0.5 mM) or vehicle control (DMSO) and then stimulated with SEB (2 µg/mL). The effect of 4-MU on SEB-induced apoptosis was determined 48 h later by Annexin V/PI (**A**) and TUNEL (**B**) assays, respectively. The level of apoptosis in individual immune cell subsets was determined by staining the spleen cells with phenotype-specific fluorescently-labeled mAbs followed by TUNEL staining (**C**).

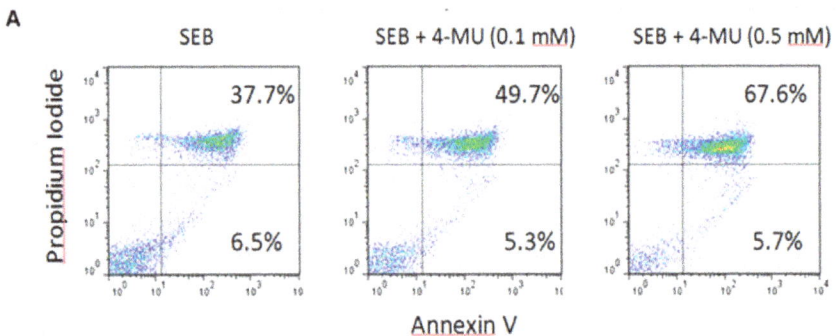

Figure 2. *Cont.*

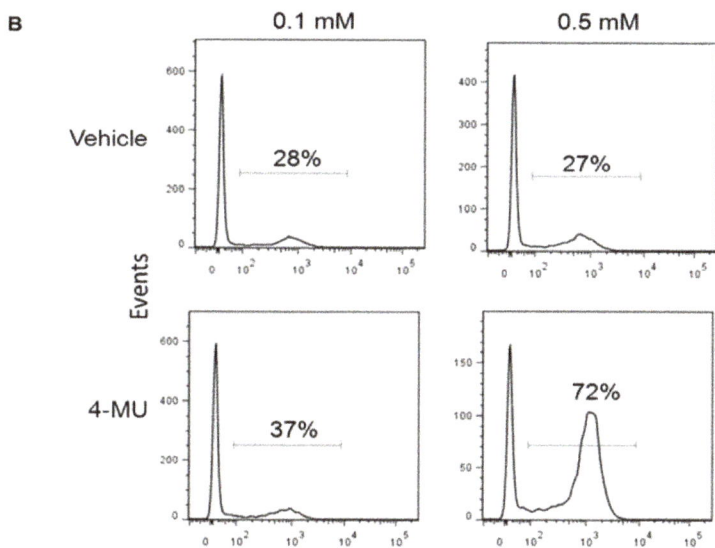

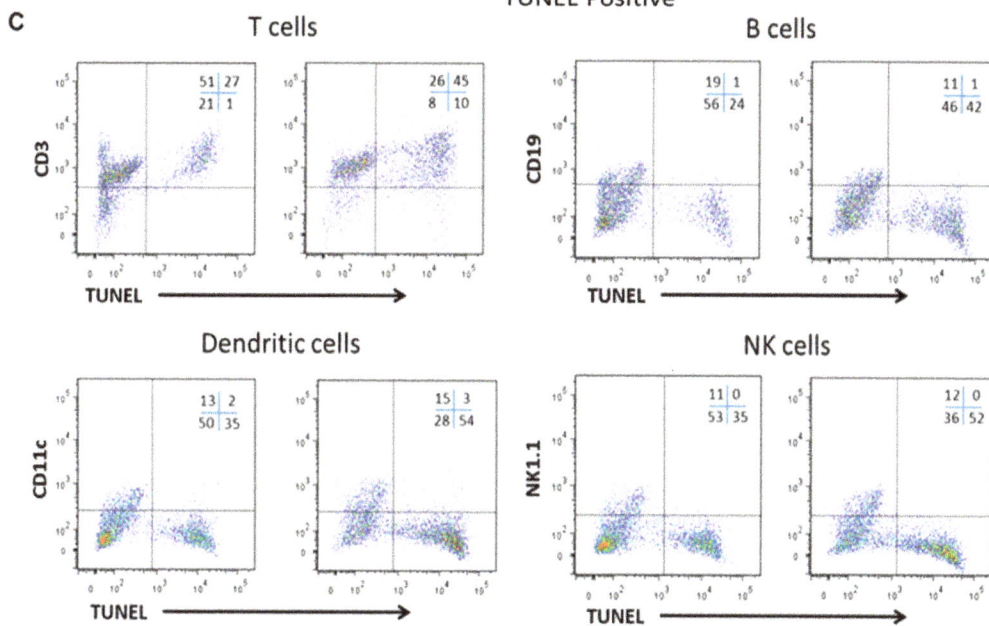

Figure 3. 4-MU treatment suppresses SEB-induced hyaluronic acid synthase expression and accumulation of soluble HA in the lungs. The effect of 4-MU on soluble HA levels in the lungs of SEB-exposed mice was determined by treating the mice with 4-MU (450 mg/mouse i.p.) or vehicle control (5% gum arabic) one day prior and on the day of SEB exposure (20 μg/50 μL PBS). The levels of lung hyaluronic acid and mRNA levels of HAS were determined 24 h later. HAS mRNA levels in whole lung extracts were determined by real-time RT-PCR (**A**). Hyaluronic acid levels in bronchoalveolar lavage fluid (BALF) were determined by ELISA (**B**). Asterisks indicate statistically significant difference when compared with the levels from vehicle-exposed mice, $p \leq 0.05$. Number sign indicates statistically significant difference when compared to PBS exposed mice, $p \leq 0.05$.

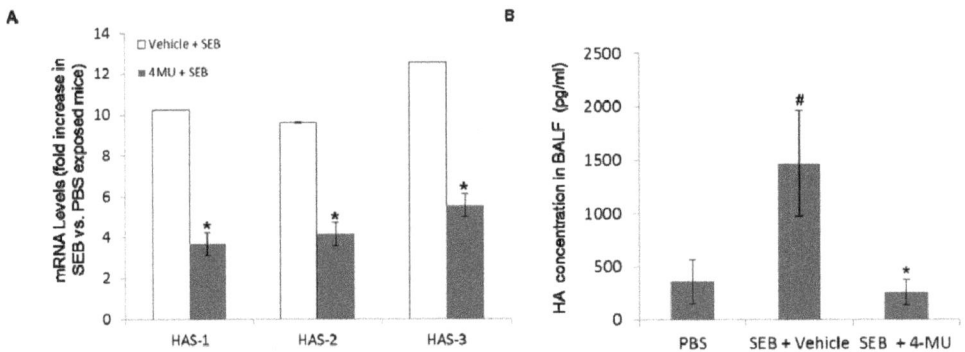

2.4. 4-MU Treatment Protects Mice from SEB-Induced Acute Lung Injury

SEB exposure can lead to acute lung injury, which is characterized by an increase in vascular permeability [16,23]. After demonstrating that treatment with 4-MU was effective at reducing the levels of HA in the lungs of SEB-exposed mice, we examined the effectiveness of targeting soluble hyaluronic acid synthesis using 4-MU as a mean to prevent SEB-induced increase in vascular permeability *in vivo*. Mice were treated one day prior to, and on the day of, SEB exposure with 4-MU (450 mg/kg) or vehicle control and then exposed to PBS or SEB (20 μg/50 μL PBS). The effect of 4-MU on the SEB-induced inflammatory response was examined by determining levels of vascular permeability in the lungs (Figure 4). The results demonstrate that vascular permeability is significantly greater in SEB-exposed mice treated with vehicle control than in corresponding PBS-exposed mice. The degree of vascular permeability in the SEB-exposed treated with 4-MU was significantly less than that seen in SEB-exposed mice treated with vehicle. Specifically, the OD reading increased from 0.082 ± 0.026 in the PBS-exposed vehicle treated mice to 0.168 ± 0.030 in the SEB-exposed vehicle treated mice which represented a 105% increase in vascular permeability following SEB exposure. In contrast the OD reading in the 4-MU treated mice increased from 0.079 ± 0.007 in the PBS-exposed mice to 0.100 ± 0.002 in the SEB-exposed mice representing a 24% increase in vascular permeability compared to the PBS exposed mice. Together, these results suggest that inhibition of hyaluronic acid synthesis by treating mice with 4-MU can lead to significant protection from SEB-induced lung injury. Furthermore, these results demonstrate that 4-MU treatment alone did not lead to a significant toxicities in the lungs as the vascular permeability in the PBS-exposed mice was not significantly increased following treatment with 4-MU when compared to vehicle-treated mice.

Figure 4. The effect of 4-MU on SEB-induced vascular permeability *in vivo* was determined by exposing mice to SEB, as described in the Material and Methods section, and treating the mice with 4-MU (450 mg/mouse i.p.) or Vehicle (5% gum arabic i.p.) one day prior, and on the day of, SEB exposure. Vascular permeability was determined as described Materials and Methods. Asterisks indicate statistically significant difference when compared with the Vehicle-treated controls, $p \leq 0.05$.

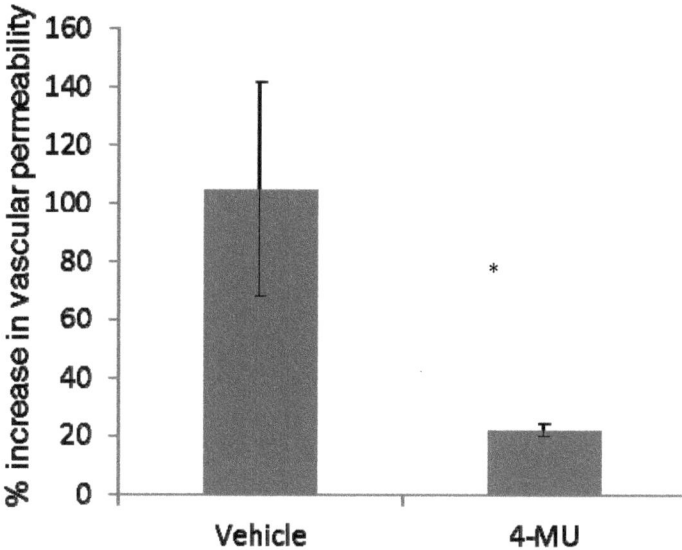

2.5. 4-MU Treatment Suppresses SEB-Induced Inflammatory Cytokine Production in the Lungs

A hallmark feature of SEB-induced acute lung injury is the activation of immune cells leading to a cytokine storm characterized by the release of large quantities of pro-inflammatory cytokines, such as IL-1β, IL-2, IL-6, IFN-γ, and TNF-α [5,16,23]. Therefore, the ability to prevent or treat the pathologies associated with SEB exposure relies on controlling the levels of SEB-induced cytokines. Experiments were set up to explore the potential use of 4-MU to reduce the SEB-induced increase in cytokine levels in the lung. To this end, groups of mice were treated with either vehicle control or 4-MU (450 mg/mouse i.p.) one day prior, and on the day of, SEB exposure. Following 4-MU treatment the mice were exposed to either PBS or SEB (20 μg/50 μL PBS i.n.). BALF and whole lung tissue were harvested 24 h later. Cytokine mRNA levels in whole lung extract were determined by real-time RT-PCR (Figure 5A). Cytokine protein levels in the BALF were determined by cytometric bead array (CBA) (Figure 5B). The results demonstrate that exposure of vehicle-treated mice to SEB led to an increase in the expression of a number of cytokines, including IL-1β, IL-2, IL-6, IFN-γ, and TNF-α, which are all reported to play a role in ALI/ARDS [13,23]. In comparison, treatment of mice with 4-MU led to a significant reduction in the SEB-induced increase in all cytokines measured.

Figure 5. 4-MU treatment suppresses SEB-induced inflammatory cytokine production in the lungs. The effect of 4-MU on SEB-induced inflammatory cytokine production *in vivo* was determined by treating the mice with 4-MU (450 mg/mouse i.p.) or vehicle (300 μL 5% gum arabic/mouse i.p.) one day prior to, and on the day of, SEB exposure. Following 4-MU treatment mice were exposed to SEB or PBS, as described in the Material and Methods section. The levels of BALF cytokine protein levels and total lung cytokine mRNA and were determined 24 h later. Cytokine mRNA levels in whole lung extracts were determined by real-time RT-PCR (**A**). The protein levels of cytokines in BALF were determined using a cytokine bead array (**B**). Asterisks indicate statistically significant difference when compared to the cytokine levels from SEB-exposed vehicle-treated mice, $p < 0.05$.

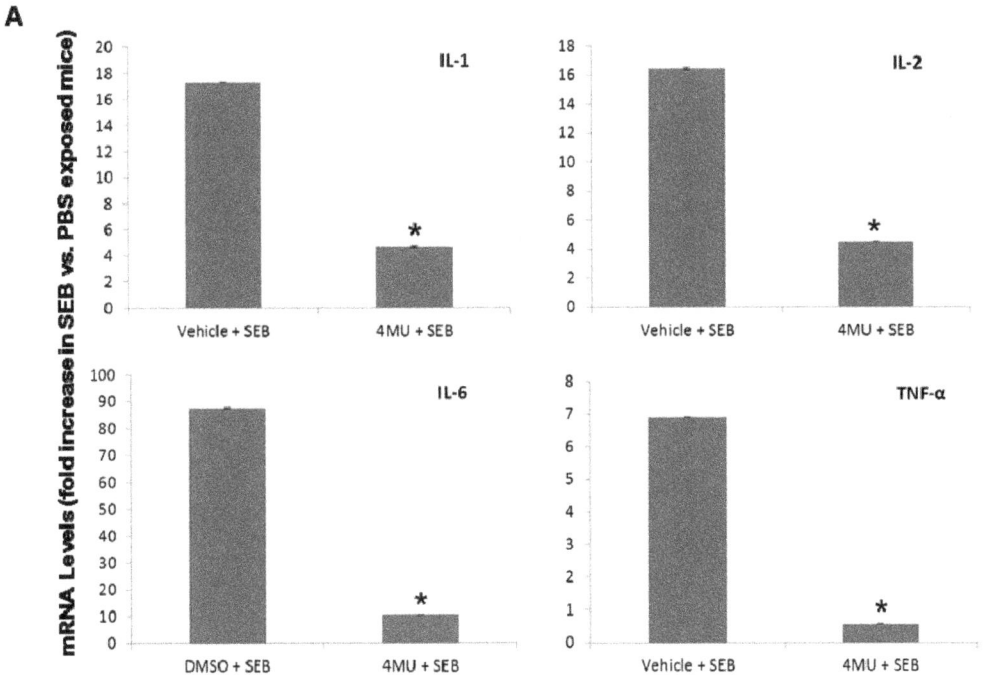

Figure 5. *Cont.*

B

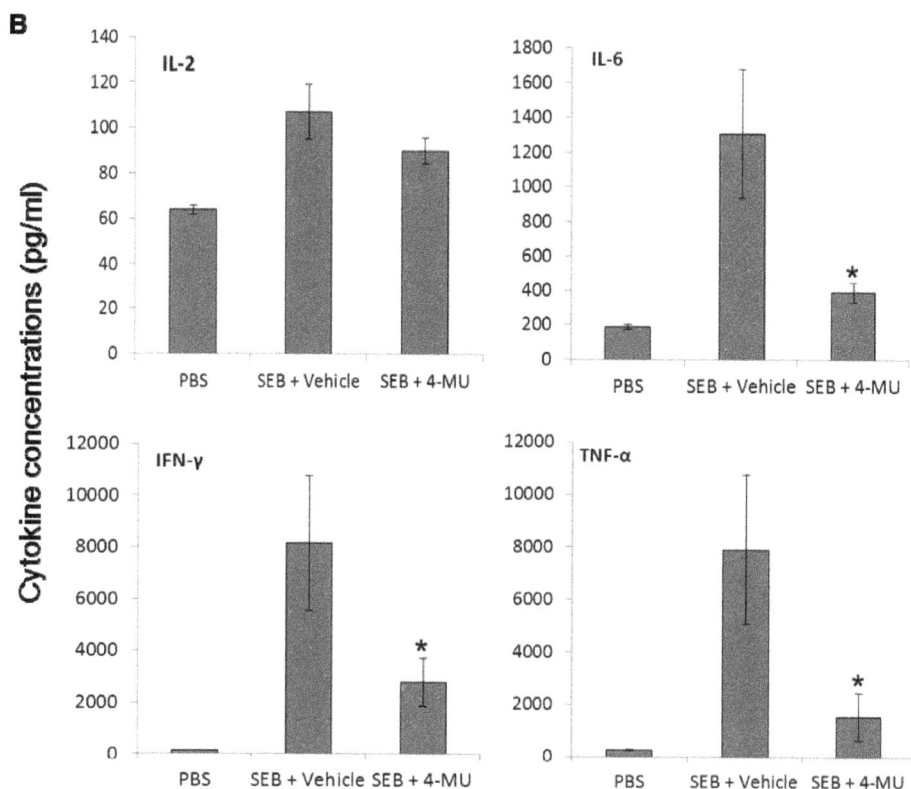

3. Experimental Section

3.1. In Vitro *Proliferation Assay*

The spleens were harvested from euthanized C57BL/6 mice and placed into 10 mL of RPMI 1640 (Gibco Laboratories, Grand Island, NY, USA) supplemented with 5% FCS, 10 mM HEPES, 1 mM glutamine, 40 µg/mL gentamicin sulfate, and 50 µM 2-mercaptoethanol, referred to as complete medium. The spleens were prepared into a single cell suspension using a laboratory homogenizer, washed twice, and adjusted to 5×10^6 /mL in complete medium. The splenocytes (5×10^5 in 100 µL/well) were cultured in 96 well flat-bottomed plates in the presence of various concentrations of 4-MU and either left unstimulated or stimulated with 2 µg/mL SEB for 48 h. The relative number of viable cells was determined using the MTT assay following the manufacturer's (Trevigen, Gaithersburg, MD, USA) instructions. Briefly, 10 µL of MTT reagent was added to each well and the plates were incubated at 37 °C for 4 h. Next, 100 µL of detergent was added to each well and the plates were incubated in the dark for 4 h, after which the absorbance at 570 nm was determined using a microplate reader.

3.2. Bead Array Analysis of Cytokine Levels

Cytokine assessment was carried out using the BD Cytometric Bead Array (CBA) (BD Bioscience, San Jose, CA, USA) for simultaneous detection of multiple cytokines (IL-2, IL-6, IFN-γ, and TNF-α) in cell supernatants and bronchoalveolar lavage fluid (BALF) of PBS or SEB exposed mice. Briefly, test samples and PE detection antibody were incubated with capture beads for 2 h, in the dark, at room temperature. After which, all unbound antibodies were washed and the beads were resuspended in 300 μL PBS. The cytokine levels were analyzed using a BD FACSAriaII cell sorter. Cytokine levels were calculated using the FCAP Array Software v3.0 (BD Bioscience, San Jose, CA, USA).

3.3. Quantification of Apoptosis

Spleen cells (5.0×10^6 cells/well) were cultured in 24-well plates in the presence of various concentrations of 4-MU (0.1 or 0.5 mM) or vehicle control and then stimulated for 48 h with SEB (2 μg/mL). Next, the cells were harvested, washed twice in PBS, and analyzed for the induction of apoptosis using the TUNEL, and Annexin V/PI methods. To detect apoptosis using the TUNEL method the cells were washed twice with PBS and fixed with 4% *P*-formaldehyde for 30 min on ice. The cells were next washed with PBS, permeabilized by adding 70% EtOH for 20 minutes, and incubated with FITC-dUTP and TdT (Promega, Madison, WI, USA) for 1 h at 37 °C and 5% CO_2. To detect apoptosis using Annexin/PI, the samples were stained with FITC labeled Annexin V and PI. The cells were incubated for 15 min at room temperature and subsequently analyzed by flow cytometric analysis. The samples were analyzed using a flow cytometer (FACSAria II, BD Biosciences, San Jose, CA, USA). In all experiments, 10,000 cells were analyzed using forward/side-scatter gating. The level of apoptosis in the individual immune cell subsets spleen cells was analyzed by flow cytometric analysis using fluorescently-labelled mAb specific for T cells (CD3), B cells (CD19), dendritic cells (CD11c), and NK (NK1.1). The level of apoptosis in activated *vs.* naïve cells was analyzed using mAbs specific for CD69. The level of apoptosis in these subsets was determined by TUNEL staining.

3.4. Hyaluronic Acid Quantification

HA levels in BALF of PBS or SEB exposed mice were measured by an ELISA-like assay according to the manufacturer's protocol (Echelon Biosciences, Salt Lake City, UT, USA).

3.5. RNA Isolation and Real-Time RT-PCR Analysis

Total RNA was isolated from a single cell suspension of splenocytes or from whole lung using the RNeasy Mini Kit (Qiagen, Valencia, CA, USA). RNA concentration and integrity was determined spectrophotometrically. cDNA was synthesized by reverse transcription of 50 ng total RNA using the High Capacity cDNA Reverse Transcriptase Kit (Applied Biosystems, Carlsbad, CA, USA). Real-time PCR was performed using a SYBR Green PCR kit (Applied Biosystems).

Amplifications were performed and monitored using an ABI 7300 real-time PCR system (Applied Biosystems). The gene-specific primers for β-actin have been previously described (17). In addition the following primers were used: IL-1β primers 5'-GAAATGCCACCTTTTGACAGTG-3' and 5'-CTGGATGCTCTCATCAGGACA-3'; IL-2 primers 5'-TGATGGACCTACAGGAGCTCCTGAG-3' and 5'-GAGTCAAATCCAGAACATGCCGCAG-3'; TNF-α primers 5'-CCAGTGTGGG AAGCTGTCTT-3' and 5'-AAGCAAAAGAGGAGGCAACA-3'; HAS-1 primers 5'-ACCTCA CCAACCGAATGCTT-3' and 5'-GAAGGAAGGAGGAGGGCG-3'; HAS-2 primers 5'-TGAGT ACAAAGAGGTTCGTTCAAGTT-3' and 5'-ATTGTCAGGGTGTGTTTGTTTCC-3'; HAS-3 primers 5'- CTACTTTGTAGCTGCCCAGAATACTG-3' and 5'-GAGTAC AAAAAACAGCA CCGGAAT-3'; HYAL-1 primers 5'-GGCCTACCTAGGACTTCCTCAA-3' and 5'-CTATTCCC GTGACTGTGCCTAT-3'; HYAL-2 primers 5'-GACCTCAACTACCTGCAGAAGC-3' and 5'-CCTTATAGTTCCG TTGGCACTG-3'; HYAL-3 primers 5'-GGAGCATTCACATCCTCTACCT-3' and 5'-GTCGTCCAGAGACAGGAATCTC-3'. The threshold cycle (CT) method was used for relative quantification of gene transcription in relation to expression of the internal standard β-actin. Fold changes of mRNA levels in SEB-stimulated immune cells relative to unstimulated cells was determined using the $2^{-\Delta\Delta Ct}$ method [24].

3.6. Quantification of Vascular Permeability

Vascular leakiness was studied by measuring the extravasation of Evan's blue, which when given i.v. binds to plasma proteins, particularly albumin, and following extravasation can be detected in various organs as described previously [25]. Vascular leak was induced by injection of SEB, as previously described [13,23]. Briefly, groups of five mice were injected i.n. with SEB (20 μg/50 μL PBS) or PBS (50 μL). The mice were exposed to SEB for 24 h. Two hours prior to harvesting the lungs, the mice were injected i.v. with 0.1 mL of 1% Evan's blue in PBS. After two hours, the mice were exsanguinated under anesthesia, and the heart was perfused with heparin in PBS as described previously [7]. The lungs were harvested and placed in formamide at 37 °C for 16 h. The Evan's blue in the organs was quantified by measuring the absorbance of the supernatant at 650 nm with a spectrophotometer. In experiments examining the effect of 4-MU on SEB-induced vascular permeability, mice were treated one day prior to, and on the day of, SEB exposure with vehicle or 4-MU (450 mg/mouse, i.p.). The vascular permeability seen in SEB-exposed mice was expressed as percent increase in extravasation when compared with that of PBS-treated controls and was calculated as: (optical density of dye in the lungs of SEB-exposed mice)-(optical density of dye in the lungs of PBS-treated controls))/(optical density of dye in the lungs of PBS-treated control) × 100. Each mouse was individually analyzed for vascular permeability, and data were expressed as mean ± SEM percent increase in vascular permeability in SEB-exposed mice when compared to that seen in PBS-treated controls [13,23].

3.7. Statistical Analysis

Student's t-test or ANOVA was used to determine statistical significance and $p < 0.05$ was considered to be statistically significant.

4. Conclusions

Exposure to bacterial superantigens, such as SEB, can lead to the induction of acute lung injury/acute respiratory distress syndrome (ALI/ARDS). Currently, there are no effective treatments for the resulting inflammatory response. Results from our previous study demonstrated an important role of hyaluronic acid in development of SEB-induced ARDS/ALI [13]. Specifically, we showed that SEB exposure leads to increased levels of hyaluronic acid in the lungs and that targeting hyaluronic acid using Pep-1 led to a significantly attenuated response to SEB *in vitro* and *in vivo*. In the current study, we explored an alternative approach to reduce hyaluronic acid levels in the lungs following SEB-exposure by targeting hyaluronic acid synthesis using 4-MU. The protective effects of 4-MU treatment were characterized by reduced expression of HAS-1, HAS-2, and HAS-3, and reduced levels of HA in the lungs of SEB-exposed mice. Furthermore, 4-MU treatment led to a reduction in SEB-induced lung permeability, and reduced cytokine production. Together, the findings of this study suggest that targeting hyaluronic acid synthesis might be a novel target for the treatment or reduction of SEB-induced lung injury.

Acknowledgments

We thank the Mercer University Seed grant program and the MEDCEN Community Health Foundation for their generous support

Conflicts of Interest

The authors declare no conflict of interest.

References

1. Marrack, P.; Kappler, J. The staphylococcal enterotoxins and their relatives. *Science* **1990**, *248*, 705–711.
2. McKallip, R.J.; Do, Y.; Fisher, M.T.; Robertson, J.L.; Nagarkatti, P.S.; Nagarkatti, M. Role of CD44 in activation-induced cell death: CD44-deficient mice exhibit enhanced T cell response to conventional and superantigens. *Int. Immunol.* **2002**, *14*, 1015–1026.
3. Rajagopalan, G.; Sen, M.M.; Singh, M.; Murali, N.S.; Nath, K.A.; Iijima, K.; Kita, H.; Leontovich, A.A.; Gopinathan, U.; Patel, R.; *et al.* Intranasal exposure to staphylococcal enterotoxin B elicits an acute systemic inflammatory response. *Shock* **2006**, *25*, 647–656.
4. Hayworth, J.L.; Mazzuca, D.M.; Maleki Vareki, S.; Welch, I.; McCormick, J.K.; Haeryfar, S.M. CD1d-independent activation of mouse and human iNKT cells by bacterial superantigens. *Immunol. Cell Biol.* **2012**, *90*, 699–709.
5. Schramm, R.; Thorlacius, H. Staphylococcal enterotoxin B-induced acute inflammation is inhibited by dexamethasone: important role of CXC chemokines KC and macrophage inflammatory protein 2. *Infect. Immun.* **2003**, *71*, 2542–2547.

6. Strandberg, K.L.; Rotschafer, J.H.; Vetter, S.M.; Buonpane, R.A.; Kranz, D.M.; Schlievert, P.M. Staphylococcal superantigens cause lethal pulmonary disease in rabbits. *J. Infect. Dis.* **2010**, *202*, 1690–1697.

7. McKallip, R.J.; Fisher, M.; Do, Y.; Szakal, A.K.; Gunthert, U.; Nagarkatti, P.S.; Nagarkatti, M. Targeted deletion of CD44v7 exon leads to decreased endothelial cell injury but not tumor cell killing mediated by interleukin-2-activated cytolytic lymphocytes. *J. Biol. Chem.* **2003**, *278*, 43818–43830.

8. Hayworth, J.L.; Kasper, K.J.; Leon-Ponte, M.; Herfst, C.A.; Yue, D.; Brintnell, W.C.; Mazzuca, D.M.; Heinrichs, D.E.; Cairns, E.; Madrenas, J.; *et al.* Attenuation of massive cytokine response to the staphylococcal enterotoxin B superantigen by the innate immunomodulatory protein lactoferrin. *Clin. Exp. Immunol.* **2009**, *157*, 60–70.

9. Bao, A.; Liang, L.; Li, F.; Zhang, M.; Zhou, X. Consequences of acute ozone exposure imposed on the culminated allergic pulmonary inflammation in an established murine model of asthma. *Front. Biosci.* **2013**, *18*, 838–851.

10. Liang, J.; Jiang, D.; Jung, Y.; Xie, T.; Ingram, J.; Church, T.; Degan, S.; Leonard, M.; Kraft, M.; Noble, P.W. Role of hyaluronan and hyaluronan-binding proteins in human asthma. *J. Allergy Clin. Immunol.* **2011**, *128*, 403–411.

11. Yoon, J.S.; Lee, H.J.; Choi, S.H.; Chang, E.J.; Lee, S.Y.; Lee, E.J. Quercetin inhibits IL-1beta-induced inflammation, hyaluronan production and adipogenesis in orbital fibroblasts from Graves' orbitopathy. *PLoS One* **2011**, *6*, e26261.

12. Stern, R.; Asari, A.A.; Sugahara, K.N. Hyaluronan fragments: An information-rich system. *Eur. J. Cell Biol.* **2006**, *85*, 699–715.

13. Uchakina, O.N.; Castillejo, C.M.; Bridges, C.C.; McKallip, R.J. The role of hyaluronic acid in SEB-induced acute lung inflammation. *Clin. Immunol.* **2013**, *146*, 56–69.

14. Piccioni, F.; Malvicini, M.; Garcia, M.G.; Rodriguez, A.; Atorrasagasti, C.; Kippes, N.; Piedra Buena, I.T.; Rizzo, M.M.; Bayo, J.; Aquino, J.; *et al.* Antitumor effects of hyaluronic acid inhibitor 4-methylumbelliferone in an orthotopic hepatocellular carcinoma model in mice. *Glycobiology* **2012**, *22*, 400–410.

15. Urakawa, H.; Nishida, Y.; Wasa, J.; Arai, E.; Zhuo, L.; Kimata, K.; Kozawa, E.; Futamura, N.; Ishiguro, N. Inhibition of hyaluronan synthesis in breast cancer cells by 4-methylumbelliferone suppresses tumorigenicity *in vitro* and metastatic lesions of bone *in vivo*. *Int. J. Cancer* **2012**, *130*, 454–466.

16. Uchakina, O.N.; Ban, H.; McKallip, R.J. Targeting hyaluronic acid production for the treatment of leukemia: Treatment with 4-methylumbelliferone leads to induction of MAPK-mediated apoptosis in K562 leukemia. *Leuk. Res.* **2013**, *37*, 1294–1301.

17. Baaten, B.J.; Li, C.R.; Deiro, M.F.; Lin, M.M.; Linton, P.J.; Bradley, L.M. CD44 regulates survival and memory development in Th1 cells. *Immunity* **2010**, *32*, 104–115.

18. Fedorchenko, O.; Stiefelhagen, M.; Peer-Zada, A.A.; Barthel, R.; Mayer, P.; Eckei, L.; Breuer, A.; Crispatzu, G.; Rosen, N.; Landwehr, T.; *et al.* CD44 regulates the apoptotic response and promotes disease development in chronic lymphocytic leukemia. *Blood* **2013**, *121*, 4126–4136.

19. Kultti, A.; Pasonen-Seppanen, S.; Jauhiainen, M.; Rilla, K.J.; Karna, R.; Pyoria, E.; Tammi, R.H.; Tammi, M.I. 4-Methylumbelliferone inhibits hyaluronan synthesis by depletion of cellular UDP-glucuronic acid and downregulation of hyaluronan synthase 2 and 3. *Exp. Cell Res.* **2009**, *315*, 1914–1923.

20. Kakizaki, I.; Kojima, K.; Takagaki, K.; Endo, M.; Kannagi, R.; Ito, M.; Maruo, Y.; Sato, H.; Yasuda, T.; Mita, S.; *et al.* A novel mechanism for the inhibition of hyaluronan biosynthesis by 4-methylumbelliferone. *J. Biol. Chem.* **2004**, *279*, 33281–33289.

21. Vigetti, D.; Rizzi, M.; Viola, M.; Karousou, E.; Genasetti, A.; Clerici, M.; Bartolini, B.; Hascall, V.C.; De Luca, G.; Passi, A. The effects of 4-methylumbelliferone on hyaluronan synthesis, MMP2 activity, proliferation, and motility of human aortic smooth muscle cells. *Glycobiology* **2009**, *19*, 537–546.

22. Itano, N.; Sawai, T.; Yoshida, M.; Lenas, P.; Yamada, Y.; Imagawa, M.; Shinomura, T.; Hamaguchi, M.; Yoshida, Y.; Ohnuki, Y.; *et al.* Three isoforms of mammalian hyaluronan synthases have distinct enzymatic properties. *J. Biol. Chem.* **1999**, *274*, 25085–25092.

23. Sun, J.; Law, G.P.; Bridges, C.C.; McKallip, R.J. CD44 as a novel target for treatment of staphylococcal enterotoxin B-induced acute inflammatory lung injury. *Clin. Immunol.* **2012**, *144*, 41–52.

24. Schmittgen, T.D.; Livak, K.J. Analyzing real-time PCR data by the comparative C_T method. *Nat. Protoc.* **2008**, *3*, 1101–1108.

25. Udaka, K.; Takeuchi Y.; Movat H.Z. Simple method for quantitation of enhanced vascular permeability. *Proc. Soc. Exp. Biol. Med.* **1970**, *133*, 1384–1387.

Sulfasalazine Attenuates Staphylococcal Enterotoxin B-Induced Immune Responses

Teresa Krakauer

Abstract: Staphylococcal enterotoxin B (SEB) and related exotoxins are important virulence factors produced by *Staphylococcus aureus* as they cause human diseases such as food poisoning and toxic shock. These toxins bind directly to cells of the immune system resulting in hyperactivation of both T lymphocytes and monocytes/macrophages. The excessive release of proinflammatory cytokines from these cells mediates the toxic effects of SEB. This study examined the inhibitory activities of an anti-inflammatory drug, sulfasalazine, on SEB-stimulated human peripheral blood mononuclear cells (PBMC). Sulfasalazine dose-dependently inhibited tumor necrosis factor α, interleukin 1 (IL-1) β, IL-2, IL-6, interferon γ (IFNγ), and various chemotactic cytokines from SEB-stimulated human PBMC. Sulfasalazine also potently blocked SEB-induced T cell proliferation and NFκB activation. These results suggest that sulfasalazine might be useful in mitigating the toxic effects of SEB by blocking SEB-induced host inflammatory cascade and signaling pathways.

Reprinted from *Toxins.* Cite as: Krakauer, T. Sulfasalazine Attenuates Staphylococcal Enterotoxin B-Induced Immune Responses. *Toxins* **2015**, *7*, 553-559.

1. Introduction

Staphylococcal enterotoxin B (SEB) and structurally related bacterial exotoxins are etiological agents that cause a variety of diseases in humans, ranging from food poisoning, autoimmune diseases, and toxic shock [1–3]. These exotoxins potently stimulate host immune responses by binding directly to the major histocompatibility complex (MHC) class II molecules on antigen-presenting cells and specific Vβ regions of the T-cell receptors [4,5]. The staphylococcal exotoxins are also known as superantigens because of their ability to polyclonally activate T cells at picomolar concentrations [1,5]. Their interactions with cells of the immune system result in a massive release of proinflammatory cytokines and chemokines [6–11]. These proinflammatory mediators enhance leukocyte migration, promote tissue injury, and coagulation [12,13]. The cytokines, interleukin 1 (IL-1), tumor necrosis factor α (TNFα), and interferon gamma (IFNγ) are pivotal mediators in animal models of superantigen-induced toxic shock [7,11].

Currently, there are no specific drugs available for treating superantigen-induced shock. However, intravenous immunoglobulin is protective if it is administered soon after SEB intoxication [14]. Targeting superantigen-induced host responses with anti-inflammatory drugs is an attractive strategy as some of these compounds block key signaling pathways induced by superantigens and the cytokines induced [15]. Sulfasalazine (SFZ) is a FDA-approved anti-inflammatory drug used clinically in the treatment of rheumatoid arthritis and Crohn's disease [16]. The mechanism underlying the biological effects of SFZ *in vivo* is complex and not completely understood. SFZ has immune-modulatory effects including inhibition of cyclooxygenase- and lipoxygenase-dependent pathways, enhancing anti-inflammatory adenosine release from sites of inflammation, and reducing leukocyte adhesion to endothelial cells [17,18]. This brief report presents the inhibitory activities of SFZ on SEB-activated human peripheral blood mononuclear cells (PBMC).

2. Results and Discussion

2.1. Effect of Sulfasalazine on Proinflammatory Mediators Release

The potency of SFZ in blocking cytokines and chemokines in SEB-stimulated human PBMC was investigated since proinflammatory mediators play key roles in superantigen-induced toxic shock. Figure 1 shows that SFZ attenuated the production of IL-1β, TNFα, IL-6, IL-2 and IFNγ in SEB-stimulated PBMC in a dose-dependent manner. The production of the chemokines, monocyte chemotactic protein 1 (MCP-1), macrophage inflammatory protein (MIP)-1α, and MIP-1β was also reduced. Reduction of these mediators were statistically significant ($p < 0.05$) between SEB and SEB plus SFZ samples at concentrations of 0.25 to 1.25 mM of SFZ. SFZ did not affect the viability of the cells over the concentration range used in these studies (0.025–1.25 mM), as demonstrated by trypan blue dye exclusion test. Lactate dehydrogenase assay also confirmed the lack of cytotoxic effects of SFZ in the concentrations used (data not shown).

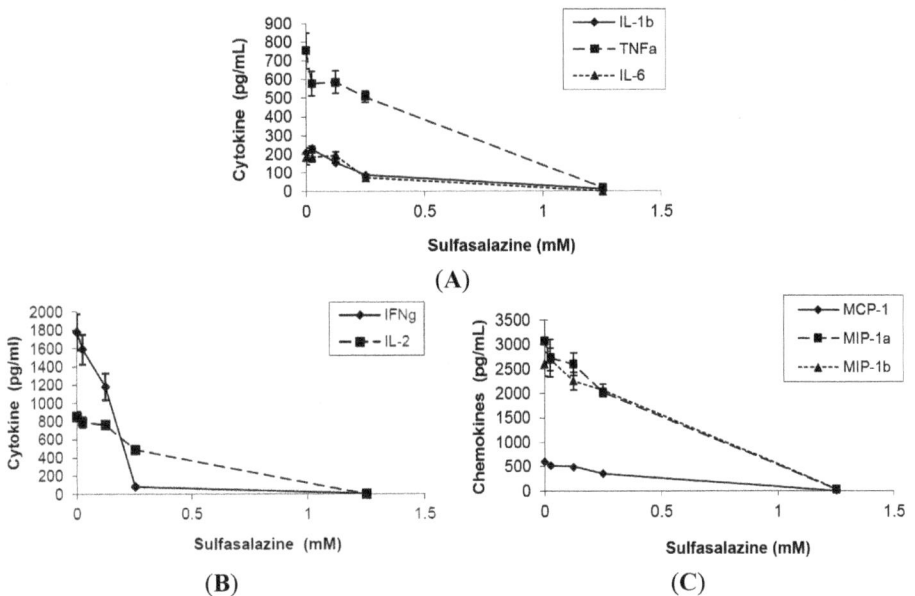

Figure 1. Dose-response inhibition of (**A**) interleukin 1β (IL-1β), tumor necrosis factor α (TNFα), and IL-6; (**B**) interferon γ (IFNγ) and IL-2; (**C**) monocyte chemotactic protein 1 (MCP-1), macrophage inflammatory protein (MIP)-1α, MIP-1β production by peripheral blood mononuclear cells (PBMC) stimulated with 200 ng/mL of staphylococcal enterotoxin B (SEB) in the presence of various concentrations of sulfasalazine (SFZ). Values represent the mean ± SD of duplicate samples and results represent three experiments. Results are statistically significant ($p < 0.05$) between SEB and SEB plus SFZ samples at concentrations of 0.25 and 1.25 mM.

2.2. Effect of Sulfasalazine on T-Cell Proliferation

Since SEB polyclonally activates T-cells, the effect of SFZ on SEB-stimulated T-cell proliferation was next examined. Figure 2 shows that SFZ effectively blocked T-cell proliferation, achieving 65% and 98% inhibition at 0.25 mM and 1.25 mM, respectively ($p < 0.05$).

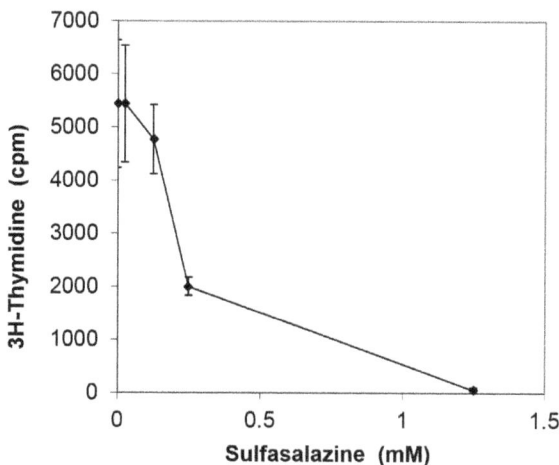

Figure 2. Inhibition of T-cell proliferation in PBMC stimulated with 200 ng/mL of SEB. Values represent the mean ± SD of triciplate samples and results represent three experiments. Results are statistically significant ($p < 0.05$) between SEB and SEB plus SFZ samples at concentrations of 0.25 and 1.25 mM.

2.3. Effect of Sulfasalazine on NFκB

The transcription factor NF-κB is a key regulator of inflammation and acts downstream of many cell surface receptors including MHC class II molecules and cytokine receptors [19,20]. Cell extracts from SEB-stimulated PBMC in the presence of SFZ indicated that SFZ reduced NF-κB activation to 3% of control cultures of SEB-stimulated cells without drug treatment ($p < 0.05$, Figure 3).

The anti-inflammatory compound SFZ has been used clinically for decades to treat various inflammatory diseases such as rheumatoid arthritis and Crohn's disease [16]. Its principal mode of action is inhibition of NF-κB, thereby down-regulating inflammation [21,22]. NF-κB is a key transcription factor involved in the regulation of a number of inflammatory cytokines, growth factors, and adhesion molecules [19]. A previous report indicated the activation of NF-κB in a human monocytic cell line treated with superantigens [23]. This study shows for the first time that SFZ reduces SEB-induced inflammatory mediators, T-cell proliferation and NF-κB.

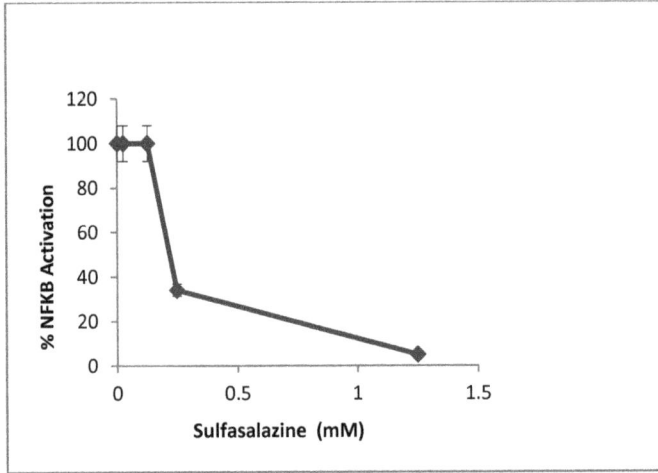

Figure 3. Inhibition of NF-κB activation in PBMC stimulated with 200 ng/mL of SEB. Values represent the mean ± SD of duplicate samples and results represent two experiments.

3. Experimental Section

3.1. Materials

Purified SEB was obtained from Toxin Technology (Sarasota, FL, USA). The endotoxin content of these preparations was <1 ng of endotoxin/mg protein as determined by the Limulus amoebocyte lysate gelation test (BioWhittaker, Walkersville, MD, USA). Human (h) recombinant (r) TNFα, antibodies against hTNFα, peroxidase-conjugated anti-rabbit IgG, and peroxidase-conjugated anti-goat IgG were obtained from Boehringer-Mannheim (Indianapolis, IN, USA). Human rIFNγ and rIL-6 were obtained from Collaborative Research (Boston, MA, USA). Antibodies against IFNγ, IL-2, and MCP-1 were obtained from BDPharMingen (San Diego, CA, USA). Recombinant IL-2, MCP-1, MIP-1α, MIP-1β; antibodies against IL-1β, IL-6, MIP-1α, and MIP-1β were purchased from R&D Systems (Minneapolis, MN, USA). SFZ and all other common reagents were purchased from Sigma (St. Louis, MO, USA).

3.2. Cell Culture

Human PBMC were isolated by Ficoll-Hypaque density gradient centrifugation of heparinized blood from normal human donors. PBMC (10^6 cells/mL) were cultured at 37 °C in RPMI 1640 medium supplemented with 10% inactivated fetal bovine serum in 24-well plates as previously described [24]. Cells were stimulated with SEB (200 ng/mL) for 16 h. Varying concentrations (0.025, 0.125, 0.25, 1.25 mM) of SFZ were added simultaneously with SEB. Culture supernatants were collected and analyzed for IL-1β, TNFα, IL-6, IFNγ, IL-2, MCP-1, MIP-1α, and MIP-1β. Cell viability was determined by the trypan blue dye exclusion method. At the end of the experiments, cells were recovered and the number of trypan blue-positive cells was counted. Cells were 93%–98% viable in the presence or absence of SFZ with SEB using concentrations described above. Additionally, cell-free supernatants were also tested for the presence of lactate dehydrogenase, an enzyme that is released from dead cells.

T-cell proliferation was assayed with PBMC (10^6 cells/well), which were plated in triplicate with SEB (200 ng/mL), with or without SFZ, for 48 h at 37 °C in 96-well microtiter plates. Cells were pulsed with 1 µCi/well of [^{3}H]thymidine (New England Nuclear, Boston, MA, USA) during the last 5 h of culture as described previously [24]. Cells were harvested onto glass fiber filters, and incorporation of [^{3}H]thymidine was measured by liquid scintillation.

3.3. Measurement of Cytokines and Chemokines

Cytokines and chemokines were measured by an enzyme-linked immunosorbent assay (ELISA) with cytokine- or chemokine-specific antibodies according to the manufacturer's instructions, as previously described [24]. Human recombinant cytokines and chemokines (20–1000 pg/mL) were used as standards for calibration on each plate. The detection limit of each assay was 20 pg/mL. The cytokine and chemokine data were expressed as the mean concentration (pg/mL) ± SD of duplicate samples.

3.4. NF-κB Activation Assay

NF-κB activation was measured with a Trans-AM NF-κB kit (Active Motif, Carlsbad, CA, USA) according to the manufacturer's instructions. Nuclear extracts (10 µg) containing NF-κB protein from PBMC with SEB in the absence or presence of SFZ were added to the wells, followed by the primary antibody against p65 subunit of NF-κB and the horseradish peroxidase-conjugated secondary antibody. Optical density was determined on an absorbance plate reader at 450 nm.

3.5. Data Analysis

Data were expressed as the mean ± SD and were analyzed for significant differences by the Student's t-test with Stata (Stata Corp., College Station, TX, USA). Differences between SFZ-treated and untreated control groups were considered significant if P was < 0.05.

4. Conclusions

Development of medical countermeasures for preventing SEB-induced toxic shock is urgently needed to improve the high morbidity and mortality associated with complications from systemic shock resulting from bacterial superantigen exposure. Receptor blockade and signal transduction pathway inhibition represent different approaches to block superantigen-induced effects with various degrees of effectiveness [15]. Anti-inflammatory drugs are potentially useful as they target many downstream signaling pathways affecting multiple cytokines and chemokines. Repurposing FDA-approved drugs represent a fast approach for discovery of therapies against SEB and other biodefense related agents as safety concerns, tolerability, bioavailability and mechanism of action of these drugs are known. The new use of a FDA-approved anti-inflammatory compound, SFZ, against the biological effects of SEB is shown in this report. A logical extension of the inhibitory effects of

SFZ on other staphylococcal superantigens may reveal broader applicability of its use and clinical potential. Further studies are underway to test the therapeutic efficacy of SFZ in various mouse models of superantigen-induced toxic shock.

Acknowledgments

This research was funded by DTRA under USAMRIID project number 1321180. Opinions, interpretations, conclusions, and recommendations are those of the author and are not necessarily endorsed by the U.S. Army.

Conflicts of Interest

The author declares no conflict of interest.

References

1. Kotzin, B.L.; Leung, D.Y.M.; Kappler, J.; Marrack, P. Superantigens and their potential role in human disease. *Adv. Immunol.* **1993**, *54*, 99–166.
2. Langley, R.; Patel, D.; Jackson, N.; Clow, F.; Fraser, J.D. Staphylococcal superantigen super-domains in immune evasion. *Crit. Rev. Immunol.* **2010**, *30*, 149–165.
3. DeVries, A.S.; Lesher, L.; Schlievert, P.M.; Rogers, T.; Villaume, L.G.; Danila, R.; Lynfield, R. Staphylococcal toxic shock syndrome 2000–2006: Epidemiology, clinical features, and molecular characteristics. *PLoS One* **2011**, *6*, doi:10.1371/journal.pone.0022997.
4. Carlsson, R.; Fischer, H.; Sjogren, H.O. Binding of staphylococcal enterotoxin A to accessory cells is a requirement for its ability to activate human T cells. *J. Immunol.* **1988**, *140*, 2484–2488.
5. Choi, Y.; Kotzin, B.; Hernon, L.; Callahan, J.; Marrack, P.; Kappler, J. Interaction of *Staphylococcus aureus* toxin "superantigens" with human T cells. *Proc. Natl. Acad. Sci. USA* **1989**, *86*, 8941–8945.
6. Jupin, C.; Anderson, S.; Damais, C.; Alouf, J.E.; Parant, M. Toxic shock syndrome toxin 1 as an inducer of human tumor necrosis factors and gamma interferon. *J. Exp. Med.* **1988**, *167*, 752–761.
7. Miethke, T.; Wahl, C.; Heeg, K.; Echtenacher, B.; Krammer, P.H.; Wagner, H. Superantigen mediated shock: A cytokine release syndrome. *Immunobiology* **1993**, *189*, 270–284.
8. Neumann, B.; Engelhardt, B.; Wagner, H.; Holzmann, B. Induction of acute inflammatory lung injury by staphylococcal enterotoxin B. *J. Immunol.* **1997**, *158*, 1862–1871.
9. Fraser, J.D.; Proft, T. The bacterial superantigen and superantigen-like proteins. *Immunol. Rev.* **2008**, *225*, 226–243.
10. Lappin, E.; Ferguson, A.J. Gram-positive toxic shock syndromes. *Lancet Infect. Dis.* **2009**, *9*, 281–290.
11. Krakauer, T.; Buckley, M.; Fisher, D. Proinflammatory mediators of toxic shock and their correlation to lethality. *Mediators Inflamm.* **2010**, doi:10.1155/2010/517594.

12. Krakauer, T.; Vilcek, J.; Oppenheim, J.J. Proinflammatory cytokines: TNF and IL-1 families, chemokines, TGFß and others. In *Fundamental Immunology*, 4th ed.; Paul, W., Ed.; Lippincott-Raven: Philadelphia, PA, USA, 1998; pp. 775–811.

13. Mattsson, E.; Herwald, H.; Egsten, A. Superantigen from *Staphylococcus aureus* induce procoagulant activity and monocyte tissue factor expression in whole blood and mononuclear cells via IL-1β. *J. Thromb. Haemost.* **2003**, *1*, 2569–2575.

14. Darenberg, J.; Soderquist, B.; Normark, B.H.; Norrby-Teglund, A. Differences in potency of intravenous polyspecific immunoglobulin G against streptococcal and staphylococcal superantigens: Implications for therapy of toxic shock syndrome. *Clin. Infect. Dis.* **2004**, *38*, 836–842.

15. Krakauer, T. Update on staphylococcal superantigen-induced signaling pathways and therapeutic interventions. *Toxins* **2013**, *5*, 1629–1654.

16. Rains, C.P.; Noble, S.; Faulds, D. Sulfasalazine. A review of its pharmacological properties and therapeutic efficacy in the treatment of rheumatoid arthritis. *Drugs.* **1995**, *50*, 137–156.

17. Hoult, J.R. Pharmacological and biochemical actions of sullphasalazine. *Drugs* **1986**, *32*, 18–26.

18. Gadangi, P.; Longaker, M.; Naime, D.; Levin, R.I.; Recht, P.A.; Montesinos, M.C.; Buckley, M.T.; Carlin, G.; Cronstein, B.N. The anti-inflammatory mechanism of sulfasalazine is related to adenosine release at inflamed sites. *J. Immunol.* **1996**, *156*, 1937–1941.

19. Vallabhapurapu, S.; Karin, M. Regulation and function of NFκB transcription factors in the immune system. *Annu. Rev. Immunol.* **2009**, *27*, 693–733.

20. Krakauer, T. Nuclear factor-κB: Fine-tuning a central integrator of diverse biologic stimuli. *Int. Rev. Immunol.* **2008**, *27*, 286–292.

21. Lappas, M.; Yee, K.; Permezel, M.; Rice, G.E. Sulfasalazine and BAY 11–7082 interfere with the nuclear factor-κB and IκB kinase pathway to regulate the release of proinflammatory cytokines from human adipose tissue and skeletal muscle *in vitro*. *Endocrinology* **2005**, *146*, 1491–1497.

22. Barnes, B.J.; Karin, M. Nuclear factor-κB: A pivotal transcription factor in chronic inflammatory disease. *N. Engl. J. Med.* **1997**, *336*, 1066–1071.

23. Trede, N.S.; Castigli, E.; Geha, R.S.; Chatila, T. Microbial superantigens induce NF-kappa B in the human monocytic cell line THP-1. *J. Immunol.* **1993**, *150*, 5604–5610.

24. Krakauer, T. Suppression of endotoxin- and staphylococcal exotoxin-induced cytokines and chemokines by a phospholipase C inhibitor in human peripheral blood mononuclear cells. *Clin. Diagn. Lab. Immunol.* **2001**, *8*, 449–453.

Assessment of the Functional Regions of the Superantigen Staphylococcal Enterotoxin B

Lily Zhang and Thomas J. Rogers

Abstract: The functional activity of superantigens is based on capacity of these microbial proteins to bind to both the β-chain of the T cell receptor (TcR) and the major histocompatibility complex (MHC) class II dimer. We have previously shown that a subset of the bacterial superantigens also binds to a membrane protein, designated p85, which is expressed by renal epithelial cells. This binding activity is a property of SEB, SEC1, 2 and 3, but not SEA, SED, SEE or TSST. The crystal structure of the tri-molecular complex of the superantigen staphylococcal enterotoxin B (SEB) with both the TcR and class II has previously been reported. However, the relative contributions of regions of the superantigen to the overall functional activity of this superantigen remain undefined. In an effort to better define the molecular basis for the interaction of SEB with the TcR β-chain, we report studies here which show the comparative contributions of amino- and carboxy-terminal regions in the superantigen activity of SEB. Recombinant fusion proteins composed of bacterial maltose-binding protein linked to either full-length or truncated toxins in which the 81 N-terminal, or 19 or 34 C-terminal amino acids were deleted, were generated for these studies. This approach provides a determination of the relative strength of the functional activity of the various regions of the superantigen protein.

Reprinted from *Toxins*. Cite as: Zhang, L.; Rogers, T.J. Assessment of the Functional Regions of the Superantigen Staphylococcal Enterotoxin B. *Toxins* **2013**, *5*, 1859-1871.

1. Introduction

The staphylococcal enterotoxins are members of a family of gram-positive pyrogenic exotoxins possessing superantigen activity. The staphylococcal enterotoxins, toxic syndrome shock toxin-1 (TSST-1), and streptococcal pyrogenic exotoxins are structurally related proteins [1–3]. These bacterial superantigens possess two common properties. First, they bind with moderate affinity to major histocompatibility complex (MHC) class II dimers [4,5], and second, these toxins are recognized by the T cell receptor (TcR) in a β-chain variable region allele-selective fashion [6,7]. It is now understood that the bacterial superantigens possess additional functional activities which are likely to involve other binding sites. For example, emetic activity which is dependent on the disulfide loop [8], and at least two binding sites which have been identified that are involved in epithelial cell binding activity [9,10].

The structural basis for the superantigen activity of these toxins has been the subject of intense research. Several approaches have been used to determine the regions of the bacterial superantigens involved in the mitogenic activity of these toxins. These include the use of synthetic peptides corresponding to regions of staphylococcal enterotoxin A (SEA) to block the activity of the native toxin [11,12], the analysis of the activity of proteolytic digestion fragments of several of the toxins including SEA, SEB, SEC1, SEC2, and TSST-1 [13–19], the use of monoclonal antibodies specific for identified epitopes to neutralize superantigen activity [16–18], the characterization of recombinant mutant SEA, SEB, TSST-1, and streptococcal pyrogenic exotoxin A (SPEA) containing amino acid substitutions [20–24], and the analysis of toxin chimeras to localize regions involved in the TcR Vβ allele selectivity [25,26]. Results from experiments carried out to determine the location of epitopes responsible for the biological activities of the superantigens have frequently appeared contradictory.

Several of the bacterial superantigens have now been crystallized [27–30], and it has been suggested that the staphylococcal and streptococcal superantigens can be grouped evolutionarily [31,32]. Analysis of the SEB crystal structure indicates that this protein consists of two domains with predominant β sheet structure, and the amino- and carboxy-termini of SEB are in close proximity. The crystal structure data of the SEB/class II complex [33], and the complex of SEB with the TcR [34–36] have provided evidence that the both amino- and carboxy-terminal residues of SEB may participate in interaction with both the MHC class II and the TcR. However, the relative contributions of residues in these regions to the binding interactions of these toxins have not been entirely clarified.

An additional approach utilized to identify regions of the toxins which contribute to superantigen activity involves the generation of recombinant truncations or internal deletions [37–41]. Hedlund *et al.* [37] have reported results showing that an N-terminal 106-amino acid SEA truncation possesses normal MHC class II-binding activity, suggesting that the region 107–233 possesses the epitopes strongly involved in this function. On the other hand, additional work [41] has shown that deletion of as few as 60 N-terminal amino acids results in greatly impaired SEB activity. It is clear, however, from the solution of the crystal structures of SEB [27] that the C-terminus of these bacterial superantigens folds back onto the N-terminal residues.

One limitation in the use of a superantigen with altered structure is that the mutation may lead to unexpected conformational changes, or decreased stability. Several investigators have employed the fusion protein approach in an effort to stabilize the structurally altered proteins. Buelow *et al.* [39] have shown that certain C- and N-terminal truncations of SEB fused to protein A possess mitogenic activity. They also found, however, that the use of protein A resulted in significantly impaired fusion protein activity for both the full-length SEB and SEB truncations.

In the present report, we describe a characterization of the functional activity of N- and C-terminal truncations of SEB using the recombinant truncation fused to bacterial maltose-binding protein (MBP). In an effort to assess superantigen activity of truncated SEB and the full-length SEB as fusion proteins, we have measured mitogenic activity, MHC class II-binding activity, and TcR Vβ allele selectivity. Our results provide further information regarding the participation of both N-terminal and C-terminal residues in the superantigen activity of this toxin, and demonstrate that MBP fusions can be utilized to assess the functional activities of these microbial toxins.

2. Results and Discussion

2.1. Proliferative Responses of Murine Splenocytes to MBP-SEB Fusion Proteins

We compared the mitogenic activity of the fusion proteins with that of SEB. The results of a representative experiment (Figure 1) show that the full-length SEB fusion protein (SEB-MBP) induced a strong response, and the mitogenic activity of native full length SEB was approximately twice that of SEB-MBP (SEB ED$_{50}$: 70 pmol; SEB-MBP ED$_{50}$: 170 pmol). Surprisingly, the proliferative response induced by the truncated-SEB fusion proteins nΔ81SEB-MBP was also substantial, and was not significantly different from the full-length fusion protein (ED$_{50}$ for nΔ81SEB-MBP: 180 pmol). In contrast, the C-terminal truncations exhibited significantly reduced mitogenic activity relative to the full-length fusion protein. The cΔ19SEB-MBP fusion induced a response which was less than 20% of that observed with the full-length fusion (ED$_{50}$ for cΔ19SEB-MBP: 900 pmol). Control experiments show that MBP alone does not exhibit detectable mitogenic activity (data not shown). We have observed essentially identical results in experiments carried out with BALB/c and B10.BR mice. The results of these experiments suggest that the 81 amino-terminal and 19 carboxy-terminal amino acids of SEB are not mandatory for significant mitogenic activity. It is clear, however, that the potency of the cΔ19SEB-MBP fusion protein is greatly reduced relative to the full-length fusion protein or to native SEB alone. On the other hand, the deletion of 34 carboxy-terminal amino acids (cΔ34SEB-MBP) appears to eliminate virtually all of the mitogenic activity of the superantigen.

2.2. Analysis of Fusion Protein Binding to HLA Class II Antigens

We attempted to analyze the HLA class II-binding characteristics of the MBP-SEB fusion proteins. Our experiments were carried out with fibroblasts transfected with HLA-DR1 (DAP.3-DR1). The results of binding experiments (Figure 2) show that these cells bind SEB with a dissociation constant (kd) of 141 nM. The Scatchard analysis is consistent with a binding density of 40,000 binding sites per cell. We then carried out an analysis of the capacity of each of the fusion proteins to compete with SEB for binding to the HLA-DR-bearing cell line. The results of a representative experiment show (Figure 2) that SEB-MBP and cΔ19SEB-MBP bind to the transfected fibroblasts in a manner which is essentially equivalent to that of SEB. The average concentrations of SEB-MBP and cΔ19SEB-MBP required to achieve 50% competition with SEB for binding to DAP.3-DR1 (determined from the mean of four experiments) are approximately 32 and 35 pmol, respectively, compared with about 25 pmol for SEB. The binding by nΔ81SEB-MBP appears to be somewhat weaker, and the concentration for 50% of binding of SEB to DAP.3-DR1 is 85 pmol. These results suggest that the binding affinities of SEB, SEB-MBP, and cΔ19SEB-MBP for HLA-DR1 are essentially equivalent. It is clear that the cΔ34SEB-MBP fusion protein fails to exhibit any detectable competition for the binding of SEB to these cells. In addition, control experiments have shown that non-transfected fibroblasts fail to exhibit detectable binding by SEB, and finally, the MBP carrier protein does not exhibit detectable competition for the binding of SEB to DAP.3-DR1 (data not shown).

Figure 1. Proliferative response of murine C3H/HeJ splenocytes to staphylococcal enterotoxin B (SEB) or SEB-MBP fusion proteins. The proliferative response to various concentrations of SEB, SEB-MBP, nΔ81SEB-MBP, cΔ19SEB-MBP, and cΔ34SEB-MBP is shown. The response to MBP alone was not detectable (data not shown). Results show the mean of quadruplicate values ± standard deviation. The control responses (no mitogen added) were 6584 ± 890 cpm.

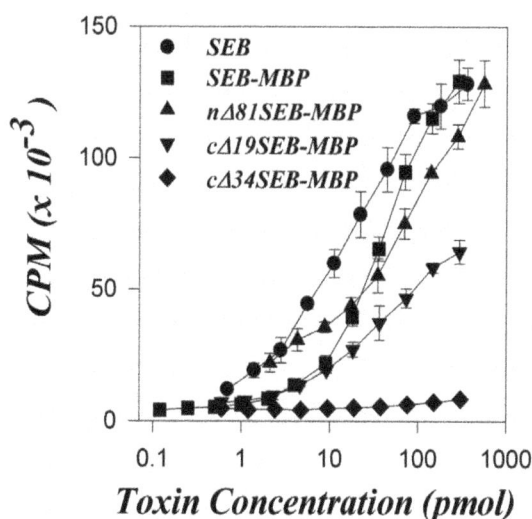

2.3. Growth of Murine T Cells Following Stimulation with Fusion Proteins

It is well established that T cells stimulated with bacterial superantigens expand in a TcR Vβ-allele selective manner in the presence of IL-2 [42]. We attempted to characterize the Vβ-allele selectivity following stimulation with the MBP-SEB fusion proteins. Following three days of stimulation with SEB or the fusion proteins SEB-MBP, nΔ81SEB-MBP or cΔ19SEB-MBP, purified murine T cells were cultured for two days with interleukin-2 (IL-2), and the surface TcR Vβ-allele expression was determined. Our results (Table 1) show the expected expansion of cells bearing the responsive TcR Vβ 8.1 and 8.2 alleles following stimulation with either SEB or the fusion proteins. Responding cells which bear CD25 and Vβ8.1 and 8.2 increase from 26.7% to between 53.4% and 68.3% of the total population following SEB or fusion protein stimulation. T cells which bear the nonresponsive TcR Vβ6 allele are reduced from 8.7% in the concanavalin A (Con A) control group, to between 1.5% and 3.3% in the SEB and fusion protein groups. Analysis of cΔ34SEB-MBP or MBP alone was not included in these studies, because these agents are not mitogenic and do not yield detectable levels of CD25-bearing T cells at the termination of culture.

Figure 2. Analysis of fusion protein binding to HLA-DR1-bearing cells. Binding of radiolabelled SEB to DAP.3-DR1 was carried out in competition with unlabelled SEB, SEB-MBP, nΔ81SEB-MBP, cΔ19SEB-MBP, and cΔ34SEB-MBP. The insert shows a representative Scatchard analysis for binding of SEB to DAP.3-DR1 cells. The inserts represent plots of bound/free (ordinate) vs. bound (abscissa).

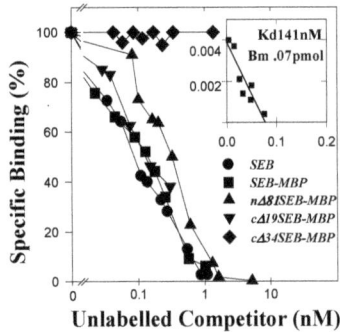

Table 1. Frequency of murine TcR Vβ alleles following stimulation of C3H/HeJ mice with Con A, SEB, SEB-MBP, nΔ81SEB-MBP, or cΔ19SEB-MBP. Results are expressed as the percentage of T cells co-expressing CD25 and the respective TcR Vβ allele, and are the means (±SEM) of four independent experiments.

Mitogen	% of Total T cells				
	Vβ6	Vβ8.1	Vβ8.2	Vβ8.3	Vβ7
Con A	9.3 ± 0.7	6.4 ± 1.0	20.6 ± 3.3	6.0 ± 0.4	10.7 ± 2.3
SEB	4.8 ± 1.6	11.0 ± 0.6	41.7 ± 4.0	20.9 ± 4.1	15.0 ± 1.2
SEB-MBP	3.9 ± 2.2	11.9 ± 0.6	42.8 ± 2.8	20.2 ± 4.3	14.6 ± 0.5
nΔ81SEB-MBP	3.0 ± 2.6	13.1 ± 0.9	44.4 ± 3.9	23.5 ± 3.8	16.1 ± 2.4
cΔ19SEB-MBP	3.9 ± 2.9	13.9 ± 0.8	49.5 ± 9.3	22.4 ± 3.7	17.0 ± 2.1

A variety of approaches have been utilized to identify the structural basis of bacterial superantigen activity. We have employed the recombinant truncation/deletion method in an effort to characterize the role of amino-terminal and carboxy-terminal amino acids in superantigen function. Our results strongly suggest that the amino-terminal 81 and carboxy-terminal 19 residues are not mandatory for substantial superantigen activity. These results may or may not be consistent with the experimental findings of some of the investigators who have examined the role of the amino-terminal region of the bacterial superantigens. Most noteworthy are the results of investigators who have generated mutant toxins by site-specific mutagenesis [20–24]. Using this approach, Kappler et al. [42] have identified three regions within the amino-terminal 61 amino acids of SEB which appeared to be involved in either the interaction with MHC class II or with the TcR. Harris et al. [22] generated mutant SEA toxins with amino acid substitutions at positions 25, 47, and 48 which failed to exhibit normal mitogenic activity. On the other hand, a number of investigators have shown by amino acid substitution analysis with SEA, SEB, SEE, SPEA and

TSST-1 that numerous residues in both the amino- and carboxy-terminal regions of these toxins appear to be involved in superantigen function [20,22,23,26]. Based on the crystal structure of both SEB and TSST-1 [27,28], the residues identified by these investigators appear to reside on at least three separate faces of the superantigen.

The crystal structure data of the complex of SEB with HLA-DR1 [33] provides evidence that the both amino- and carboxy-terminal residues of SEB participate in the interaction with the MHC class II. These sites are composed primarily of residues in the regions 43–47, 65–78, 92–96 and 211–215. It is apparent from the crystal structure that roughly half of the residues of SEB which interact with class II are located in the amino-terminal 81 amino acids, and our studies with the truncation of these N-terminal amino acids retains substantial class II binding activity. However, we do observe a measureable reduction in class II binding activity with this truncation. On the other hand, our results suggest that the amino acids in the carboxy-terminal 158 residues contribute significantly to the class II interaction, and this is likely due to the contribution of the remaining class II-binding residues. Moreover, a comparison of our results with the two carboxy-terminal truncations shows substantial activity with the 1–220 region, but no detectable activity with the 1–205 region. This suggests that very critical MHC binding residues are located in the 206–220 region, and this would be fully in agreement with published crystal structure data showing MHC II contact sites in the 211–215 region [33].

The crystal structure of the complex of SEB with the TcR β-chain reveals contact sites on the SEB surface that are distributed over both N-terminal and C-terminal regions [34–36]. These contact residues include T18, G19, L20, E22, N23, N60, Y91, F177, E210, and L214. Our results here show that substantial mitogenic activity is retained following deletion of the N-terminal 81 amino acids, and this suggests that the loss of the TcR contact residues T18, G19, L20, E22, N23, and N60 are not mandatory for interaction with the TcR, both in terms of the proliferative response, and the TcR Vβ-allele specificity of the response. In contrast, the mitogenic activity of the C-terminal truncations is much more substantially reduced, suggesting that the contact sites in this region have more substantial TcR interaction activity. Analysis of the results with the carboxy-terminal truncations suggests that residues in the 221–239 region contribute substantially to the interaction with the TcR, since the proliferative response is clearly reduced with this truncation, yet the interaction of this truncation with MHC-II remains essentially intact. It should be appreciated that while the interaction of the 1–220 region with TcR may be reduced (based on reduced mitogenic activity), the Vβ-selectivity of this truncation remains unaltered.

An additional approach to the question of the location of structural epitopes involved in superantigen function has been the generation of toxin chimeras. Results using SEA/SEE chimeras [25,26] have suggested that residues corresponding to residues 208 and 209 of SEB participate in TcR β-chain allele selectivity. The crystal structure of SEB shows that residues in this region form a part of the surface of one face of the toxin. The location of the TcR interaction site in the region which includes the carboxy-terminal residues is consistent with the results reported here. The mitogenic activity of the fusion protein cΔ19SEB-MBP is significantly reduced relative to full-length SEB, while the class II-binding activity of this fusion protein is essentially normal. These results suggest that the altered mitogenic activity of cΔ19SEB-MBP is due primarily to an altered ability to bind to the TcR.

We have examined the TcR Vβ-allele selectivity of the fusion proteins by fluorescence-activated cell sorter analysis. Our results show similar expansion of the responsive Vβ alleles in each of the fusion proteins when compared to the wild-type toxin. These results suggest that the reduced mitogenic activity of cΔ19SEB-MBP is not due to an apparent failure to selectively activate T cells. It should be pointed out, however, that analysis of this kind is not extremely sensitive to minor changes in selectivity, and it is possible that certain Vβ alleles may be more or less affected by loss of the 19 carboxy-terminal residues.

The superantigen truncation/MBP fusion protein approach described in these studies should have value in identifying regions of the superantigen responsible for toxin activities which are dependent on other binding interactions. For example, human colon carcinoma cells bind SEB in a class II-independent manner [43]. Moreover, we have previously identified a membrane protein expressed by a renal epithelial cell line, designated p85, which binds to SEB, SEC1, 2, and 3 [44,45]. Our previous studies, using this truncation/fusion protein approach, showed that the binding of SEB to p85 was dependent on residues in the C-terminal 19 amino acids since deletion of these residues eliminates all detectable binding activity. Our attempts to identify this protein have been unsuccessful at this point, and we have not been able to identify a membrane protein with this molecular mass which would be a likely candidate [44,45]. Indeed, we have previously eliminated the possibility that this protein is either MHC-II or MHC-I, and we established that SEB bound to p85 on epithelial cells does not allow for a productive interaction with the TcR on T cells [44,45]. The most likely interpretation of this result is that the complex of SEB with p85 does not result in access to critical TcR binding residues (most likely within the carboxy-terminal 19 amino acids). More recently, studies have shown that a dodecapeptide region of SEB (SEB152–161) is involved in the transcytosis of enterotoxins across intestinal epithelial cell monolayers [10]. The precise nature of the functional activity of this region is not clear at this time, but additional binding studies with epithelial cell populations may reveal more detailed information about the interaction of SEB with these cell types. We suggest that the deletion/MBP fusion method may be particularly useful as these studies progress.

Taken together, the results suggest that the activation of T cells by SEB involves interactions between multiple cell populations. It is generally accepted that for superantigens like SEB and SEC, the binding of superantigen to the MHC-II expressed by antigen-presenting cells (predominantly dendritic cells and macrophages) occurs first, and this binding stabilizes and concentrates superantigen for presentation to the appropriate T cells [36]. Depending on the anatomical site, it is possible that epithelial cells in the environment may also bind (via p85) these superantigens, and essentially "block" productive interaction with T cells. However, for the antigen-presenting cells that bind SEB via MHC-II, the final step is for the complex of MHC-II-SEB to initiate a productive binding interaction with the TcR. This then leads to induction of TcR signaling cascades that can lead to T cell activation and proliferation in a TcR Vβ allele-specific manner. One could view the activation of T cells with SEB as a part of a dynamic interaction of T cells with SEB-bound to epithelial cells (non-productive for the T cell), and SEB-bound to antigen-presenting cells (productive for the T cell). This suggests that competition between epithelial cells and antigen-presenting cells for superantigen binding may dictate successful T cell activation and proliferation.

3. Experimental Section

3.1. Bacterial Strains and Plasmids

Vectors used to generate the MBP fusion proteins, and for the expression of SEB, have been described in detail previously [45]. Briefly, the MBP fusion protein vector pMAL-C2 (New England Biolabs, Inc., Beverly, MA, USA) contains the *malE* gene upstream of a multiple cloning site, and the *entB* sequence, and truncations of the *entB* gene, were inserted to allow for the generation of MBP-fusion proteins with the carrier protein at the N-terminus of the inserted protein. The pMAL-C2 vector was engineered to remove the *malE* signal sequence in order to prevent the transport of MBP to the periplasmic space.

3.2. SEB Constructs

The construction of MBP-SEB expression vectors carrying full length, and truncated, *entB* sequences, including the procedures for cloning and PCR amplification, have been described previously in detail [45]. The constructs utilized for the present studies were pTR6532 (full-length SEB), pTR65816 (N-terminal 81 amino acid truncation), pTR65192 (C-terminal 19 amino acid truncation), and pTR65342 (C-terminal 34 amino acid truncation).

3.3. Production of Fusion Proteins

Fusion proteins were prepared from cultures of transformed *E. coli* as previously described [45]. Briefly, bacterial cultures grown to mid-log phase were treated with 1 mM isopropyl-β-D-thiogalactoside, grown for an additional 3–4 h at 37 °C, and the bacteria were harvested and lysed with 25 mM Tris, pH 7.4, 10 mM EDTA, and 0.3% lysozyme. The treated bacteria were then subjected to rapid freeze-thaw, sonicated for 2 min, 0.5 M NaCl was added, followed by centrifugation at 10,000g for 30 min at 4 °C. The lysate was collected and purified by chromatography on an amylose column (New England Biolabs). Fusion proteins were eluted from the amylose column with 10 mM maltose. The purified proteins (>95% pure) were subjected to SDS-PAGE analysis and the identity of proteins was confirmed based on electrophoretic mobility and reactivity by western blot analysis (the relative molecular mass values: SEB-MBP 70.6 kDa, nΔ81SEB-MBP 62.5 kDa, cΔ19SEB-MBP 68.0 kDa, and cΔ34SEB-MBP 66.0 kDa). The western blot analysis was conducted using both polyclonal anti-MBP antibody (New England Biolabs, Beverly, MA, USA) and a monoclonal anti-SEB antibody, 2GD9, which recognizes a determinant in the C-terminal 140-amino acid region [18,46].

3.4. Proliferative Response Assay

The proliferative response of murine splenocytes to SEB and the MBP fusion proteins was carried out as described previously [15], using endotoxin-unresponsive C3H/HeJ splenocytes.

Cultures of splenocytes (8×10^5 cells in a volume of 0.2 mL in Dulbecco's modified Eagle's medium (DMEM) supplemented with 10% FCS and 50 µg mL^{-1} gentamicin). After 48 h, cultures were given 1 µCi of ^{3}H-thymidine, and the cells were harvested after an additional 18 h. The proliferative response was assessed by measuring thymidine uptake. Mitogenic activity was assessed in part by determining the effective dose to elicit 50% of the maximal response (ED$_{50}$).

3.5. Radiolabelled Cell-Binding Assay

The binding assay for the HLA-DR1-transfected murine DAP.3 clone DAP.3-DR1 [47] was carried out by a standard binding assay as described previously [48,49]. Briefly, the DAP.3-DR1 cells at a density of 2×10^6 in 100 µL of cold competitor were diluted in a binding medium composed of DMEM containing 1% bovine serum albumin (BSA), 25 mM HEPES, and 0.05% azide. An additional 100 µL of ^{125}I-SEB was added immediately, and the cells were incubated at 37 °C for 4 h. The cells were washed with binding medium, and then treated with 1 N NaOH. The radioactivity of the dissolved cells was determined with a gamma counter. SEB was radio-iodinated using the iodo-bead method as described previously [45]. A specific activity of $5–6 \times 10^5$ DPM/pmol of SEB was normally achieved.

3.6. Flow Cytometry Analysis of Superantigen-Induced Murine T Cells

Using a minor modification of a standard protocol [7,49], murine T cells obtained from lymph nodes were purified by nylon wool and cultured at a density of 2×10^6/mL with 4×10^6 irradiated splenocytes and either Con A (10 µg mL^{-1}), SEB (1 µg mL^{-1}), MBP-SEB (2 µg mL^{-1}), or nΔ81SEB-MBP (8 µg mL^{-1}), or cΔ19SEB-MBP (10 µg mL^{-1}) in DMEM supplemented with 0.1 mM non-essential amino acids, 1 mM sodium pyruvate, 50 µg mL^{-1} gentamicin, 2 mM L-glutamine, 10% FCS, 0.05 mM β-mercaptoethanol, and 10 µg mL^{-1} each of adenosine, uridine, cytosine and guanosine. The cells were harvested after 3 days, and viable T cells were returned to culture with IL-2 (25 U mL^{-1}), cultured for an additional 2 days, and staining with antibodies for FACS analysis. Fluorescent antibody staining was carried out as described previously [50,51]. All monoclonal antibodies were obtained from Becton-Dickinson Laboratories (San Diego, CA, USA), and the cytometry was conducted using an EPICS Elite analyzer (Coulter Corporation, Hialeah, FL, USA).

4. Conclusions

We describe an approach to the localization of functional regions of superantigens in which truncations and/or deletions can be studied as fusion proteins using the maltose-binding protein. Our studies with SEB suggest that this approach allows for determination of T cell activation activity, as well as binding interactions with MHC class II or the alternative superantigen binding protein p85 since MBP does not possess functional activity on its own for either of these interactions. Finally, the studies presented here suggest that the superantigen functional activity of SEB is dominated by the contributions of residues in the C-terminal region of the protein.

Acknowledgments

This work was supported by grants AI23828 from the National Institutes of Health. We wish to thank Gregory Harvey for his editorial assistance.

Conflicts of Interest

The authors declare no conflict of interest.

References

1. Bergdoll, M.S. Enterotoxins. In *Staphylococci and Staphylococcal Infections*; Easmon, C.S.F., Adlam, C., Eds.; Academic Press: New York, NY, USA, 1983; Volume 2, p. 559.
2. Marrack, P.; Kappler, J. The staphylococcal enterotoxins and their relatives. *Science* **1990**, *248*, 705–711.
3. Herman, A.; Kappler, J.W.; Marrack, P.; Pullen, A.M. Superantigens: Mechanism of T cell stimulation and role in immune responses. *Ann. Rev. Immunol.* **1991**, *9*, 745–772.
4. Fraser, J.D. High affinity binding of staphylococcal enterotoxins A and B to HLA DR. *Nature* **1989**, *339*, 221–223.
5. Mollick, J.A.; Cook, R.G.; Rich, R.R. Class II molecules are specific receptors for staphylococcus enterotoxin A. *Science* **1989**, *244*, 817–820.
6. Kappler, J.; Kotzin, B.; Herron, L.; Gelfand, E.W.; Bigler, R.D.; Boylston, A.; Carrel, S.; Posnett, D.N.; Choi, Y.; Marrack, P. Vβ specific stimulation of human T cells by staphylococcal toxins. *Science* **1989**, *244*, 811–813.
7. White, J.; Herman, A.; Pullen, A.M.; Kubo, R.; Kappler, J.W.; Marrack, P. The V beta specific superantigen staphylococcal enterotoxin B: Stimulation of mature T cells and clonal deletion in neonatal mice. *Cell* **1989**, *56*, 27–35.
8. Hovde, C.; Marr, J.C.; Hoffmann, M.L.; Hackett, S.P.; Chi, Y.-I.; Crum, K.K.; Stevens, D.L.; Stauffacher, C.V.; Bohach, G.S. Investigation of the role of the disulphide bond in the activity and structure of staphylococcal enterotoxin C1. *Mol. Microbiol.* **1994**, *13*, 897–909.
9. Arad, G.; Levy, R.; Hillman, D.; Kaempfer, R. Superantigen antagonist protects against lethal shock and defines a new domain for T-cell activation. *Nat. Med.* **2000**, *6*, 414–421.
10. Shupp, J.W.; Jett, W.; Pontzer, C.H. Identification of a transcytosis epitope on staphylococcal enterotoxins. *Infect. Immun.* **2002**, *70*, 2178–2186.
11. Pontzer, C.H.; Russell, J.K.; Johnson, H.M. Structural basis for differential binding of staphylococcal enterotoxin A and toxic shock syndrome toxin 1 to class II major histocompatibility molecules. *Proc. Natl. Acad. Sci. USA* **1991**, *88*, 125–128.
12. Griggs, N.D.; Pontzer, C.H.; Jarpe, M.A.; Johnson, H.M. Mapping of multiple binding domains of the superantigen staphylococcal enterotoxin A for HLA. *J. Immunol.* **1992**, *148*, 2516–2521.
13. Spero, L.; Morlock, B.A. Biological activities of the peptides of staphylococcal enterotoxin C formed by limited tryptic hydrolysis. *J. Biol. Chem.* **1978**, *253*, 8787–8791.

14. Ezepchuk, Y.V.; Noskov, A.N. NH2 terminal localization of that part of the staphylococcal enterotoxins polypeptide chain responsible for binding with membrane receptor and mitogenic effect. *Int. J. Biochem.* **1986**, *18*, 485–488.

15. Edwin, C.; Parsonnet, J.; Kass, E.H. Structure activity relationship of Toxic Shock Syndrome Toxin 1: Derivation and characterization of immunologically and biologically active fragments. *J. Infect. Dis.* **1988**, *158*, 1287–1295.

16. Edwin, C.; Kass, E.H. Identification of functional antigenic segments of Toxic Shock Syndrome Toxin 1 by differential immunoreactivity and by differential mitogenic responses of human peripheral blood mononuclear cells, using active toxin fragments. *Infect. Immun.* **1989**, *57*, 2230–2236.

17. Bohach, G.A.; Handley, J.P.; Schlievert, P.M. Biological and immunological properties of the carboxyl terminus of staphylococcal enterotoxin C1. *Infect. Immun.* **1989**, *57*, 23–28.

18. Binek, M.; Newcomb, J.R.; Rogers, C.M.; Rogers, T.J. Localisation of the mitogenic epitope of staphylococcal enterotoxin B. *J. Med. Microbiol.* **1992**, *36*, 156–163.

19. Alakhov, V.; Klinsky, E.; Kolosov, M.I.; Maurer-Fogy, I.; Moskaleva, E.; Sveshnikov, P.G.; Pozdnyakova, L.P.; Shemchukova, O.B.; Severin, E.S. Identification of functionally active fragments of staphylococcal enterotoxin B. *Eur. J. Biochem.* **1992**, *209*, 823–828.

20. Grossman, D.; Van, M.; Mollick, J.A.; Highlander, S.K.; Rich, R.R. Mutation of the disulfide loop in staphylococcal enterotoxin A. Consequences for T cell recognition. *J. Immunol.* **1991**, *147*, 3274–3281.

21. Kappler, J.W.; Herman, A.; Clements, J.; Marrack, P. Mutations defining functional regions of the superantigen staphylococcal enterotoxin B. *J. Exp. Med.* **1992**, *175*, 387–396.

22. Harris, T.O.; Grossman, D.; Kappler, J.W.; Marrack, P.; Rich, R.R.; Betley, M.J. Lack of complete correlation between emetic and T cell stimulatory activities of staphylococcal enterotoxins. *Infect. Immun.* **1993**, *61*, 3175–3183.

23. Hartwig, U.F.; Fleischer, B. Mutations affecting MHC class II binding of the superantigen streptococcal erythrogenic toxin A. *Int. Immunol.* **1993**, *5*, 869–875.

24. Murray, D.L.; Prasad, G.S.; Earhart, C.A.; Leonard, B.A.; Kreiswirth, B.N.; Novick, R.P.; Ohlendorf, D.H.; Schlievert, P.M. Immunobiologic and biochemical properties of mutants of Toxic Shock Syndrome Toxin 1. *J. Immunol.* **1994**, *152*, 87–95.

25. Irwin, M.J.; Hudson, K.R.; Fraser, J.D.; Gascoigne, N.R. Enterotoxin residues determining T cell receptor V beta binding specificity. *Nature* **1992**, *359*, 841–843.

26. Mollick, J.A.; MCmasters, R.L.; Grossman, D.; Rich, R.R. Localization of a site on bacterial superantigens that determines T cell receptor beta chain specificity. *J. Exp. Med.* **1993**, *177*, 283–293.

27. Swaminathan, S.; Furey, W.; Pletcher, J.; Sax, M. Crystal structure of staphylococcal enterotoxin B, a superantigen. *Nature* **1992**, *359*, 801–806.

28. Acharya, K.R.; Passalacqua, E.F.; Jones, E.Y.; Karlos, K.; Stuart, D.I.; Brehm, R.D.; Tranter, H.S. Structural basis of superantigen action inferred from crystal structure of toxic shock syndrome toxin 1. *Nature* **1994**, *367*, 94–97.

29. Schad, E.M.; Zaitseva, I.; Zaitsev, V.N.; Dohlsten, M.; Kalland, T.; Schlievert, P.; Ohlendorf, D.H.; Svensson, L.A. Crystal structure of the superantigen staphylococcal enterotoxin type A. *EMBO J.* **1995**, *14*, 3292–3301.

30. Papageorgiou, A.C.; Tranter, H.S.; Acharya, K.R. Crystal structure of microbial superantigen staphylococcal enterotoxin B at 1.5 A resolution: Implications for superantigen recognition by MHC class II molecules and T-cell receptors. *J. Mol. Biol.* **1998**, *277*, 61–79.

31. Kozono, H.; Parker, D.; White, J.; Marrack, P.; Kappler, J. Multiple binding sites for bacterial superantigens on soluble class II molecules. *Immunity* **1995**, *3*, 187–196.

32. Sundberg, E.J.; Li, Y.; Mariuzza, R.A. So many ways of getting in the way: Diversity in the molecular architecture of superantigen-dependent T-cell signaling complexes. *Curr. Opin. Immunol.* **2002**, *14*, 36–44.

33. Jardetzky, T.S.; Brown, J.H.; Gorga, J.C.; Stern, L.J.; Urban, R.G.; Chi, Y.-I.; Stauffacher, C.; Strominger, J.L.; Wiley, D.C. Three-dimensional structure of a human class II histocompatibility molecule complexed with superantigen. *Nature* **1994**, *368*, 711–718.

34. Li, H.; Liera, A.; Tsuchiya, D.; Leder, L.; Ysern, X.; Schlievert, P.M.; Karjalainen, K.; Mariuzza, R.A. Three-dimensional structure of the complex between a T cell receptor β chain and the superantigen staphylococcal enterotoxin B. *Immunity* **1998**, *9*, 807–816.

35. Sundberg, E.J.; Deng, L.; Mariuzza, R.A. TCR recognition of peptide/MHC class II complexes and superantigens. *Semin. Immunol.* **2007**, *19*, 262–271.

36. Fraser, J.D.; Proft, T. The bacterial superantigen and superantigen-like proteins. *Immunol. Rev.* **2008**, *225*, 226–243.

37. Hedlund, G.; Dohlsten, M.; Herrmann, T.; Buell, G.; Lando, P.A.; Segren, S.; Schrimsher, J.; MacDonald, H.R.; Sjogren, H.O.; Kalland, T. A recombinant C terminal fragment of staphylococcal enterotoxin A binds to human MHC class II products but does not activate T cells. *J. Immunol.* **1991**, *147*, 4082–4085.

38. Hufnagle, W.O.; Tremaine, M.T.; Betley, M.J. The carboxyl terminal region of staphylococcal enterotoxin type A is required for a fully active molecule. *Infect. Immun.* **1991**, *59*, 2126–2134.

39. Buelow, R.; O'Hehir, R.E.; Schreifels, R.; Kummerehl, T.J.; Riley, G.; Lamb, J.R. Localization of the immunologic activity in the superantigen staphylococcal enterotoxin B using truncated recombinant fusion proteins. *J. Immunol.* **1992**, *148*, 1–6.

40. Harris, T.O.; Hufnagle, W.O.; Betley, M.J. Staphylococcal enterotoxin type A internal deletion mutants: Serological activity and induction of T cell proliferation. *Infect. Immun.* **1993**, *61*, 2059–2068.

41. Metzroth, B.; Marx, T.; Linnig, M.; Fleischer, B. Concomitant loss of conformation and superantigenic activity of staphylococcal enterotoxin B deletion mutant proteins. *Infect. Immun.* **1993**, *61*, 2445–2452.

42. Kappler, J.W.; Pullen, A.; Callahan, J.; Choi, Y.; Herman, A.; White, J.; Potts, W.; Wakeland, E.; Marrack, P. Consequences of self and foreign superantigen interaction with specific V beta elements of the murine TCR alpha beta. Cold Spring Harbor Sympos. *Quantit. Biol.* **1989**, *1*, 401–407.

43. Dohlsten, M.; Hedlund, G.; Segren, S.; Lando, P.A.; Herrmann, T.; Kelly, A.P.; Kalland, T. Human histocompatibility complex class II-negative colon carcinoma cells present staphylococcal superantigens to cytotoxic T lymphocytes: Evidence for a novel enterotoxin receptor. *Eur. J. Immunol.* **1991**, *21*, 1229–1233.

44. Rogers, T.J.; Guan, L.; Zhang, L. Characterization of an alternative superantigen binding site expressed on a renal fibroblast cell line. *Int. Immunol.* **1995**, *7*, 1721–1727.

45. Rogers, T.J.; Zhang, L. Structural basis for the interaction of superantigen with the alternate superantigen-binding receptor p85. *Mol. Immunol.* **1997**, *34*, 263–272.

46. Lin, Y.S.; Largen, M.T.; Newcomb, J.R.; Rogers, T.J. Production and characterization of monoclonal antibodies specific for staphylococcal enterotoxin B. *J. Med. Microbiol.* **1988**, *27*, 263–270.

47. Long, E.O.; Rosen-Bronson, S.; Karp, D.R.; Malnati, M.; Sekaly, R.P.; Jaraquemada, D. Efficient cDNA expression vectors for stable and transient expression of HLA-DR in transfected fibroblast and lymphoid cells. *Hum. Immunol.* **1991**, *31*, 229–235.

48. Donigan, A.M.; Cavalli, R.C.; Pena, A.A.; Savage, C.R.; Soprano, D.R.; Soprano, K.J. Epidermal growth factor receptors lose ligand binding ability as WI 38 cells progress from short term to long term quiescence. *J. Cell. Physiol.* **1993**, *155*, 164–170.

49. Julius, M.; Simpson, E.; Herzenberg, L. A rapid method for the isolation of functional thymus-derived lymphocytes. *Eur. J. Immunol.* **1973**, *3*, 645–649.

50. Briggs, C.; Garcia, C.; Zhang, L.; Guan, L.; Gabriel, J.L.; Rogers, T.J. Mutations affecting the superantigen activity of staphylococcal enterotoxin B. *Immunology* **1997**, *90*, 169–175.

51. Garcia, C.; Briggs, C.; Zhang, L.; Guan, L.; Gabriel, J.L.; Rogers, T.J. Molecular characterization of the putative T-cell receptor cavity of the superantigen staphylococcal enterotoxin B. *Immunology* **1998**, *94*, 160–166.

Toxic Shock Syndrome Toxin-1-Mediated Toxicity Inhibited by Neutralizing Antibodies Late in the Course of Continual *in Vivo* and *in Vitro* Exposure

Norbert Stich, Nina Model, Aysen Samstag, Corina S. Gruener, Hermann M. Wolf and Martha M. Eibl

Abstract: Toxic shock syndrome (TSS) results from the host's overwhelming inflammatory response and cytokine storm mainly due to superantigens (SAgs). There is no effective specific therapy. Application of immunoglobulins has been shown to improve the outcome of the disease and to neutralize SAgs both *in vivo* and *in vitro*. However, in most experiments that have been performed, antiserum was either pre-incubated with SAg, or both were applied simultaneously. To mirror more closely the clinical situation, we applied a multiple dose (over five days) lethal challenge in a rabbit model. Treatment with toxic shock syndrome toxin 1 (TSST-1) neutralizing antibody was fully protective, even when administered late in the course of the challenge. Kinetic studies on the effect of superantigen toxins are scarce. We performed *in vitro* kinetic studies by neutralizing the toxin with antibodies at well-defined time points. T-cell activation was determined by assessing T-cell proliferation (3H-thymidine incorporation), determination of IL-2 release in the cell supernatant (ELISA), and IL-2 gene activation (real-time PCR (RT-PCR)). Here we show that T-cell activation occurs continuously. The application of TSST-1 neutralizing antiserum reduced IL-2 and TNFα release into the cell supernatant, even if added at later time points. Interference with the prolonged stimulation of proinflammatory cytokines is likely to be *in vivo* relevant, as postexposure treatment protected rabbits against the multiple dose lethal SAg challenge. Our results shed new light on the treatment of TSS by specific antibodies even at late stages of exposure.

Reprinted from *Humanities*. Cite as: Stich, N.; Model, N.; Samstag, A.; Gruener, C.S.; Wolf, H.M.; Eibl, M.M. Toxic Shock Syndrome Toxin-1-Mediated Toxicity Inhibited by Neutralizing Antibodies Late in the Course of Continual *in Vivo* and *in Vitro* Exposure. *Toxins* **2014**, *6*, 1724-1741.

1. Introduction

Sepsis is the most common cause of death in critically ill patients, and it results from the overwhelming inflammatory response of the host [1], as well as from the inability of the immune system to limit bacterial spread during an ongoing infection. Numerous clinical trials have been conducted with the aim of blocking the uncontrolled inflammatory cascade, but with limited success [2–7]. The most common causative organisms in patients with sepsis are *Staphylococcus aureus*, *Pseudomonas aeruginosa* and *Escherichia coli* [8]. Whereas in the 1980s the most frequently identified organisms were Gram-negative bacteria, Gram-positive bacteria have accounted for the greatest proportion of hospital admissions with sepsis in the last decade [8,9]. This might be a consequence of the increasing prevalence of multiresistant organisms such as methicillin-resistant *S. aureus* [10] and the wider use of prostheses and invasive vascular devices [11]. *S. aureus* causes significant illnesses, including pneumonia, acute kidney injury, infective endocarditis, and toxic shock syndrome (TSS) [12]. Major contributors to these diseases are superantigens (SAgs), such as toxic shock syndrome toxin- 1 (TSST-1) and staphylococcal enterotoxin B (SEB), both of which remarkably hyperactivate the host's inflammatory response.

Numerous efforts have been undertaken to develop a specific therapy for TSS [13,14]. Therapies of sepsis have included the application of intravenous immunoglobulin (IVIG), which has been only partially useful [15]. Hyperimmune IVIG could be produced by vaccination with a recombinant attenuated SAg vaccine. These immunoglobulins could offer the advantage of both neutralizing SAgs and modulating the inflammatory reaction, e.g., by lowering the levels of circulating cytokines [16–18]. Since there is a strong relation between toxicity and increased serum levels of cytokines, many therapeutic approaches in animal models aimed at blocking these proinflammatory mediators [19–21]. However, anticytokine treatments have not been successful in clinical trials since sepsis is a complex process involving excessive and suppressed inflammatory and immune responses [22].

In studies using staphylococcal enterotoxin B (SEB), it has been shown that mouse and non-human primates were protected from SEB-induced TSS by the use of antibodies up to 4 h after toxin exposure [23]. Larkin *et al.* investigated the effect of monoclonal Fab fragments and whole monoclonal antibodies against SEB in an extensive kinetic study. Some of these antibodies bound their targets with very high affinity, and the protective effect in the mouse toxic shock model reached 68% [24]. In a rabbit model, IVIG reduced the toxic effects of exotoxins, but mainly when immunoglobulins and toxins were injected simultaneously and not when the application of antiserum was delayed [25]. Similar results were obtained *in vitro* when human PBMCs were stimulated with SEB, and T-cell responses could be inhibited by antibodies up to 12 h after SAg exposure [24].

In a rabbit infection model of TSS using *S. aureus* producing TSST-1, fatal disease could be inhibited by application of TSST-1-neutralizing monoclonal antibodies [26]. Notably, for protection in this model, the antibodies had to be given constantly before and during the challenge (on days−1, 0, 1). We chose the rabbit model, since the sensitivity and the susceptibility of humans and rabbits to SAgs is comparable. Furthermore, the pathological effects of SAgs are highly similar in humans and rabbits [12,27–29]. In the latter publication it was shown that rabbits could be protected from lethal pneumonia after having been challenged with SAg (SEB) followed by delayed administration of IVIG (up to 48 h). Moreover, it was shown that rabbit immune serum was protective when given prior to challenge.

Another strategy to limit the overproduction of cytokines and to elicit a powerful antibody response against SAgs such as TSST-1 is achieved by vaccination. Rabbits which received TSST-1 toxoids developed strong antibody titers that neutralized TSST-1 in TSS models *in vitro* and *in vivo* [30]. Mice vaccinated with mutant TSST-1 could be protected against *S. aureus*-induced septic death by neutralizing antibodies and downregulation of IFNγ production [16]. Anti-SAg antibodies are widespread among the human population, and there is a good correlation between antibody titers and the inhibition of superantigenic effects of these toxins [31]. SAg-specific antibodies from pooled sera could suppress T-cell proliferation *in vitro* and protect mice against SAg-induced TSS [31]. In these previous studies, the toxin-neutralizing effect of antibodies was mostly analyzed by *in vivo* and *in vitro* systems in which antibodies were present before toxin challenge (e.g., through vaccination): Antibodies and toxins were applied either simultaneously or after pre-incubation, or antibodies were given after a single challenge with toxin. However, patients usually receive clinical treatment several hours (if not days or weeks) after exposure to pathogens and their toxins, and during this lag period they are continuously exposed to an ongoing production of bacterial toxins. Continual exposure was achieved by inserting a pump, which showed that lethal doses were much lower under these conditions than with a bolus injection [27,30]. In the present study we applied defined amounts of recombinant TSST-1 wild-type (rTSST-1 wt) within a five-day period in a rabbit multiple dose challenge model, thus mimicking the clinical situation of systemic Gram-positive bacterial infection with continuous exposure to toxins over a longer time period. In this model, treatment with neutralizing antibodies given late in the course of toxin challenge fully protected from toxin-induced lethality.

A major component involved in the pathogenesis of Gram-positive sepsis is the ability of bacterial antigens to induce an exaggerated release of proinflammatory and immune cell-activating cytokines, e.g., TNFα, IFNγ, IL-6 and IL-2. The superantigenic characteristic to hyperactivate T cells might play a role in this context [12,32,33]. The effect of postexposure antibody treatment on T-cell activation, cytokine gene expression and the release of inflammatory cytokines following TSST-1 stimulation was assessed in an *in vitro* system. Toxin-neutralizing antiserum was added to PBMC cultures at different time points after superantigen exposure. Cytokine mRNA expression and protein secretion were analyzed in parallel in order to monitor gene expression patterns and cytokine release. Our results demonstrate that the release of cytokines into the supernatant decreases following neutralization of the stimulus, as long as the maximum cytokine concentration in the supernatant has not been reached. Cytokine mRNA is continuously produced upon stimulation with SAg toxin and is labile. Neutralization of the toxin stops cytokine gene transcription completely, if given within the first hours after toxin exposure.

2. Results and Discussion

2.1. Multiple Dose Lethal Challenge and Protection by Neutralizing Antibody

We first examined whether rabbits could be protected against multiple dose lethal challenge with rTSST-1 by pre-incubation of toxin with antiserum before application. The control group (two rabbits) received 30 μg of toxin per dose as explained in the legend to Table 1 (in this model the lethal dose range was between 20 and 40 μg of rTSST-1 wt: number of rabbits that survived/number of animals that were challenged: 20 μg per dose: 0/4; 30 μg/dose: 1/11; 40 μg per dose: 0/2; 100 μg per dose: 0/4). The rabbits died at days 5 and 6. The second group (three rabbits) received neutralized toxin (as described in the Experimental Section) for each injection. All rabbits survived until nine days *post* challenge. The third group (three rabbits) received toxin until day 3 (six injections) and neutralized toxin on days 4 and 5. All rabbits survived, were free of severe symptoms on day 5 and day 7, as determined by a veterinarian observing behavior, vital signs, mucosa of the nose, eyes and anus, and food and water uptake. The rabbits were euthanized 9 days *post* challenge.

We then examined whether rabbits could be protected against lethal challenge by passive transfer of TSST-1 antibodies injected at a site different from the one used for toxin administration. Multiple doses of 40 µg or 30 µg rTSST-1 were given (see legend to Table 1). One group of rabbits was injected with 2 mL of antiserum as described in the Experimental Section at a different site on days 2 and 3, while another group received 2 mL of antiserum on day 3 (after the 6th toxin dose), day 4 (after dose 7) and day 5 (after dose 9) (three doses). For timeline and detailed description of treatment see experimental section. The survival rates of all rabbits were monitored during these five days of continual challenge and for a further 7–9 days (Table 1).

Table 1. Survival of rabbits challenged with multiple doses rTSST-1 and effect of anti-TSST-1 antiserum.

Antiserum raised against	Days of treatment	Survival (No. of animals that survived/total No. of animals challenged)	
-	-	0/4 *	
Negative/irrelevant serum	1–5	0/3 [++]	
TSST-1 variant	2, 3	5/5 **	$p = 0.02$ ***
TSST-1 variant	3–5	5/5 **	$p = 0.02$
TSST-1wt	3– 5	5/5 **	$p = 0.02$

Notes: * Of 4 rabbits, 2 rabbits received 2 doses of 30 µg rTSST-1 (days 1–4) and 1 dose of 30 µg rTSST-1 (day 5), and 2 rabbits received 2 doses of 40 µg rTSST-1 (days 1–4) and 1 dose of 40 µg (day 5); [++] Two rabbits were treated with antiserum raised against *S. aureus* alpha toxin, one rabbit was treated with pre-immune serum (days 1–5); ** Of 5 rabbits, 3 received 2 doses of 30 µg rTSST-1 (days 1–4) and 1 dose of 30 µg rTSST-1 (day 5), and 2 rabbits received 2 doses of 40 µg rTSST-1 (days 1–4) and 1 dose of 40 µg rTSST-1 (day 5); *** Statistical significance of the difference between treatment and control group determined by chi-square analysis.

In these experiments, rabbits survived the rTSST-1 challenge when treated with antiserum either on days 2 and 3, or on days 3, 4 and 5. Both types of antiserum (gained via immunization with TSST-1-wild-type or TSST-1-variant) were equally protective. Rabbits without protective antiserum died between four days and one week after the first challenge. Our data demonstrate the efficacy of anti-TSST-1 antibodies for the protection of rabbits, even when given delayed during continual exposure to bacterial toxin in an animal model that more closely resembles the clinical situation in patients suffering from systemic gram-positive bacterial infection.

Multiple dose lethal challenge continual exposure, during which toxin concentrations are accumulated (over several days of treatment), leads to death. Since rTSST-1 challenge is not lethal for three days, treatment with immunoglobulin at that time point may have a prophylactic effect by neutralizing and preventing further accumulation of the toxin.

2.2. Inhibition of T-Cell Proliferation

As T-cell hyperactivation is a crucial step in the pathogenesis of Gram-positive systemic inflammatory syndrome, and given the positive effect of antiserum treatment late in the course of toxemia as described above, we examined whether the antisera used in our animal model of continual toxin exposure could also inhibit superantigen activation of human T cells *in vitro* when applied after toxin stimulation of the cells. As a first readout for T-cell activation, T-cell proliferation assays were performed. As described in the experimental section, human PBMCs were stimulated with rTSST-1 wt, and antiserum was added at different times thereafter. Figure 1 shows that T-cell proliferation was nearly completely inhibited when the antiserum was added at 2 h and 4 h after TSST-1 stimulation. Addition at 7 h resulted in 20% inhibition of T-cell proliferation. Application of antiserum at later times did not have any influence on the TSST-1-induced proliferation of T cells. In the course of these experiments, two different neutralizing antibodies (raised against rTSST-1 wt or rTSST-1 variant) were used, but the inhibitory effect obtained using antiserum against rTSST-1 variant (one of the mutations is at the T-cell receptor binding site) or rTSST-1 wt was comparable: ^{3}H thymidin incorporation at the end of the four-day culture period in the presence of negative serum was 60,186 ccpm ± 28,002 (mean ± SD, n = 4). When antiserum against rTSST-1 wt was added at 4 h, it was 25,110 ccpm (mean, n = 2). When antiserum raised against rTSST-1 mutant was added at 4 h, ^{3}H thymidin incorporation was 19,783 ccpm (mean, n = 2). Therefore, results from both antisera were pooled for analysis.

These results suggest that shortly after stimulation of the cells, before certain activation steps, e.g., IL-2 induction and IL-2 release into the culture supernatant have reached a threshold, antisera can still significantly inhibit superantigen-induced T-cell activation. Thereafter, T-cell activation is self-perpetuating, and it cannot be inhibited, even by complete neutralization of the stimulus still present in the cell culture.

2.3. Kinetics of Inhibition of Cytokine Expression after rTSST-1 wt Stimulation for Defined Time Periods

There is a strong association between superantigen toxicity and an exaggerated release of proinflammatory and immune cell activating cytokines [19–21]. Several therapeutic approaches in animal models aimed at blocking the action of these proinflammatory mediators [23–25]. To examine the effect of postexposure immune prophylaxis by antiserum treatment on the levels of superantigen-induced cytokines, we applied an *in vitro* system for monitoring the kinetics of the cytokine expression profile in human PBMC following stimulation with recombinant rTSST-1 wild-type (rTSST-1 wt) and sequential neutralization of the toxin. We chose the expression of two cytokines produced early *in vitro* and also early after applying the toxin *in vivo*. We selected IL-2 as an indicator of T-cell activation and determined TNFα as one of the cytokines produced by T cells and non-T cells that play a pivotal role in superantigen-mediated toxicity. The appropriate concentration of rTSST-1 wt (1 ng/mL) had been chosen by previous titration experiments (data not shown).

Figure 1. Inhibition of rTSST-1-induced T-cell proliferation by antiserum. Human PBMC were stimulated with 0.3 ng/mL rTSST-1 wt and cultured for four days as described in the experimental section. Antiserum generated against rTSST-1 was added at 2 h (n = 8), 4 h (n = 8), 7 h (n = 4), 9 h (n = 4), 24 h (n = 4), 2d (n = 4) or 3d (n = 4) after stimulation in a final dilution of 1:100. Each experiment was carried out in triplicate. Inhibition of proliferation is reported in percentage in relation to negative control serum, values represent the mean, and error bars indicate the standard deviation of multiple experiments. PBMC proliferative response following stimulation with rTSST-1 wt in the presence of negative serum was 60,186 ccpm $\pm$ 28,002 (mean $\pm$ SD, n = 4). The term "negative serum" refers to a serum derived from animals not immunized against TSST-1, *i.e.*, preimmune serum. PBMC proliferative response following stimulation with rTSST-1 wt without antiserum was 43,388 ccpm $\pm$ 6510 (mean $\pm$ SD, n = 4). Background proliferative response of cells cultured in medium alone was 1020 ccpm $\pm$ 462 (mean $\pm$ SD, n = 4). % inhibition was calculated as described in the experimental section using the following formula: % inhibition = 100 – ((ccpm ^{3}H thymidin incorporation in the presence of antiserum/ccpm ^{3}H thymidin incorporation in control cultures containing negative serum) × 100). An asterisk indicates a statistically significant ($p < 0.05$) inhibition (paired Student's *t*-test).

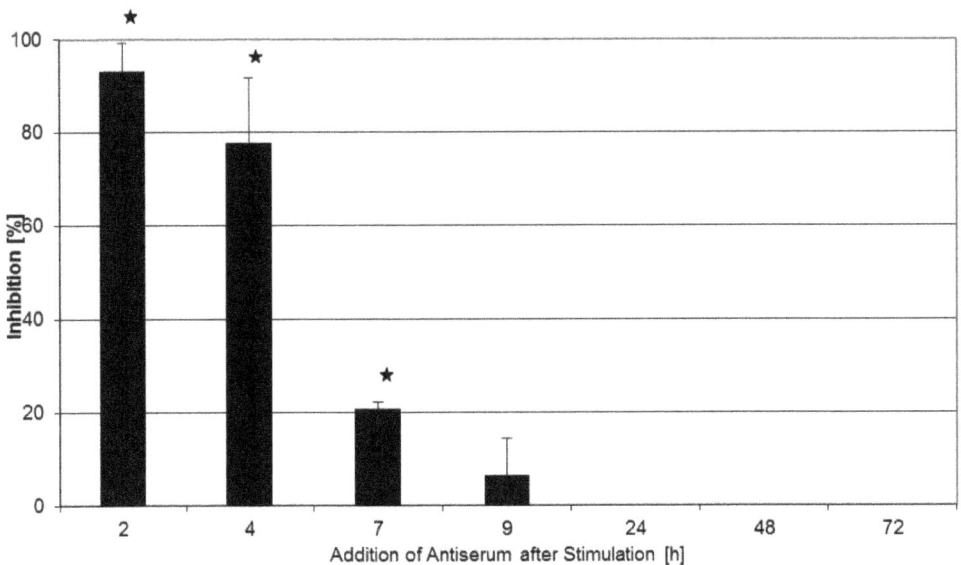

First, we examined human T-cell cytokine induction after rTSST-1 wt stimulation by studying the kinetics of IL-2 mRNA production applying RT PCR (Figure 2a,b). rTSST-1 wt led to a strong induction of IL-2 mRNA at 2 h, 3 h, 4 h, and 5 h with the maximum induction at 5 h (more than 200-fold) (Figure 2a). At 24 h, IL-2 mRNA production returned close to baseline. We then studied the kinetics of toxin neutralization by adding neutralizing antibody to the culture of toxin-stimulated PBMC at 4 time points (1, 2, 3 and 4 h after addition of TSST-1), and analyzed IL-2 mRNA expression after 5 h of stimulation (Figure 2b). Induction of IL-2 mRNA at 5 h was completely blocked when antiserum was added after 1 h and 2 h of TSST-1 stimulation. Addition of antiserum after 3 h or later was too late to completely inhibit induction of IL-2 message, but even if given as late as 4 h, it had a slight inhibitory effect on cytokine induction.

Moreover, the data depicted in Figure 2 indicate that the amount of superantigen-induced cytokine mRNA present in the cell culture results from a balance between ongoing production and continuous degradation, and that antiserum can significantly affect this steady state equilibration. Without antiserum, IL-2 mRNA levels were continuously increasing during five hours after toxin stimulation (Figure 2a). Neutralizing the toxin at 2, 3 and 4 h after toxin challenge led to a stop in newly produced IL-2 mRNA, with already induced cytokine mRNA being degraded. This process effectively led to decreased IL-2 mRNA levels when analyzed at 5 h (Figure 2b). Therefore, it appears likely that after neutralization of the stimulus, subsequent production of mRNA is abrogated and IL-2 mRNA decay seems to initiate immediately afterwards. The initiation of this effect could be observed immediately after neutralization. After 4 h of stimulation, newly produced mRNA accounts for more than 50% of the total IL-2 mRNA in the 5 h culture, and this increase is abolished by neutralization of the stimulus.

Studying the kinetics of IL-2 secretion after rTSST-1 wt exposure, we found that the highest concentration of IL-2 in our system was detected at 24 h in the supernatant of the cultures (22,043 pg/mL) (Figure 2c). We then examined whether IL-2 release could be inhibited by neutralizing TSST-1 antiserum if given as postexposure immune prophylaxis at defined time points. In the first set of experiments, IL-2 protein was analyzed in the supernatant after 5 h (Figure 2d). Subsequently, we followed IL-2 secretion over a period of 24 h, and added antiserum at 1 h, 3 h, 5 h, 7 h, 9 h, 20 h and 22 h (Figure 2e). The results depicted in Figure 2d,e show that early application of antiserum could efficiently inhibit the release of IL-2. An addition of antiserum at 1 h postexposure led to immediate and complete abrogation of IL-2 release. IL-2 concentration present at the respective time points of SAg neutralization did not significantly increase during the further culture period. Addition of neutralizing antibodies stopped protein secretion within the next hour. By comparing Figure 2c,d, one can see that protein levels with antiserum added at, e.g., 2 h (Figure 2d, IL-2 conc. 509 pg/mL), did not significantly exceed the levels produced in cultures at 3 h without antiserum (Figure 2c, IL-2 conc. 631 pg/mL).

However, even at late time points, such as neutralization after 7 h, a statistically significant inhibition could be observed, and the addition of antiserum at 20 h still showed a tendency towards reduced IL-2 secretion.

Figure 2. (**a**) Kinetics of IL-2 mRNA expression in human PBMC after stimulation with TSST-1 for designated time intervals. At the corresponding time point, cells were harvested for IL-2 mRNA fold induction analysis via RT PCR as indicated in the experimental section. The supernatant of these cultures served for determination of IL-2 concentration (see Figure 2c); (**b**) IL-2 gene activation in human PBMCs at 5 h after TSST-1 stimulation and inhibition at different time points with anti-TSST-1 antiserum. Antiserum was added at 1 h, 2 h, 3 h and 4 h followed by PBMC harvesting at 5 h. IL-2 mRNA expression was analyzed with real-time PCR as described in the experimental section. Stimulation with TSST-1 for 5 h without antiserum served as a control. Simultaneous addition of rTSST-1 wt and antiserum to PBMCs resulted in blocking of IL-2 mRNA expression (fold induction of 1; 1–2; median; interquartile range; n = 6). Simultaneous stimulation with rTSST-1 and negative serum led to a fold induction of 151, 109–167; (median, interquartile range) (n = 5); (**c**) IL-2 concentration in the supernatant of cultured human PBMCs stimulated with 1 ng/mL TSST-1 for indicated periods. At the designated time points, cells were harvested and IL-2 protein concentration was assessed by ELISA; (**d**) Amount of IL-2 after 5 h of stimulation with rTSST-1 wt and addition of antiserum at 1 h, 2 h, 3 h, and 4 h. When antiserum and TSST-1 were applied simultaneously for 5 h, we detected 367 pg/mL, 292–379 pg/mL; (median, interquartile range) IL-2 in the supernatant (n = 3) at this time point. PBMCs in medium alone for 5 h secreted 6 pg/mL IL-2, 0–43 pg/mL; (median, interquartile range) (n = 6). When negative serum and rTSST-1 were given simultaneously for 5 h, 3655 pg/mL, 2677–3944 (median, interquartile range) of IL-2 was detected in the supernatant (n = 3); (**e**) IL-2 secretion 24 h after stimulation. Antiserum was added to the culture at 1 h, 3 h, 5 h, 7 h, 9 h, 20 h, and 22 h and IL-2 secretion was determined at 24 h. The data presented are representative of 6 separate experiments. PBMCs in medium alone for 24 h secreted 27 pg/mL, 15–282 (median, interquartile range) IL-2 (n = 6). ** Statistically significant difference compared with stimulated PBMCs without antiserum using the Wilcoxon signed-ranks test ($p < 0.001$; n = 6). Box plot diagrams indicate the median (+), interquartile range (box) and minimum and maximum values (whiskers).

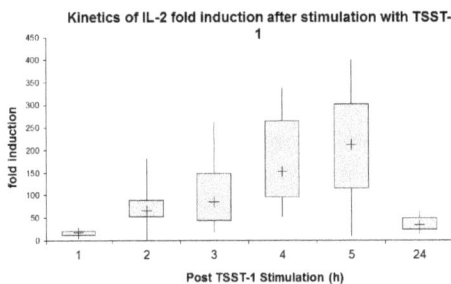

(**a**)

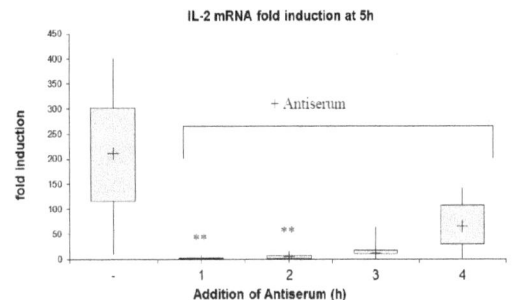

(**b**)

Figure 2. *Cont.*

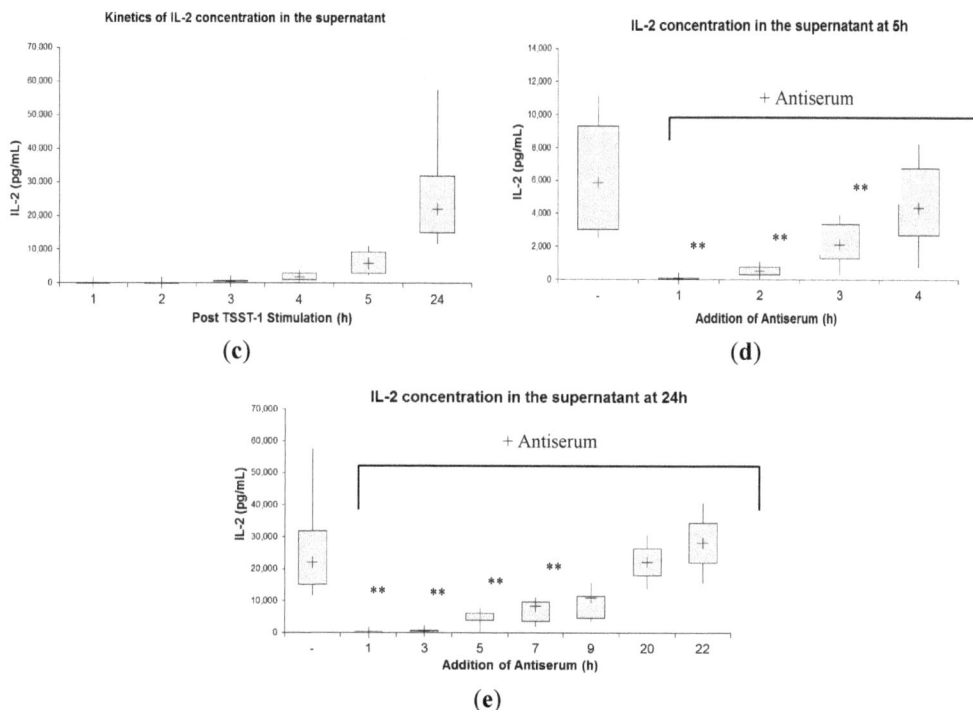

(c)

(d)

(e)

In summary, T-cell activation in human PBMCs occurred quickly after stimulation with rTSST-1 wt (as measured by IL-2 mRNA induction and secretion) and could be inhibited by postexposure antibody treatment at both, early (complete inhibition is given at 1 and 2 h after stimulation) and late, (partial inhibition) time points. After inhibition by antiserum, T-cell activation stopped immediately and IL-2 mRNA already induced was continuously degraded. On a protein level, IL-2 secretion continued slightly after inhibition through antiserum for a short period of time (~1 h), and remained at this level in the 5 h culture period.

Induction of the inflammatory cytokine TNFα followed different kinetics (Figure 3a). When TNFα mRNA induction was examined over a period of 24 h post rTSST-1 wt stimulation, TNFα mRNA was rapidly induced at 1 h to maximum levels remaining relatively constant up to 5 h. At 24 h, induction of TNFα mRNA was largely over. Our results indicate that TNFα was induced more rapidly than IL2 and maximum levels were reached as early as one hour after stimulation. Even in view of the different kinetics of mRNA-induction, addition of antiserum to the culture after stimulation could inhibit TNFα mRNA induction, although in contrast to IL-2, mRNA inhibition was not complete when antibody was added early (one hour) after stimulation (Figure 3b). Levels of TNFα mRNA produced were significantly reduced as antiserum inhibited newly induced mRNA, while mRNA already present was gradually degraded.

With respect to protein secretion, maximum levels of TNFα in the cell supernatant were detected at 24 h (Figure 3c) (6008 pg/mL). At an early time point (*i.e.*, 1 h), there was already a slight but statistically significant increase in TNFα protein release detectable. Addition of antiserum hampered protein synthesis at early time points, but also at later times (e.g., 5 h), though TNFα inhibition was less efficient than the one observed at IL-2 (Figure 3d,e). More than 9 h after stimulation, no inhibitory effect was observed.

Figure 3. (**a**) Kinetics of TNFα mRNA expression in human PBMC after stimulation with TSST-1 for indicated periods. Cells were harvested and TNFα mRNA was analyzed via real-time PCR. The supernatant of these cultures served for analysis of TNFα protein secretion (see Figure 3c); (**b**) TNFα gene activation at 5 h after TSST-1 stimulation. Antiserum was added to PBMCs at 1 h, 2 h, 3 h, and 4 h and TNFα mRNA expression was analyzed at 5 h. Simultaneous addition of TSST-1 and antiserum to PBMCs resulted in blocking of TNFα mRNA expression (fold induction of 1.5; 1–2.5, median, interquartile range; n = 6). Simultaneous addition of rTSST-1 and negative serum led to TNFα fold induction of 4 5, 38–50 (median, interquartile range) (n = 6); (**c**) TNFα concentration in the supernatant of cultured human PBMCs stimulated with 1 ng/mL TSST-1 for indicated periods. TNFα concentration in the supernatant at 5 h (**d**) and 24 h (**e**) and the effect of antiserum. When antiserum and TSST-1 were applied simultaneously for 5 h, we detected 2 pg/mL, 1–13 (median, interquartile range) of TNFα in the supernatant (n = 3) at this time point. Negative serum applied simultaneously with rTSST-1 did not influence the accumulation of TNFα in the supernatant compared to rTSST-1 alone (4890 pg/mL, 4869–4911; median, interquartile range) (n = 3). PBMCs in medium alone secreted 0.5 pg/mL, 0–4; median, interquartile range TNFα at 5 h (n = 6) and 26 pg/mL, 15–74 (median, interquartile range) TNFα at 24 h (n = 6). ** Statistically significant difference compared with stimulated PBMCs without antiserum using the Wilcoxon signed-ranks test ($p < 0,001$; n = 6). Box plot diagrams indicate the median (+), interquartile range (box) and minimum and maximum values (whiskers).

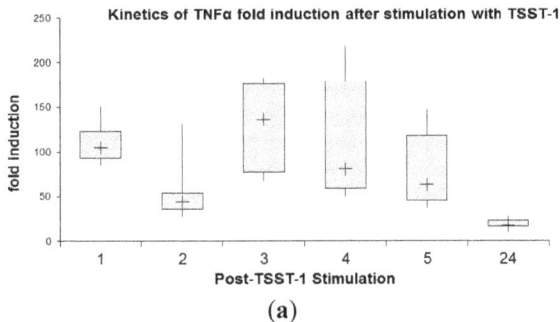

(**a**)

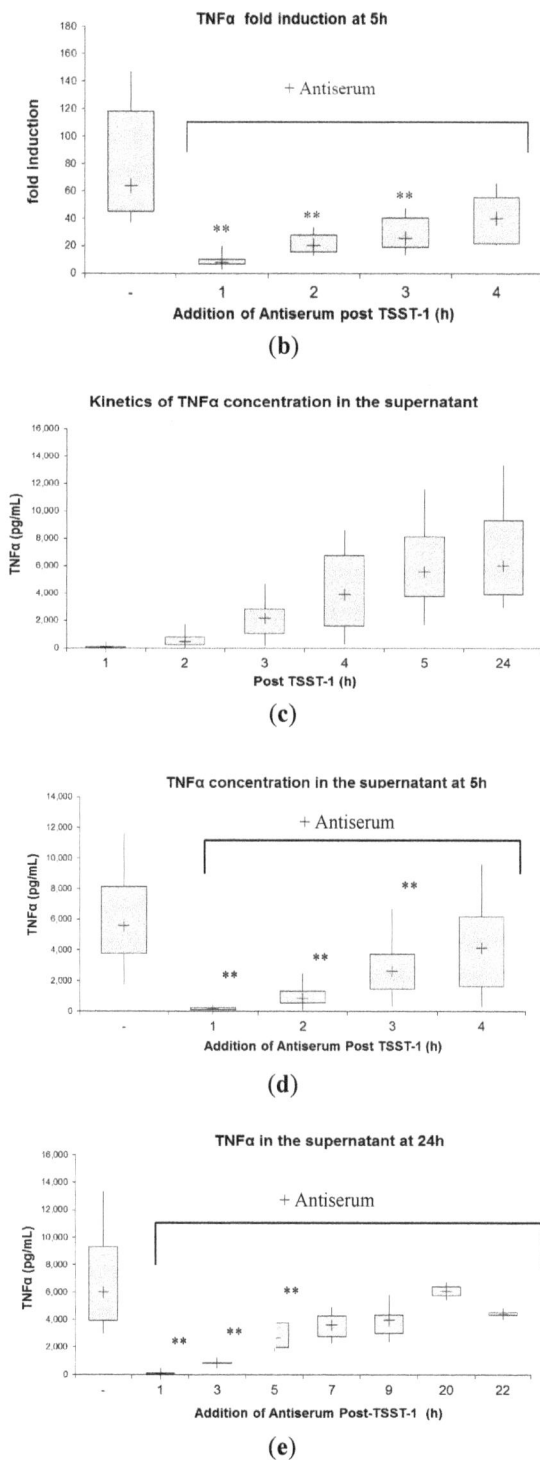

Figure 3. *Cont.*

(b)

(c)

(d)

(e)

3. Experimental Section

3.1. In Vitro *Cytokine Gene Expression Assay*

Peripheral blood mononuclear cells (PBMC) were isolated from heparinized blood of healthy human adults using density gradient centrifugation with Lymphoprep™ (Axis-Shield PoC, Oslo, Norway) as previously described [34]. PBMC were cultured in complete medium (RPMI 1640 medium (Gibco), 10% FCS (HyClone, Logan, UK), 2 mM L-glutamine (Invitrogen, Paisley, UK), 100 U/mL penicillin, and 100 μg/mL streptomycin (Invitrogen)) at a concentration of 5×10^6/mL in 24-well flat-bottom tissue culture plates (Sarstedt, Newton, NC, USA) and stimulated with a final concentration of 1 ng/mL recombinant TSST-1 wildtyp (rTSST-1 wt) in humidified atmosphere (37 °C, 5% CO_2) for time periods indicated in the text. Antiserum was added to the *in vitro*-stimulated PBMC at final dilutions of 1:100 at the time points indicated in the text.

At the end of stimulation, cells were resuspended in culture medium and transferred to Eppendorf tubes followed by centrifugation at 4 °C and $2000 \times g$ for 5 min. Cell pellets were frozen at −20 °C before extracting RNA. The supernatants were saved for analysis of protein concentration via ELISA.

3.2. RNA Isolation and Reverse Transcription

RNA was extracted from frozen PBMC pellets using the *High pure RNA Isolation kit* from Roche. RNA was then reversely transcribed into cDNA using the *2xRT Kit* from Invitrogen (Paisley, UK) following the protocol of the manufacturer.

3.3. Primer Design

Gene-specific oligonucleotide primers were designed by hand and by primer design Primer Express® v2.0 software from Applied Biosystems (Foster City, CA, USA). Primer pairs were synthesized at MWG/Eurofins Biotech (Heidelberg, Germany) and at Invitrogen. cDNA standards were prepared as previously described [34]. The primer sequences used for amplification of human cytokine cDNAs by real-time PCR were as follows:

IL-2-Forward:	5'- AAACCTCTGGAGGAAGTG-3';
IL-2-Reverse:	5'- GTTCAGAAATTCTACAATGG-3';
TNFα-Forward:	5'- CTGTACCTCATCTACTCCC-3';
TNFα-Reverse:	5'- GAGAGGAGGTTGACCTTG-3';
HPRT-Forward:	5'- AGGCCATCACATTGTAGCCC-3';
HPRT-Reverse:	5'- GTTGAGAGATCATCTCCACCG-3'.

3.4. Quantitative Real-Time PCR (QRT-PCR) and Quantification

QRT-PCR was previously described [34]. In short, cDNA from samples and standards were simultaneously amplified on the same plate (MicroAmp, Applied Biosystems, Vienna, Austria) using an ABI Prism 7500-FAST (Applied Biosystems, Vienna, Austria) with the KAPA SYBR

FAST Super Mix from PEQLAB (Erlangen, Germany) and ROX as reference dye. At the end of the amplification, a melting curve analysis was performed. The number of target cDNA copy numbers in the cellular samples was calculated by creating a standard curve where the cycles at threshold (CT) were plotted against the logarithmic values of the cDNA standard copy number. The housekeeping gene HPRT served as an internal standard. Fold induction of mRNA expression was assessed from values normalized for the expression of HPRT and then related to the mean values derived from unstimulated PBMCs of three human donors.

3.5. Lymphocyte Proliferation Assay

PBMC were isolated as described elsewhere [34]. In 96-well round-bottom tissue culture plates (Sarstedt, Newton, NC, USA), 1×10^5 cells/well were then cultured in complete medium consisting of RPMI 1640 medium (Gibco), 10% FCS (HyClone, Logan, UK), 2 mM L-glutamine (Invitrogen, Paisley, UK), 100 U/mL penicillin, and 100 µg/mL streptomycin (Invitrogen). PBMC were stimulated in triplicate with rTSST-1 wt in final concentrations of 0.3 ng/mL. Sera of rabbits, immunized with rTSST-1 wt or rTSST-1 variant, were added at different times in a final dilution of 1:100. As a positive control, phytohaemagglutinin (PHA, Sigma–Aldrich, St. Louis, MO, USA) was used in a final dilution of 1:160. As a negative control, cells were grown in culture medium alone. Stimulated cells were cultured for four days in a humidified atmosphere (37 °C, 5% CO_2). On day 3, 0.5 µCi/well 3H-thymidine (GE Healthcare, Chalfont St Giles, UK) was added and 18 h later plates were frozen and stored at −20 °C until harvesting onto glass fiber filters. Incorporated radioactivity was determined on a MicroBeta Trilux 1450 scintillation counter (Wallac, Turku, Finland) and expressed as ccpm. Percentage (%) inhibition was determined by calculating $100 - (($ccpm ^{3}H thymidin incorporation in the presence of antiserum/ccpm ^{3}H thymidin incorporation in control cultures containing negative serum) $\times 100)$.

3.6. Animals

New Zealand White male and female rabbits weighing between 1.5 and 2 kg were purchased from Charles River Laboratories (Sulzfeld, Germany). Animals were kept in standard care facilities according to the guidelines of the Austrian Ministry of education, science and culture, and had free access to food and water (Ssniff®, Alleindiaet fuer Kaninchen, Ssniff Soest, Germany). The animal experiments had been approved and controlled by the municipal Veterinary Department of the City of Vienna (Austria).

3.7. Substances and Production of Antiserum

rTSST-1 wt and its mutant form rTSST-1 variant were produced in our laboratory. The rTSST-1 variant bears two mutations affecting the MHC binding site and the T-cell receptor binding site: G31R and H135A. The expression, purification and characterization of rTSST-1 wt and rTSST-1 variant were extensively described elsewhere [18]. All substances were tested for their chemical and biological properties in our laboratory and were proven to be below the detection limit of the Limulus test for LPS (0.01 EU/mL). Antisera were obtained in rabbits after four immunizations.

Antiserum raised against wild-type TSST-1 had an ELISA titer of 46,330 and showed 99% inhibition of T-cell proliferation at a dilution of 1:100, when human PBMCs were stimulated with 0.3 ng/mL rTSST-1 wt. Antisera raised against TSST-1 variant had ELISA titers of 49,921 and 60,714 in two different rabbits and both sera showed 98% inhibition of T-cell proliferation when stimulated with 0.1 ng/mL rTSST-1 wt and an application of antiserum diluted 1:300.

3.8. Multiple Dose Lethal Challenge

Different quantities of TSST-1 wt (30 and 40 µg) in 1 mL sterile PBS were given subcutaneously twice a day for four days and once on day 5. Lethality was monitored over a period of 7–9 days.

3.9. Neutralization of rTSST-1 wt by Pre-Incubation with Antiserum

Thirty µg rTSST-1 wt were incubated in 1 mL undiluted antiserum at 37 °C overnight (neutralized toxin). Experimental design: two rabbits were challenged via multiple dose lethal challenge. Three rabbits were challenged with neutralized toxin at the same schedule. Three rabbits were challenged for three days (six injections) with toxin and, on days 4 and 5, with neutralized toxin.

3.10. Passive Immunization with Antiserum

One of two doses of rTSST-1 wt (30 µg; 40 µg) was dissolved in 1 mL PBS and given subcutaneously twice a day (at intervals of 6–8 h) for four days and once on day five. Two mL of antiserum was given subcutaneously either on days 2 and 3 or on days 3, 4, and 5 at different application sites. Lethality was monitored over a period of 7–9 days.

4. Conclusions

Taken together, our results demonstrate that treatment with neutralizing antibodies (hyperimmune serum) is protective late in the course of an ongoing exposure to TSST-1 in a multiple dose rabbit TSS model. In an *in vitro* system for monitoring the kinetics of cytokine release of human PBMCs after exposure to TSST-1 for defined periods, we show that IL-2 concentration in the PBMC cell supernatant increases up to 24 h after stimulation. Neutralization of the toxin at early as well as late time points during stimulation (e.g., at 7 h) leads to a significant reduction of IL-2 release in the supernatant. In contrast, TNFα secretion reached its maximum very early, e.g., at 5 h, and plateaued at this level up to 24 h. Therefore, neutralization of the toxin given after 5 h only had a marginal effect on the release of this cytokine.

It is well known that during T-cell activation, mRNA stability contributes significantly to changes in gene expression [35,36]. Large-scale gene expression profiling in activated T cells revealed that a large proportion of genes is regulated by mRNA stability [35,37]. Cytokine transcripts are labile [38] and regulated at the level of mRNA decay in a stimulus-specific manner. Activation of T cells by stimulation of CD28 led to selective stabilization of IL-2 and TNFα transcripts. In the absence of anti-CD28, mRNA degradation of these cytokines was rapidly induced [39]. We followed the kinetics of cytokine gene expression in the course of prolonged toxin exposure. The results of this study clearly indicate that in the presence of toxin, transcription of IL-2 and TNFα mRNA is ongoing and that the mRNAs of these cytokines are highly labile. It is likely that the gene expression of IL-2 and TNFα after exposure to TSST-1 is mainly regulated via mRNA turnover as it was shown for other modes of T-cell activation [39]. Both transcripts were rapidly induced upon stimulation with TSST-1. Neutralization of the toxin via addition of toxin-neutralizing antibodies immediately stopped cytokine gene transcription and the degradation of (already expressed) mRNA could be observed.

Interference with the prolonged stimulation of proinflammatory and immune activating cytokines, e.g., by inhibiting continual cytokine gene transcription, is likely to participate in the protective effect of neutralizing antibodies given as a postexposure treatment in animal models mimicking the clinical situation of prolonged toxin exposure such as observed in our multiple dose lethal rabbit model of TSS. At this stage, it is premature to present the mode of action of TSST-1-neutralizing antibodies in our animal model. Dissociation of superantigen bound to immune cells by neutralizing antibodies is one possible explanation whereby antibodies could protect late in the course of TSST-1 exposure. However, it is more likely that multiple activating hits over time, *i.e.* repeated activation of T cells by superantigenic toxins, rather than a single stimulatory event at the first encounter of immune cells with superantigenic toxins, leads to the prolonged hyperactivation of inflammatory responses characteristic of TSS. Using a TCR transgenic mouse model, it has been shown that prolonged antigen stimulation (thus delivering multiple activating hits over time) fully activates CD4 T-cell response [40]. Preliminary results indicate that in either case, high affinity antibodies produced after repeated immunization are required for superantigen neutralization as opposed to toxin-binding antibodies observed relatively early in the course of rabbit vaccination (data not shown). A candidate vaccine intended for use in humans has to be detoxified, e.g., by mutations in the toxin sites responsible for binding to TCR and MHC class II. Such a double mutant toxin might theoretically be less effective in inducing neutralizing antibodies as it lacks epitopes crucial for toxin activity. This was not the case in our system, as titers of neutralizing antibodies induced by double mutant and wild-type TSST-1 were comparable. This is in good agreement with previous studies showing that nonvirulent mutant toxins were immunogenic in animals and humans, e.g., a licensed acellular pertussis vaccine [41], a recombinant toxoid vaccine against *Clostridium difficile* [42], a recombinant *Clostridium perfringens* epsilon toxin mutant vaccine [43], and a mutant recombinant SEB vaccine [44]. Further studies are required to explore the mechanism(s) by which neutralizing antibodies administered late in the course of toxin exposure interfere with a prolonged toxin-mediated stimulation of immune cells. These studies should investigate important and currently unknown characteristics of the neutralizing antibodies involved, such as toxin-binding affinity, half-life of binding to TSST-1, and linear and conformational epitopes recognized by these antibodies.

Author Contributions

Nina Model, Aysen Samstag, Corina S. Gruener and Norbert Stich conducted the experiments and analyzed the data. Norbert Stich wrote the manuscript. Martha M. Eibl designed the experiments and supervised the work. Hermann M. Wolf performed the statistical analysis. Martha M. Eibl and Hermann M. Wolf discussed interpretation of the results and edited the manuscript.

Conflicts of Interest

The authors declare no conflict of interest.

Disclosure

This work was supported by grants of the Austrian Research Promotion Agiency (FFG) (Nos. 839130 and 844049).

References

1. Hotchkiss, R.S.; Karl, I.E. The pathophysiology and treatment of sepsis. *N. Engl. J. Med.* **2003**, *348*, 138–150.
2. Fisher, C.J.; Agosti, J.M.; Opal, S.M. Treatment of septic shock with the tumor necrosis factor receptor: Fc fusion protein. *N. Engl. J. Med.* **1996**, *334*, 1697–1702.
3. Reinhart, K.; Karzai, W. Anti-tumor necrosis factor therapy in sepsis: Update on clinical trials and lessons learned. *Crit. Care Med.* **2001**, *29*, S121–S125.
4. Warren, H.S.; Suffredini, A.F.; Eichacker, P.Q.; Munford, R.S. Risks and benefits of activated protein C treatment for severe sepsis. *N. Engl. J. Med.* **2002**, *347*, 1027–1030.
5. Riedemann, N.C.; Guo, R.F.; Ward, P.A. The enigma of sepsis. *J. Clin. Investig.* **2003**, *112*, 460–467.
6. Fisher, C.J. Recombinant human interleukin 1 receptor antagonist in the treatment of patients with sepsis syndrome: Results from a randomized, double-blind, placebo-controlled trial. *JAMA* **1994**, *271*, 1836–1843.
7. Riedemann, N.C.; Guo, R.F.; Ward, P.A. Novel strategies for the treatment of sepsis. *Nat. Med.* **2003**, *9*, 517–524.
8. Vincent, J.L.; Sakyr, Y.; Sprung, C.L.; Ranieri, V.M.; Reinhart, K.; Gerlach, H.; Moreno, R.; Carlet, J.; Le Gall, J.R.; Payen, D. Sepsis in European intensive care units: Results of the SOAP study. *Crit. Care Med.* **2006**, *34*, 344–353.
9. Martin, G.S.; Mannino, D.M.; Eaton, S.; Moss, M. The epidemiology of sepsis in the United States from 1979 through 2000. *N. Engl. J. Med.* **2003**, *348*, 1546–1554.
10. Zetola, N.; Francis, J.S.; Nuermberger, E.L.; Bishai, W.R. Community-acquired methicillin-resistant *Staphylococcus aureus*: An emerging threat. *Lancet Infect. Dis.* **2005**, *5*, 275–286.
11. Shams, W.E.; Rapp, R.P. Methicillin-resistant staphylococcal infections: An important consideration for orthopaedic surgeons. *Orthopedics* **2004**, *27*, 565–568.
12. Salgado-Pabon, W.; Breshears, L.; Spaulding, A.R.; Merriman, J.A.; Stach, C.S.; Horswill, A.R.; Peterson, M.L.; Schlievert, P.M. Superantigens are critical for *Staphylococcus aureus* infective endocarditis, sepsis, and acute kidney injury. *mBio* **2013**, *4*, doi:10.1128/mBio.00494-13.
13. Krakauer, T. Therapeutic down-modulators of staphylococcal superantigen-induced inflammation and toxic shock. *Toxins* **2010**, *2*, 1963–1983.

14. Krakauer, T. Update on staphylococcal superantigen-induced signaling pathways and therapeutic interventions. *Toxins* **2013**, *5*, 1629–1654.

15. Shankar-Hari, M.; Spencer, J.; Sewel, W.A.; Rowan, K.M.; Singer, M. Bench-to-bedside review: Immunoglobulin therapy for sepsis- biological plausibility from a critical care perspective. *Crit. Care* **2012**, *16*, 206–220.

16. Hu, D.L.; Omoe, K.; Sasaki, S.; Sashinami, H.; Sakuraba, H.; Yokomizo, Y.; Shinagawa, K.; Nakane, A. Vaccination with nontoxic mutant toxic shock syndrome toxin 1 protects against *Staphylococcus aureus* infection. *J. Infect. Dis.* **2003**, *188*, 743–752.

17. Stiles, B.G.; Krakauer, T.; Bonventre, P.F. Biological activity of toxic shock syndrome toxin 1 and a site-directed mutant, H135A, in a lipopolysaccharide-potentiated mouse lethality model. *Infect. Immun.* **1995**, *63*, 1229–1234.

18. Gampfer, J.; Thon, V.; Gulle, H.; Wolf, H.M.; Eibl, M.M. Double mutant and formaldehyde inactivated TSST-1 as vaccine candidates for TSST-1- induced toxic shock syndrome. *Vaccine* **2002**, *20*, 1354–1364.

19. Krakauer, T.; Buckley, M. Dexamethasone attenuates staphylococcal enterotoxin B- induced hypothermic response and protects mice from superantigen-induced toxic shock. *Antimicrob. Agents Chemother.* **2006**, *50*, 391–395.

20. Hale, M.L.; Margolin, S.B.; Krakauer, T.; Roy, C.J.; Stiles, B.G. Pirfenidone blocks *in vitro* and *in vivo* effects of staphylococcal enterotoxin B. *Infect. Immun.* **2002**, *70*, 2989–2994.

21. Krakauer, T.; Buckley, M. Intransal rapamycin rescues mice from staphylococcal enterotoxin B-induced shock. *Toxins* **2012**, *4*, 718–728.

22. Schulte, W.; Bernhagen, J.; Bucala, R. Cytokines in sepsis: Potent immunoregulators and potential therapeutic targets- an updated view. *Mediat. Inflamm.* **2013**, 165974:1–165974:16.

23. LeClaire, R.D.; Hunt, R.E.; Bavari, S. Protection against bacterial superantigen staphylococcal enterotoxin B by passive vaccination. *Infect. Immun.* **2002**, *70*, 2278–2281.

24. Larkin, E.A.; Stiles, B.G.; Ulrich, R.G. Inhibition of toxic shock by human monoclonal antibodies against staphylococcal enterotoxin B. *PLos ONE* **2010**, *5*, e13253.

25. Perkins, S.L.; Han, D.P.; Burke, J.M.; Schlievert, P.M.; Wirostko, W.J.; Tarasewicz, D.G.; Skumatz, C.M.B. Intravitreally injected human immunoglobulin attenuates the effects of *Staphylococcus aureus* culture supernatant in a rabbit model of toxin-mediated endophthalmitis. *Arch. Ophthalmol.* **2004**, *122*, 1499–1506.

26. Bonventre, P.F.; Heeg, H.; Cullen, C.; Lian, C.J. Toxicity of recombinant toxic shock syndrome toxin 1 and mutant toxins produced by *Staphylococcus aureus* in a rabbit infection model of toxic shock syndrome. *Inf. Immun.* **1993**, *61*, 793–799.

27. Schlievert, P.M. Cytolysins, superantigens, and pneumonia due to community-associated methicillin-resistant *Staphylococcus aureus*. *J. Infect. Dis.* **2009**, *200*, 676–678.

28. Stich, N.; Waclavicek, M.; Model, N.; Eibl, M.M. Staphylococcal superantigen (TSST-1) mutant analysis reveals that T cell activation is required for the biological effects in the rabbit including the cytokine storm. *Toxins* **2010**, *2*, 2272–2288.

29. Spaulding, A.R.; Salgado-Pabon, W.; Merriman, J.A.; Stach, C.S.; Ji, Y.; Gillman, A.N.; Peterson, M.L.; Schlievert, P.M. Vaccination against *Staphylococcus aureus* Pneumonia. *J. Infect. Dis.* **2013**, doi:10.1093/infdis/jit823.

30. Spaulding, A.R.; Ying-Chi, L.; Merriman, J.A.; Brosnahan, A.J.; Peterson, M.L.; Schlievert, P.M. Immunity to *Staphylococcus aureus* secreted proteins protects rabbits from serious illnesses. *Vaccine* **2012**, *30*, 5099–5109.

31. LeClaire, R.D.; Bavari, S. Human Antibodies to bacterial superantigens and their ability to inhibit T cell activation and lethality. *Antimicrob. Agents. Chemother.* **2001**, *45*, 460–463.

32. Visvanathan, K.; Charles, A.; Bannan, J.; Pugach, P.; Kashfi, K.; Zabriskie, J.B. Inhibition of bacterial superantigens by peptides and antibodies. *Infect. Immun.* **2001**, *69*, 875–884.

33. Delsesto, D.; Opal, S.M. Future perspectives on regulating pro- and anti-inflammatory responses in sepsis. *Contrib. Microbiol.* **2011**, *17*, 137–156.

34. Waclavicek, M.; Stich, N.; Rappan, I.; Bergmeister, H.; Eibl, M.M. Analysis of the early response to TSST-1 reveals Vβ-unrestricted extravasation, compartmentalization of the response, and unresponsiveness but not anergy to TSST-1. *J. Leukoc. Biol.* **2009**, *85*, 44–54.

35. Cheadle, C.; Fan, J.; Cho-Chung, Y.S.; Werner, T.; Ray, J.; Do, L.; Gorospe, M.; Becker, K.G. Control of gene expression during T cell activation: Alternate regulation of mRNA transcription and mRNA stability. *BMC Genomics* **2005**, *6*, doi:10.1186/1471-2164-6-75.

36. Khabar, K.S.A. Rapid transit in the immune cells: The role of mRNA turnover regulation. *J. Leukoc. Biol.* **2007**, *81*, 1335–1344.

37. Raghavan, A.; Bohjanen, P.R. Microarray-based analyses of mRNA decay in the regulation of mammalian gene expression. *Brief. Funct. Genomics Proteomics* **2004**, *3*, 112–124.

38. Palanisamy, V.; Jakymiw, A.; van Tubergen, E.A.; D'Silva, N.J.; Kirkwood, K.L. Control of cytokine mRNA expression by RNA-binding proteins and microRNAs. *J. Dent. Res.* **2012**, *91*, 651–658.

39. Lindsten, T.; June, C.H.; Ledbetter, J.A.; Stella, G.; Thompson, C.B. Regulation of lymphokine messenger RNA stability by a surface-mediated T cell activation pathway. *Science* **1989**, *244*, 339–343.

40. Rabenstein, H.L.; Behrendt, A.C.; Ellwart, J.W.; Naumann, R.; Horsch, M.; Beckers, J.; Obst, R.J. Differential kinetics of antigen dependency of CD4$^+$ and CD8$^+$ T Cells. *J. Immunol.* **2014**, *192*, 3507–3517.

41. Greco, D.; Salmaso, S.; Mastrantonio, P.; Giuliano, M.; Tozzi, A.E.; Anemona, A.; Ciofi degli Atti, M.L.; Giammanco, A.; Panei, P.; Blackwelder, W.C.; Klein, D.L.; Wassilak, S.G. A controlled trial of two acellular vaccines and one whole-cell vaccine against pertussis. *N. Engl. J. Med.* **1996**, *334*, 341–348.

42. Donald, R.G.; Flint, M.; Kalyan, N.; Johnson, E.; Witko, S.E.; Kotash, C.; Zhao, P.; Megati, S.; Yurgelonis, I.; Lee, P.K.; Matsuka, Y.V.; Severina, E.; Deatly, A.; Sidhu, M.; Jansen, K.U.; Minton, N.P.; Anderson, A.S. A novel approach to generate a recombinant toxoid vaccine against Clostridium difficile. *Microbiology* **2013**, *159*, 1254–1266.

43. Bokori-Brown, M.; Hall, C.A.; Vance, C.; Fernandes da Costa, S.P.; Savva, C.G.; Naylor, C.E.; Cole, A.R.; Basak, A.K.; Moss, D.S.; Titball, R.W. Clostridium perfringens epsilon toxin mutant Y30A-Y196A as a recombinant vaccine candidate against enterotoxemia. *Vaccine* **2014**, *32*, 2682–2687.

44. Boles, J.W.; Pitt, M.L.; LeClaire, R.D.; Gibbs, P.H.; Torres, E.; Dyas, B.; Ulrich, R.G.; Bavari, S. Generation of protective immunity by inactivated recombinant staphylococcal enterotoxin B vaccine in nonhuman primates and identification of correlates of immunity. *Clin. Immunol.* **2003**, *108*, 51–59.

Superantigens Modulate Bacterial Density during *Staphylococcus aureus* Nasal Colonization

Stacey X. Xu, Katherine J. Kasper, Joseph J. Zeppa and John K. McCormick

Abstract: Superantigens (SAgs) are potent microbial toxins that function to activate large numbers of T cells in a T cell receptor (TCR) Vβ-specific manner, resulting in excessive immune system activation. *Staphylococcus aureus* possesses a large repertoire of distinct SAgs, and in the context of host-pathogen interactions, staphylococcal SAg research has focused primarily on the role of these toxins in severe and invasive diseases. However, the contribution of SAgs to colonization by *S. aureus* remains unclear. We developed a two-week nasal colonization model using SAg-sensitive transgenic mice expressing HLA-DR4, and evaluated the role of SAgs using two well-studied stains of *S. aureus*. *S. aureus* Newman produces relatively low levels of staphylococcal enterotoxin A (SEA), and although we did not detect significant TCR-Vβ specific changes during wild-type *S. aureus* Newman colonization, *S. aureus* Newman Δ*sea* established transiently higher bacterial loads in the nose. *S. aureus* COL produces relatively high levels of staphylococcal enterotoxin B (SEB), and colonization with wild-type *S. aureus* COL resulted in clear Vβ8-specific T cell skewing responses. *S. aureus* COL Δ*seb* established consistently higher bacterial loads in the nose. These data suggest that staphylococcal SAgs may be involved in regulating bacterial densities during nasal colonization.

Reprinted from *Toxins*. Cite as: Xu, S.X.; Kasper, K.J.; Zeppa, J.J.; McCormick, J.K. Superantigens Modulate Bacterial Density during *Staphylococcus aureus* Nasal Colonization. *Toxins* **2015**, *7*, 1821-1836.

1. Introduction

Staphylococcus aureus is recognized as a major human pathogen causing a range of illnesses from superficial skin infections to invasive diseases including bacteremia, sepsis, pneumonia, and endocarditis [1]. Within the healthcare setting, *S. aureus* infections are particularly serious, including infection by methicillin-resistant *S. aureus* (MRSA) strains, and this pathogen is now the most significant cause of serious infections in the United States [1–4].

Despite the massive burden of disease, asymptomatic carriage by *S. aureus* is pervasive within human populations, being found most typically on the skin and in the nasal cavity. Nasal carriers have been defined as persistent, intermittent, or non-carriers, and although the definitions for each group can vary between studies, ~50% of the population are persistent or intermittent carriers, with some studies showing even higher levels of colonization [4,5]. *S. aureus* typically resides within the vestibulum nasi of the anterior nares and has been found colonizing the cornified layer of stratified squamous epithelium, keratinized surfaces and mucous debris, as well as hair follicles of the human nose [6]. Given these anatomical findings, it is not surprising that *S. aureus* is able to bind to both keratinized cells and desquamated nasal epithelia. These act as key host cells upon which *S. aureus* initiates colonization [7,8]. Nasal carriers of *S. aureus* are generally asymptomatic and healthy, forming a commensal relationship with the bacteria. However, colonization status increases the risk of a severe infection from the carrier strain, although nasal carriers tend to have a better prognosis in the event of a staphylococcal infection [9]. This is thought to be due to specific immunity built up against the colonizing strain [10]. Bacterial components contributing to staphylococcal colonization are multifactorial and include host genetic factors that influence carrier status [11], as well as a variety of bacterial adhesins and cell-wall associated factors such as clumping factor B (ClfB) [12], wall teichoic acids [13], surface protein SasG [14], and iron-regulated surface determinant A (IsdA) [15].

Superantigens (SAgs) are a group of toxins produced by bacteria including *S. aureus* that mediate interactions between peptide-MHC class II and the CDR2 loop of the variable chain of the T cell receptor [16]. As SAg-mediated T cell activation is not dependent on the antigenic peptide presented in the MHC class II molecule, this response can activate very large numbers of the exposed T cell population and may, in rare cases, lead to a 'cytokine storm' disease known as the toxic shock syndrome (TSS). These toxins have also been implicated in many other diseases including infectious endocarditis, Kawasaki disease, atopic dermatitis, and various autoimmune diseases [16,17]. To date, more than twenty *S. aureus* SAgs have been identified including an operon of SAgs known as the enterotoxin gene cluster (*egc*) encoding staphylococcal enterotoxins (SE) G, I and SE-like (SEl) M, N, O and U [16,18].

Assessment of the humoral response from *S. aureus* colonized individuals have shown that persistent carriers produce high titer neutralizing antibodies with specificity for the SAgs produced by the carrier strain [10,19]. Additionally, assessment of antibody titers to seven different staphylococcal SAgs showed increased antibodies to toxic shock syndrome toxin-1 (TSST-1) and staphylococcal enterotoxin A (SEA) in persistent carriers compared with non-carriers [20]. Epidemiological studies of clinical isolates revealed a high prevalence of *egc* SAgs [21], as well as a negative correlation of these toxins with severe septic shock [22]. Nasal swabs from persistent carriers revealed that *sea*, *sec* and *sel-o* were actively transcribed; however, neutralizing antibodies against SEA and SEC, but not SEl-O, were detected in this cohort [23]. It was concluded that the robust antibody response against the non-*egc* SAgs was due to minor infections rather than colonization [23]. Also, vaccination of mice with SAg toxoids seems to protect only against the early phase of colonization (days 1 and 3) [24] suggesting that SAgs may be involved in initial colonization, but further implications are difficult to extrapolate. Collectively, these studies have shed light on the highly complex nature of nasal colonization and hinted at a role for SAgs in nasal carriage in humans, and mouse infection models. However, the role of SAgs during nasal colonization, either for establishing initial colonization, or involvement in dissemination, has not been experimentally addressed.

Human studies reveal low levels of bacteria in the nose, with 10^1–10^4 colony forming units (CFU) of *S. aureus* being typically isolated from nasal swabs [25]. We hypothesized that secreted SAgs may act as 'checkpoints' of colonization in order to maintain this state of commensalism and to prevent high bacterial densities through activation of the immune system, and subsequent elimination of invasive organisms. In order to test this hypothesis, we created isogenic SAg deletions of two well-characterized strains of *S. aureus*, and tested these strains against their wild-type counterparts in a SAg-sensitized murine model of staphylococcal nasal colonization.

2. Results

2.1. SAg Deletion Strains Have Reduced Superantigen Production and Activity in vitro

To assess the role of SAgs in experimental *S. aureus* nasal colonization, a mutant strain of *S. aureus* COL with a deletion of the *seb* gene was generated as described in Materials and Methods. Growth curve analysis of the SEB deletion strain compared to the wild-type counterpart showed no obvious growth defects *in vitro* (Figure 1A). As expected, *S. aureus* COL Δ*seb* did not produce SEB as shown by the exoprotein profiles and Western blot analysis (Figure 1B). Additionally, we did not detect significant levels of IL-2 from DR4-B6 splenocytes treated with cultural supernatants from COL Δ*seb* compared to wild-type COL (Figure 1C). Although *S. aureus* COL also encodes *sei*, *sel-k* and *sel-x*, these data indicate that SEB is the dominant SAg produced by *S. aureus* COL *in vitro*. The *sea* deletion mutant was generated in *S. aureus* Newman as previously described, and similarly, has been characterized as lacking superantigenic activity [26].

2.2. Lack of SEA Transiently Increases S. aureus Newman Δsea Nasal Colonization

To investigate if SEA plays a role during murine nasal colonization, DR4-B6 mice pre-treated with streptomycin (Sm) were inoculated with 1×10^8 CFUs of *S. aureus* Newman or *S. aureus* Newman Δ*sea*. *S. aureus* was detected in the nasal passages of both *S. aureus* Newman- and *S. aureus* Newman Δ*sea*-infected mice up to day 14 post-inoculation. Generally, CFU counts were higher during the first week of colonization compared to the second week (Figure 2A). Infected mice did not show overt signs of infection (lack of piloerection, conjunctivitis, skin rashes, and dehydration, with normal activity levels), had no weight loss, and were generally healthy for the duration of the experiment (data not shown). Despite the apparent lack of infection, the lungs and livers of both infection groups revealed spread of bacteria beyond the nose, although the bacterial burdens in these organs were lower than in the nasal passage and generally very low by day 14 (Figure 2B,C). Bacteria were not detected in the kidneys or spleen (data not shown). No significant differences in bacterial loads were observed between bacterial strains on days 3 or 7 in the nasal passage. However, by day 10, *S. aureus* Newman Δ*sea*-colonized mice had increased CFUs compared to wild-type Newman-colonized mice (Figure 2A); however, this phenotype reverted to no differences between treatment groups by day 14. These data suggest that SEA does not play a major role during the initial stages of colonization, but may prevent higher bacterial densities from forming in the nose. While the lack of SEA production did allow higher bacterial densities to form, this transient difference did not result in better colonization at later time points, suggesting that it does not enhance the overall colonization capabilities of *S. aureus* Newman. No significant differences were observed in the spread of infection to other organs between wild-type and *sea*-null infections indicating that SEA likely does not influence dissemination in this model.

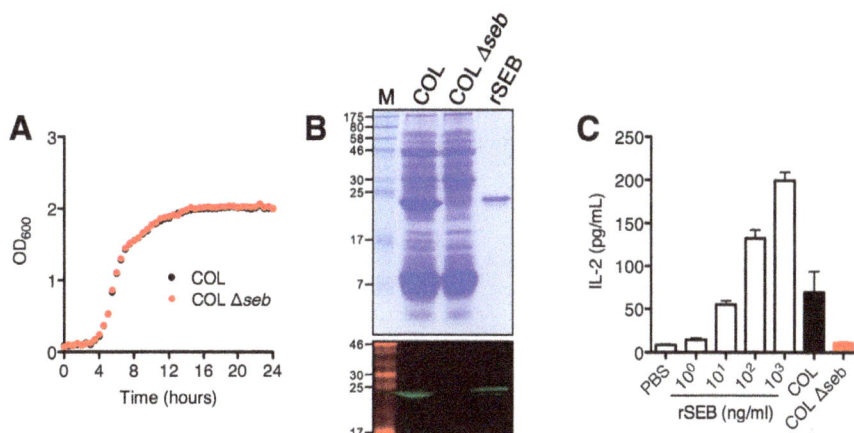

Figure 1. *S. aureus* COL Δ*seb* does not produce staphylococcal enterotoxin B (SEB) and has greatly reduced superantigenic activity. (**A**) Growth curve analysis of *S. aureus* COL (black) and COL Δ*seb* (red), grown in triplicate; (**B**) TCA-precipitated supernatants (5 OD units) showing the exoprotein and superantigen profiles of *S. aureus* COL and COL Δ*seb* with detection of SEB production by anti-SEB antibodies using Western blot; (**C**) IL-2 production from DR4-B6 splenocytes activated with increasing concentrations of recombinant SEB, and bacterial supernatants diluted 1:10 from *S. aureus* COL and COL Δ*seb*. Results shown as the mean ± SEM from a representative data set.

2.3. SEA Does Not Skew Vβ3 Subsets in vivo

We next aimed to evaluate if SEA was produced during *S. aureus* colonization by examining the Vβ profiles of infected mice. As SEA has been previously shown to skew murine Vβ3+ T cells during bacteremia [26], we analyzed the Vβ3 subset as well as levels of serum IgG against SEA in order to assess if SEA had *in vivo* activity. Analysis of the Vβ3+CD3+ lymphocytes from lymph nodes revealed no significant changes in this subset between *S. aureus* Newman or Newman Δ*sea*-inoculated mice on any of the days analyzed (Figure 3), although there was a slight trend of decreased Vβ3+ T cells in wild-type Newman-colonized mice. These data suggest that SEA may not be produced in large amounts, or is weakly active, during the length of the experiment. Additionally, no IgG against SEA could be detected in Newman and Newman Δ*sea*-inoculated mice sera (data not shown). Collectively, these data suggest that SEA was not produced in functional quantities *in vivo* during colonization. This may explain that lack of differences seen in bacterial burdens at earlier time points (Figure 2A), since the lack of SEA production by *S. aureus* Newman is functionally equivalent to infection with Newman Δ*sea*.

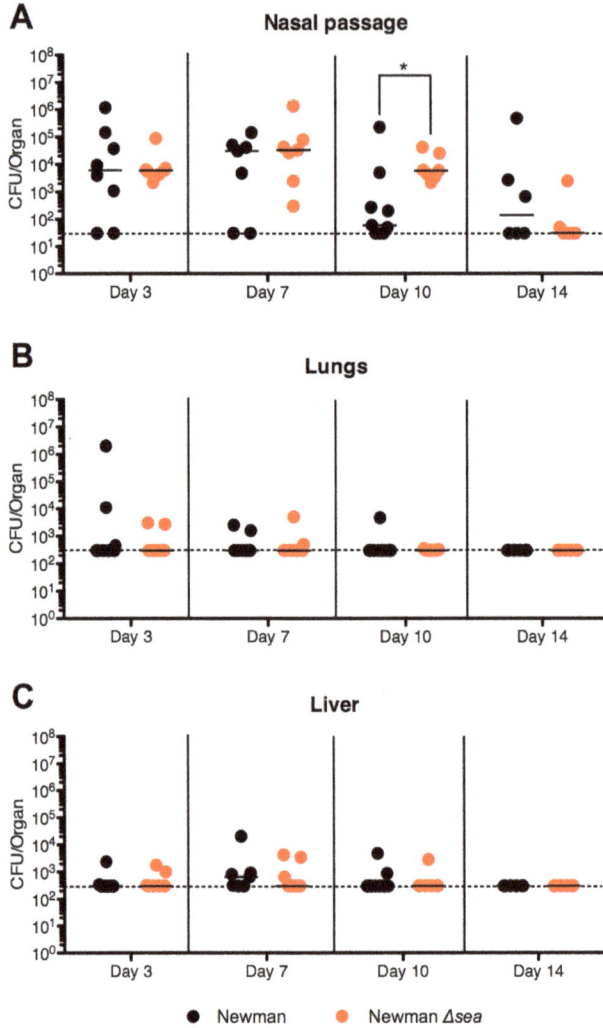

Figure 2. Nasal colonization of DR4-B6 mice with *S. aureus* Newman Δ*sea* results in a transient increase in bacterial load compared to wild-type Newman. DR4-B6 mice were infected nasally with 1×10^8 CFUs of *S. aureus* Newman or Newman Δ*sea* ($n = 6$–9). Mice were sacrificed on days 3, 7, 10 and 14 and the (**A**) nasal passage; (**B**) lungs and (**C**) livers were assessed for overall *S. aureus* burdens. Each point represents an individual mouse and the line in each treatment group represents the median. The horizontal dotted line indicates the limit of detection. Data are pooled from of at least three independent experiments. Significant differences were determined by Mann-Whitney U test (*, $p < 0.05$).

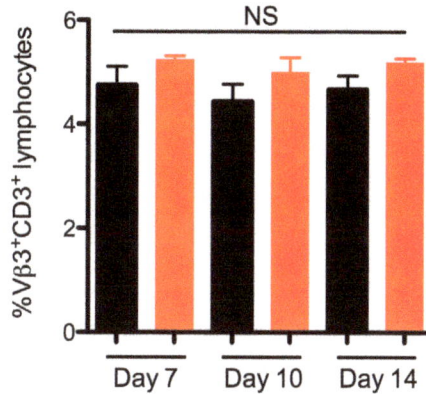

Figure 3. *S. aureus* Newman nasal colonization does not result in significant changes in the percentage of Vβ3+CD3+ T cells. Analysis of lymphocytes from lymph nodes isolated from DR4-B6 mice nasally inoculated with 1×10^8 CFU *S. aureus* Newman or Newman Δ*sea* (*n* = 3–4). Cells were stained with antibodies against CD3 and Vβ3 and gated on CD3+ lymphocytes, followed by gating on the Vβ3+CD3+ population. Data are shown as the mean ± SEM and significant differences (*p* < 0.05) were determined by unpaired student's *t*-test (NS = not statistically different).

2.4. SEB Decreases Nasal Colonization

Unlike SEA, SEB is transcriptionally activated by the accessory gene regulator (*agr*) quorum-sensing system during exponential and late stages of growth [27] and this may also result in differential expression in response to environmental cues. Similar to colonization with *S. aureus* Newman, COL was found to colonize the nasal passages of infected mice in both treatment groups; however, colonization with wild-type *S. aureus* COL persisted with higher bacterial numbers ($\sim10^3$–10^4) (Figure 4A) compared to wild-type Newman (10^2–10^3) (Figure 2A), especially at later time points, suggesting that COL may be a better nasal colonizer than Newman in DR4-B6 mice. When mice were colonized with *S. aureus* COL Δ*seb*, bacteria recovered from the nasal passages were ~100-fold higher in CFUs at all time points compared with wild-type COL colonization alone (Figure 4A). As with nasal colonization by *S. aureus* Newman, all mice were apparently healthy for the duration of the experiment with no overt signs of infection. Spread of the infection to the lungs and livers was also observed during *S. aureus* COL and COL Δ*seb* colonization, although no significant differences were observed between CFUs of the two strains (Figure 4B,C). While a complete SAg-negative strain was not assessed (*i.e.*, *sel-k*, *sel-i* and *sel-x* are still encoded within COL), these data suggest that production of SEB actually inhibited high-density colonization within the nasal passage.

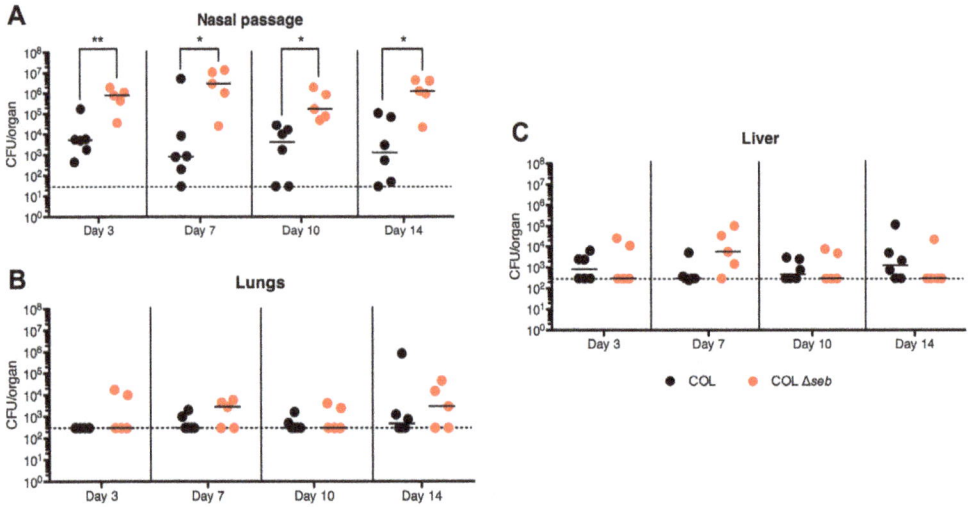

Figure 4. Murine nasal colonization with *S. aureus* COL Δ*seb* results in enhanced bacterial counts compared to wild-type COL. DR4-B6 mice infected nasally with 1×10^8 CFUs of *S. aureus* COL ($n = 6$) or COL Δ*seb* ($n = 5$) were sacrificed on days 3, 7, 10 and 14. The (**A**) nasal passage, (**B**) lungs, and (**C**) livers were assessed for overall *S. aureus* loads. Each point represents an individual mouse and the line in each treatment group represents the median. The horizontal dotted line indicates the limit of detection. Data are pooled from of at least three independent experiments. Significant differences were determined by Mann-Whitney U test (*, $p < 0.05$; **, $p < 0.01$).

2.5. SEB Induces Late Vβ8 Skewing but not Anti-SEB IgG during Nasal Colonization

To evaluate if the phenotype observed during *S. aureus* COL colonization was SEB-dependent, we assessed Vβ-skewing in mice colonized with wild-type *S. aureus* COL and COL Δ*seb* to test for functional SEB activity *in vivo*. It is well-established that SEB targets Vβ8.1/8.2+ (henceforth Vβ8+) T cells in mice [28] and Vβ3 was used as an internal control. The murine Vβ subsets targeted by SEl-K, SEl-I, and SEl-X are unknown to date and thus could not be assessed for *in vivo* activity although these strains showed no superantigenic activity *in vitro* (Figure 1C). While no differences could be detected early, by day 10 there was a trend of decreased Vβ8+ T cells, which was significantly decreased by day 14 (Figure 5). Interestingly, anti-SEB IgG antibodies were not detected from either COL- or COL Δ*seb*-colonized mice by day 14 (data not shown). The demonstrated Vβ-skewing by day 14 indicates that SEB was produced and functional during *S. aureus* COL nasal colonization. Furthermore, the difference in bacterial loads between COL and COL Δ*seb* (Figure 4A) at the early time points suggests that SEB was functioning early on during colonization, although we were not able to detect functional activity until the later time points.

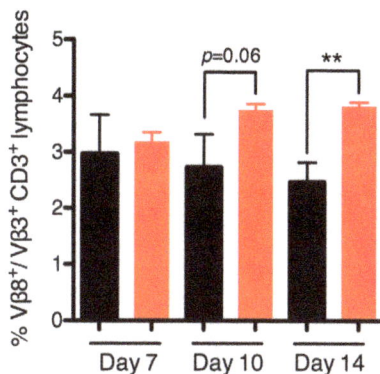

Figure 5. Staphylococcal enterotoxin B (SEB) is produced during *S. aureus* COL nasal colonization and specifically interacts with Vβ8⁺CD3⁺ lymphocytes. Lymphocytes from lymph nodes isolated from DR4-B6 mice nasally inoculated with 1×10^8 CFU *S. aureus* COL or COL Δ*seb* were analyzed using flow cytometry ($n = 3$–5). Samples were stained with antibodies against either CD3 and Vβ3 or CD3 and Vβ8. Each mouse sample was stained with both Vβ3 and Vβ8, using Vβ3 as the internal control. Samples were gated on CD3⁺ lymphocytes, followed by gating on the Vβ3⁺CD3⁺ or Vβ8⁺CD3⁺ population and expressed as a ratio of Vβ8⁺ CD3⁺ to Vβ3⁺CD3⁺ cells per mouse. Data are shown as the mean ± SEM and significant differences determined by unpaired student's *t*-test (**, $p < 0.01$).

3. Discussion

This is the first study where the role of SAgs has been directly and experimentally assessed during a controlled model of nasal colonization using SAg-sensitive, humanized transgenic mice. Our findings reveal that different SAgs may play distinctive roles during colonization as SEA only transiently altered CFUs for *S. aureus* Newman nasal colonization, while SEB production reduced *S. aureus* COL colonization throughout all experimental time points. Although *S. aureus* Newman also encodes *sel-x* and COL additionally encodes *sei*, *sel-k* and *sel-x*, the *in vitro* stimulation data suggest that in our growth conditions, these SAgs are not made in high quantities by these strains and thus may not play a role in our model. However, future studies should assess a complete SAg deletion strain in comparison to wild-type colonization.

Data from previous human studies suggest that SAgs may be involved during *S. aureus* colonization from two lines of evidence: real-time PCR analysis of nasal swabs from persistent carriers have demonstrated *in vivo* transcription of *sea* [23], and the finding that persistently-colonized individuals have increased levels of neutralizing antibodies against SEA and TSST-1 [20]. Although it has been suggested that non-*agr* regulated SAgs such as SEA may be involved during the early phases of colonization [29], this was not supported by our experimental model when we inoculated DR4-B6 mice with *S. aureus* Newman. SEA expression during Newman colonization is supported by the increase in bacterial colonization at day 10 by *S. aureus* Newman Δ*sea* despite the lack of significant Vβ-skewing. These data suggest that SEA was expressed in low amounts that transiently inhibited the formation of higher bacterial densities in the nasal cavities. Conversely, the decrease in $V\beta8^+$ T cells during colonization with *S. aureus* COL compared to COL Δ*seb* mice is indicative of SEB expression by COL, which is responsible for the difference in nasal bacterial burdens. Direct comparison of the role of SEA *versus* SEB is difficult because they are encoded by two distinct strains. However, a notable difference between SEA and SEB lies in their regulation and expression: SEA is generally not produced in large amounts, whereas SEB production can reach high concentrations *in vitro* (Figure 1B), likely due to the activation of the *agr* two-component system. Thus, the high expression of SEB by *S. aureus* COL may have resulted in colonization with lower bacterial counts due to its inflammatory properties at all time points, while lower expression of SEA by *S. aureus* Newman did not have as dramatic differences. Although we have not genetically complemented the *S. aureus* COL Δ*seb* strain, compared with wild-type COL, other than production of SEB the exoprotein profiles are virtually indistinguishable.

The absence of anti-SAg IgG antibodies by day 14 is suggestive that either the SAgs were not processed as conventional antigens and presented to B cells, or that anti-SAg antibodies were not IgG isotypes and thus could not be detected by the assay employed. Human studies have concluded that colonization by *S. aureus* does not appear to induce a strong humoral response [23,30]. Thus, the high levels of anti-SEA antibodies in healthy subjects [20] may not be a result of persistent colonization, but rather breaches of the nasal mucosa from colonizing *S. aureus*, or skin infections. It has also been noted that anti-SAg antibodies are not always produced when the immune system is subjected to wild-type SAg, whereas SAg toxoids are much more immunogenic and are capable of forming robust anti-SAg antibodies [24,31], suggesting that SAgs can dysregulate the antibody response. Furthermore, it has been shown that naïve T cells exposed to SAgs will restrict antibody production, but will not affect 'primed' T cells [32], which may explain the lack of anti-SAg IgG in our colonized mice. TSS patients that fail to seroconvert after an episode may lead to recurrence, which has been attributed to the mechanisms of TSST-1 that prevent the development of Th2 responses, and thus T-cell dependent B cell activation [33,34].

Our study was extended to 14 days to observe differences in dissemination to other organs. *agr*-regulated SAgs such as SEB and TSST-1 may be involved in dissemination from the main bacterial colony, during which many exoproteins and virulence factors are produced, as opposed to cell-surface factors such as MSCRAMMs that are primarily involved in the initial colonization phase [29]. Surprisingly, we found bacteria in the lungs and livers of colonized mice as early as after three days, even though the mice did not show any overt signs of infection. However, there were no significant differences in the bacterial loads in these extra-nasal locations between the wild-type strains and their SAg deletion counterparts, suggesting that neither SEA nor SEB were involved in dissemination from the nasal cavity.

While SAgs are generally thought to enhance virulence [16,35], including the development of toxic shock syndrome [36], the deletion of SAgs actually increased staphylococcal CFUs in the nasal cavity. Interestingly, although colonization with *S. aureus* Newman Δ*sea* resulted in higher bacterial counts at day 10, this did not translate into long-term fitness and actually decreased back to wild-type levels by day 14. This suggests that higher bacterial densities in the nose may not be beneficial for asymptomatic colonization. Extending the length of the study may further clarify this theory since COL Δ*seb* maintained a higher bacterial density throughout the duration of the experiment. Although we did not observe differences in dissemination in our model during *S. aureus* COL and COL Δ*seb* colonization, the highest bacterial counts in the lungs were mostly COL Δ*seb*-inoculated mice, suggesting increased seeding from the higher bacterial counts in the nasal cavity. Given that bacteria colonizing the anterior nares are poised for both transmission between people and dissemination within the host, the vestibulum nasi is a desirable environment for *S. aureus* to reside in. Thus, *S. aureus* may utilize SAgs to prevent nasal bacteria from overwhelming this niche and breaching the mucosa, potentially leading to elimination by the immune system, thus acting as 'checkpoints' of dissemination. Since higher densities of bacteria may result in a greater inflammatory response, maintaining a low presence in the nose may be an evolutionarily prudent tactic to maintain long-term asymptomatic colonization. This is supported by the low bacterial burdens isolated from human nasal carriers during asymptomatic colonization [25]. Thus, this work supports the clinical finding that SAgs are expressed during nasal colonization [23], and in the context of colonization, these toxins may play an important role for influencing bacterial densities during this commensal lifestyle. This provides evidence for a novel role for SAgs, contrary to the traditional role of having been associated with enhancing virulence in severe invasive diseases.

4. Experimental Section

4.1. Mice

Six-to-twelve week old male and female HLA-DR4-IE (DRB1 * 0401) humanized transgenic mice lacking endogenous mouse MHC-II on a C57BL/6 (B6) background [37] (herein referred to as DR4-B6 mice) were used for all *in vivo* infection experiments. B6 mice were purchased from Charles River. All animal experiments were performed according to protocols approved by the Animal Use Subcommittee at Western University and in accordance with the Canadian Council on Animal Care Guide to the Care and Use of Experimental Animals.

4.2. Bacterial Strains, Media and Growth Conditions

Escherichia coli DH5α was used as a cloning host, grown in Luria Bertani (LB) broth (Difco; Mississauga, ON, Canada) at 37 °C with shaking at 250 rpm and supplemented with 150 μg/mL ampicillin when necessary. Strains of *S. aureus* listed in Table 1 were grown in tryptic soy broth (TSB) (Difco) at either 30 °C or 37 °C with shaking, and supplemented with appropriate antibiotics (Sigma Aldrich; Oakville, ON, Canada). Growth curves were performed using a Bioscreen C MBR system (Thermo Labsystems; Milford, MA, USA).

4.3. Selection of a Streptomycin-Resistant S. aureus Strain

S. aureus strain Newman is a methicillin-sensitive clinical isolate from the 1950s that is commonly used in experimental studies of staphylococcal pathogenesis [38]. Initial attempts to colonize mice resulted in competition with endogenous bacterial species and poor *S. aureus* colonization. This phenomenon has been documented previously in the literature [41] and represents an additional challenge for *S. aureus* to colonize in nature. However, for the purposes of testing our hypothesis, an antibiotic dosing regime was instated with streptomycin sulfate (Sm) in order to reduce the endogenous murine microbiota, as previously described [41]. Since *S. aureus* Newman is not naturally resistant to Sm, a Sm-resistant strain was generated by plating *S. aureus* Newman on Sm gradient TSA plates and selecting for bacteria with increased resistance. *S. aureus* Newman SmR was able to be grown in TSB containing 500 µg/mL Sm. No loss of resistance was observed after daily 1% subcultures in TSB without Sm for up to 6 days. Since the growth rate was reduced with the inclusion of Sm, preparations of bacteria for inoculation into mice were cultured without Sm. *spa* genotyping [42] showed that *S. aureus* Newman SmR had the same genetic background as Sm-sensitive Newman and qRT-PCR showed normal levels of *sea* expression (data not shown). The *sea* gene was deleted in the Newman SmR background to maintain isogenicity as described [26]. For the remainder of the experiments, Newman SmR will be referred to as Newman and the isogenic *sea* deletion strain as Newman Δ*sea*.

Table 1. Bacterial strains used in this study.

Strain	Description	Source
S. aureus Newman	Early methicillin sensitive isolate from secondary infection in a patient with tubercular osteomyelitis (Sm sensitive)	[38]
S. aureus Newman SmR	*S. aureus* Newman resistant to Sm	This study
S. aureus Newman SmR Δ*sea*	*sea*-null *S. aureus* Newman (with resistance to Sm)	[26]
S. aureus RN4220	Restriction-deficient derivation of NCTC8325-4	[39]
S. aureus COL	Early methicillin-resistant strain of *S. aureus* isolated in the 1960s	[40]
S. aureus COL Δ*seb*	*seb* deletion strain of *S. aureus* COL	This study
E. coli DH5α	Cloning strain	Invitrogen

4.4. Construction of S. aureus COL Δseb

S. aureus COL is one of the earliest MRSA strains to be isolated in the 1960's and data mining of the sequenced COL genome [40] revealed the SAgs: SEB, SEl-K and SEl-I (formerly SEQ [43]) as well as SEl-X. COL was found to be inherently resistant to Sm and thus did not require a new Sm-resistant strain to be generated. A markerless deletion was created in *seb* based on previously described methods [44]. Briefly, a 524-bp fragment upstream of *seb* was amplified using the primers 5'-TAGGGATCCAGCTCGTGATATGTTGGGTAAA-3' and 5'-GGGCGGGTCGACTGAAATAAA TAATCTCTTATACA-3' along with a 505 bp region downstream of *seb* amplified by the primers 5'-CGATGTCGACTATCTTACGACAAAGAAAAAGTGAAA-3' and 5'-TCAGGAATTCGAGATGC TTTGAAAGAAGCAAA-3'. These products were directionally cloned into pMAD, creating pMAD::*seb* which only includes 54 bp of the original 801 bp encoding *seb*. This knockout construct was methylated by *S. aureus* RN4220 and electroporated into *S. aureus* COL. To create the *seb* knockout, a single-integration event was first isolated, followed by subcultures in TSB without antibiotics grown at 30 °C. Since pMAD contains β-galactosidase, patching of white colonies detected colonies that had lost resistance to erythromycin, evident of plasmid curing and screened by PCR to verify successful deletion of *seb*.

4.5. Detection of SAgs in Cultural Supernatants in vitro

Bacterial cultures were grown overnight in TSB, cells were pelleted, and cell-free supernatants equivalent to 5.0 OD_{600} units of culture were collected. Proteins were precipitated with 10% trichloroacetic acid (TCA) overnight on ice, washed twice with ice-cold 70% ethanol and resuspended in Laemmli buffer as previously described [45]. Samples were analyzed on 12% polyacrylamide gels stained with Coomassie Brilliant Blue R-250. For Western blot analysis of SEB expression, samples were transferred to polyvinylidene difluoride (PVDF) membranes (Millipore; Etobicoke, ON, Canada) at 100V for 1 h. The membrane was blocked at roomed temperature for 1 h with PBS supplemented with 10% skim milk and 5% horse serum (Gibco; Burlington, ON, Canada). Following removal of the blocking buffer, the membrane was incubated with rabbit polyclonal anti-SEB antibodies (kindly provided by Patrick Schlievert, University of Iowa, IA, USA) diluted 1:100 in PBS supplemented with 5% skim milk and 2.5% horse serum. The membrane was washed three times with PBS supplemented with 0.02% Tween-20 (Fisher Scientific; Ottawa, ON, Canada) (PBST), followed by incubation with IRDye-conjugated goat anti-rabbit secondary antibody (LI-COR Biosciences; Lincoln, NB, USA) diluted 1:10,000 in PBST supplemented with 5% skim milk and 2.5% horse serum for 1 h in the dark. The membrane was imaged using an Odyssey imager (LI-COR Biosciences).

4.6. Assessment of Superantigenic Activity of S. aureus COL Strains in vitro

Supernatants from *S. aureus* strains were tested for SAg activity using DR4-B6 splenocytes seeded into 96-well plates as described above. Titrations of recombinant SEB, and supernatants from overnight cultures of *S. aureus* COL and COL Δ*seb* were diluted 1:10 were added to splenocytes for 18 h at 37 °C, and supernatants were assayed for IL-2 by ELISA according to manufacturer's instructions (eBioscience; San Diego, CA, USA).

4.7. Staphylococcus aureus Nasal Colonization Model

Twenty-four hours prior to inoculation, mice were administered drinking water supplemented with 2.0 mg/mL of Sm *ad libitum*, which was changed every 3–4 days for the duration of the experiment. Bacteria picked from a TSA plate were grown in 5 mL TSB overnight (16–18 h), OD_{600} was adjusted to 1.0, subcultured 2% into 50 mL TSB and grown to $OD_{600} \sim 3.0$–3.5. The bacterial pellet was washed 3 times with Hank's Buffered Salt Solution (HBSS) (Hyclone; Logan, UT, USA) and suspended at a concentration of 1×10^{10} CFU/mL in HBSS. Isofluorane-anesthetized mice were nasally inoculated by slowly pipetting 5 μL into each nare and allowing the animal to breathe in the suspension naturally, resulting in a total inoculum of 1×10^8 CFU *S. aureus* per mouse. Mice were weighed and monitored daily according to animal ethics use protocol and sacrificed at days 3, 7, 10, and 14. To enumerate the amount of bacteria in the nose, euthanized mice were decapitated and the lower jaws removed. The entire snout was excised using the back of the mouth opening as an anatomical marker in order to include any bacteria in the nasal passage. The whiskers and surrounding skin were removed without touching the nose, and the remaining tissue was collected in HBSS. The kidneys, hearts, lungs, livers and spleens were also collected and all organs were homogenized and serially diluted and plated on MSA (Difco) to differentiate between *S. aureus* and any endogenous bacteria. *S. aureus* CFUs were not different between plates containing Sm and without Sm (data not shown), thus Sm was not included in plates. Plates were enumerated after being incubated at 37 °C for 24 h. Counts less than 3 CFU/100 µL were considered below the detectable limit.

4.8. Determination of SAg Function in vivo

Lymph nodes (cervical, axillary, brachial, inguinal, and popliteal) were isolated *in toto* from mice and pushed through a cell strainer to create a single cell suspension in PBS. Cells were stained with APC-conjugated anti-CD3 (clone 145-2C11) (eBioscience) and FITC-conjugated anti-Vβ3 (clone KJ25) (BD Pharmingen; Mississauga, ON, Canada) or FITC-conjugated anti-Vβ8 (clone KJ16) (eBioscience) and assayed by flow cytometry. Data were analyzed using FlowJo v.8.7. (Treestar; Ashland, OR, USA).

4.9. Statistical Analyses

All statistical analyses were performed using Prism v5.0 (GraphPad; La Jolla, CA, USA) with $p < 0.05$ being considered significant.

Acknowledgments

We thank Patrick M. Schlievert (University of Iowa) for the gift of anti-SEB rabbit antiserum. This work was supported by Canadian Institutes of Health Research (CIHR) operating grant (MOP-64176) to John K. McCormick. Stacey X. Xu and Joseph J. Zeppa were supported, in part, by Ontario Graduate Scholarships.

Author Contributions

Stacey X. Xu, Katherine J. Kasper, Joseph J. Zeppa and John K. McCormick conceived and designed the experiments; Stacey X. Xu, Katherine J. Kasper and Joseph J. Zeppa performed the experiments; Stacey X. Xu and Joseph J. Zeppa analyzed the data; Stacey X. Xu and John K. McCormick wrote the paper.

Conflicts of Interest

The authors declare no conflict of interest.

References

1. Lowy, F.D. *Staphylococcus aureus* infections. *N. Engl. J. Med.* **1998**, *339*, 520–532.
2. Rehm, S.J.; Tice, A. *Staphylococcus aureus*: Methicillin-susceptible *S. aureus* to methicillin-resistant *S. aureus* and vancomycin-resistant *S. aureus*. *Clin. Infect. Dis.* **2010**, *51*, S176–S182.

3. Boucher, H.; Miller, L.G.; Razonable, R.R. Serious infections caused by methicillin-resistant *Staphylococcus aureus*. *Clin. Infect. Dis.* **2010**, *51*, S183–S197.

4. Kluytmans, J.A.J.W.; Wertheim, H.F.L. Nasal carriage of *Staphylococcus aureus* and prevention of nosocomial infections. *Infection* **2005**, *33*, 3–8.

5. Wertheim, H.F.L.; Melles, D.C.; Vos, M.C.; van Leeuwen, W.; van Belkum, A.; Verbrugh, H.A.; Nouwen, J.L. The role of nasal carriage in *Staphylococcus aureus* infections. *Lancet Infect. Dis.* **2005**, *5*, 751–762.

6. Ten Broeke-Smits, N.J.P.; Kummer, J.A.; Bleys, R.L.A.W.; Fluit, A.C.; Boel, C.H.E. Hair follicles as a niche of *Staphylococcus aureus* in the nose; is a more effective decolonisation strategy needed? *J. Hosp. Infect.* **2010**, *76*, 211–214.

7. Bibel, D.J.; Aly, R.; Shinefield, H.R.; Maibach, H.I.; Strauss, W.G. Importance of the keratinized epithelial cell in bacterial adherence. *J. Investig. Dermatol.* **1982**, *79*, 250–253.

8. Corrigan, R.M.; Miajlovic, H.; Foster, T.J. Surface proteins that promote adherence of *Staphylococcus aureus* to human desquamated nasal epithelial cells. *BMC Microbiol.* **2009**, *9*, 22, doi:10.1186/1471-2180-9-22.

9. Wertheim, H.F.; Vos, M.C.; Ott, A.; van Belkum, A.; Voss, A.; Kluytmans, J.A.; van Keulen, P.H.; Vandenbroucke-Grauls, C.M.; Meester, M.H.; Verbrugh, H.A. Risk and outcome of nosocomial *Staphylococcus aureus* bacteraemia in nasal carriers *versus* non-carriers. *Lancet* **2004**, *364*, 703–705.

10. Kolata, J.; Bode, L.G.M.; Holtfreter, S.; Steil, L.; Kusch, H.; Holtfreter, B.; Albrecht, D.; Hecker, M.; Engelmann, S.; van Belkum, A.; *et al.* Distinctive patterns in the human antibody response to *Staphylococcus aureus* bacteremia in carriers and non-carriers. *Proteomics* **2011**, *11*, 3914–3927.

11. Brown, A.F.; Leech, J.M.; Rogers, T.R.; McLoughlin, R.M. *Staphylococcus aureus* colonization: Modulation of host immune response and impact on human vaccine design. *Front. Immun.* **2013**, *4*, doi:10.3389/fimmu.2013.00507.

12. Wertheim, H.F.; Walsh, E.; Choudhurry, R.; Melles, D.C.; Boelens, H.A.; Miajlovic, H.; Verbrugh, H.A.; Foster, T.; van Belkum, A. Key role for clumping factor B in *Staphylococcus aureus* nasal colonization of humans. *PLoS Med.* **2008**, *5*, e17, doi:10.1371/journal.pmed.0050017.

13. Weidenmaier, C.; Kokai-Kun, J.F.; Kristian, S.A.; Chanturiya, T.; Kalbacher, H.; Gross, M.; Nicholson, G.; Neumeister, B.; Mond, J.J.; Peschel, A. Role of teichoic acids in *Staphylococcus aureus* nasal colonization, a major risk factor in nosocomial infections. *Nat. Med.* **2004**, *10*, 243–245.

14. Roche, F.M.; Meehan, M.; Foster, T.J. The *Staphylococcus aureus* surface protein SasG and its homologues promote bacterial adherence to human desquamated nasal epithelial cells. *Microbiol.* **2003**, *149*, 2759–2767.

15. Clarke, S.R.; Andre, G.; Walsh, E.J.; Dufrêne, Y.F.; Foster, T.J.; Foster, S.J. Iron-regulated surface determinant protein A mediates adhesion of Staphylococcus aureus to human corneocyte envelope proteins. *Infect. Immun.* **2009**, *77*, 2408–2416.

16. Xu, S.X.; McCormick, J.K. Staphylococcal superantigens in colonization and disease. *Front. Cell Infect. Microbiol.* **2012**, *2*, doi:10.3389/fcimb.2012.00052.

17. Salgado-Pabón, W.; Breshears, L.; Spaulding, A.R.; Merriman, J.A.; Stach, C.S.; Horswill, A.R.; Peterson, M.L.; Schlievert, P.M. Superantigens are critical for *Staphylococcus aureus* infective endocarditis, sepsis, and acute kidney injury. *mBio.* **2013**, *4*, doi:10.1128/mBio.00494-13.

18. Jarraud, S.; Peyrat, M.A.; Lim, A.; Tristan, A.; Bes, M.; Mougel, C.; Etienne, J.; Vandenesch, F.; Bonneville, M.; Lina, G.; *et al.* A highly prevalent operon of enterotoxin gene, forms a putative nursery of superantigens in *Staphylococcus aureus. J. Immunol.* **2001**, *166*, 669–677.

19. Holtfreter, S.; Roschack, K.; Eichler, P.; Eske, K.; Holtfreter, B.; Kohler, C.; Engelmann, S.; Hecker, M.; Greinacher, A.; Bröker, B.M. *Staphylococcus aureus* carriers neutralize superantigens by antibodies specific for their colonizing strain: A potential explanation for their improved prognosis in severe sepsis. *J. Infect. Dis.* **2006**, *193*, 1275–1278.

20. Verkaik, N.J.; de Vogel, C.P.; Boelens, H.A.; Grumann, D.; Hoogenboezem, T.; Vink, C.; Hooijkaas, H.; Foster, T.J.; Verbrugh, H.A.; van Belkum, A.; *et al.* Anti-staphylococcal humoral immune response in persistent nasal carriers and noncarriers of *Staphylococcus aureus. J. Infect. Dis.* **2009**, *199*, 625–632.

21. Holtfreter, S.; Grumann, D.; Schmudde, M.; Nguyen, H.T.T.; Eichler, P.; Strommenger, B.; Kopron, K.; Kolata, J.; Giedrys-Kalemba, S.; Steinmetz, I.; *et al.* Clonal distribution of superantigen genes in clinical *Staphylococcus aureus* isolates. *J. Clin. Microbiol.* **2007**, *45*, 2669–2680.

22. Ferry, T.; Thomas, D.; Genestier, A.L.; Bes, M.; Lina, G.; Vandenesch, F.; Etienne, J. Comparative prevalence of superantigen genes in *Staphylococcus aureus* isolates causing sepsis with and without septic shock. *Clin. Infect. Dis.* **2005**, *41*, 771–777.

23. Burian, M.; Grumann, D.; Holtfreter, S.; Wolz, C.; Goerke, C.; Bröker, B.M. Expression of staphylococcal superantigens during nasal colonization is not sufficient to induce a systemic neutralizing antibody response in humans. *Eur. J. Clin. Microbiol. Infect. Dis.* **2012**, *31*, 251–256.

24. Narita, K.; Hu, D.L.; Tsuji, T.; Nakane, A. Intranasal immunization of mutant toxic shock syndrome toxin 1 elicits systemic and mucosal immune response against *Staphylococcus aureus* infection. *FEMS Immunol. Med. Microbiol.* **2008**, *52*, 389–396.

25. Krismer, B.; Peschel, A. Does *Staphylococcus aureus* nasal colonization involve biofilm formation? *Future Microbiol.* **2011**, *6*, 489–493.

26. Xu, S.X.; Gilmore, K.J.; Szabo, P.A.; Zeppa, J.J.; Baroja, M.L.; Haeryfar, S.M.M.; McCormick, J.K. Superantigens subvert the neutrophil response to promote abscess formation and enhance Staphylococcus aureus survival *in vivo. Infect. Immun.* **2014**, *82*, 3588–3598.

27. Gaskill, M.E.; Khan, S.A. Regulation of the enterotoxin B gene in *Staphylococcus aureus. J. Biol. Chem.* **1988**, *263*, 6276–6280.

28. Janeway, C.A., Jr. Selective elements for the V beta region of the T cell receptor: Mls and the bacterial toxic mitogens. *Adv. Immunol.* **1991**, *50*, 1–53.

29. Bohach, G.; Schlievert, P.M. Staphylococcal and streptococcal superantigens: An update. In *Superantigens: Molecular Basis for the Role in Human Diseases*; Fraser, J.D., Kotb, M., Eds.; ASM Press: Washington, DC, USA, 2007; pp. 21–36.

30. Holtfreter, S.; Nguyen, T.T.H.; Wertheim, H.; Steil, L.; Kusch, H.; Truong, Q.P.; Engelmann, S.; Hecker, M.; Völker, U.; van Belkum, A.; *et al.* Human immune proteome in experimental colonization with *Staphylococcus aureus*. *Clin. Vaccine Immunol.* **2009**, *16*, 1607–1614.

31. Spaulding, A.R.; Lin, Y.C.; Merriman, J.A.; Brosnahan, A.J.; Peterson, M.L.; Schlievert, P.M. Immunity to *Staphylococcus aureus* secreted proteins protects rabbits from serious illnesses. *Vaccine* **2012**, *30*, 5099–5109.

32. Lussow, A.R.; MacDonald, H.R. Differential effects of superantigen-induced "anergy" on priming and effector stages of a T cell-dependent antibody response. *Eur. J. Immunol.* **1994**, *24*, 445–449.

33. Lappin, E.; Ferguson, A.J. Gram-positive toxic shock syndromes. *Lancet Infect. Dis.* **2009**, *9*, 281–290.

34. Hofer, M.F.; Newell, K.; Duke, R.C.; Schlievert, P.M.; Freed, J.H.; Leung, D.Y. Differential effects of staphylococcal toxic shock syndrome toxin-1 on B cell apoptosis. *Proc. Natl. Acad. Sci. USA* **1996**, *93*, 5425–5430.

35. Llewelyn, M.; Cohen, J. Superantigens: Microbial agents that corrupt immunity. *Lancet Infect. Dis.* **2002**, *2*, 156–162.

36. McCormick, J.K.; Yarwood, J.M.; Schlievert, P.M. Toxic shock syndrome and bacterial superantigens: An update. *Annu. Rev. Microbiol.* **2001**, *55*, 77–104.

37. Ito, K.; Bian, H.J.; Molina, M.; Han, J.; Magram, J.; Saar, E.; Belunis, C.; Bolin, D.R.; Arceo, R.; Campbell, R.; *et al.* HLA-DR4-IE chimeric class II transgenic, murine class II-deficient mice are susceptible to experimental allergic encephalomyelitis. *J. Exp. Med.* **1996**, *183*, 2635–2644.

38. Duthie, E.; Lorenz, L.L. Staphylococcal coagulase: Mode of action and antigenicity. *Microbiology* **1952**, *6*, 95–107.

39. Novick, R. Properties of a cryptic high-frequency transducing phage in *Staphylococcus aureus*. *Virology* **1967**, *33*, 155–166.

40. Gill, S.R.; Fouts, D.E.; Archer, G.L.; Mongodin, E.F.; DeBoy, R.T.; Ravel, J.; Paulsen, I.T.; Kolonay, J.F.; Brinkac, L.; Beanan, M.; *et al.* Insights on evolution of virulence and resistance from the complete genome analysis of an early methicillin-resistant *Staphylococcus aureus* strain and a biofilm-producing methicillin-resistant *Staphylococcus epidermidis* strain. *J. Bacteriol.* **2005**, *187*, 2426–2438.

41. Kiser, K.B.; Cantey-Kiser, J.M.; Lee, J.C. Development and characterization of a *Staphylococcus aureus* nasal colonization model in mice. *Infect. Immun.* **1999**, *67*, 5001–5006.

42. Fenner, L.; Widmer, A.F.; Dangel, M.; Frei, R. Distribution of *spa* types among meticillin-resistant *Staphylococcus aureus* isolates during a 6 year period at a low-prevalence university hospital. *J. Med. Microbiol.* **2008**, *57*, 612–616.

43. Yarwood, J.M.; McCormick, J.K.; Paustian, M.L.; Orwin, P.M.; Kapur, V.; Schlievert, P.M. Characterization and expression analysis of *Staphylococcus aureus* pathogenicity island 3. Implications for the evolution of staphylococcal pathogenicity islands. *J. Biol. Chem.* **2002**, *277*, 13138–13147.

44. Arnaud, M.; Chastanet, A.; Debarbouille, M. New vector for efficient allelic replacement in naturally nontransformable, low-GC-content, gram-positive bacteria. *Appl. Environ. Microbiol.* **2004**, *70*, 6887–6891.

45. Arsic, B.; Zhu, Y.; Heinrichs, D.E.; McGavin, M.J. Induction of the staphylococcal proteolytic cascade by antimicrobial fatty acids in community acquired methicillin resistant *Staphylococcus aureus*. *PLoS One* **2012**, *7*, e45952.

Clostridium perfringens Epsilon Toxin: A Malevolent Molecule for Animals and Man?

Bradley G. Stiles, Gillian Barth, Holger Barth and Michel R. Popoff

Abstract: *Clostridium perfringens* is a prolific, toxin-producing anaerobe causing multiple diseases in humans and animals. One of these toxins is epsilon, a 33 kDa protein produced by *Clostridium perfringens* (types B and D) that induces fatal enteric disease of goats, sheep and cattle. Epsilon toxin (Etx) belongs to the aerolysin-like toxin family. It contains three distinct domains, is proteolytically-activated and forms oligomeric pores on cell surfaces via a lipid raft-associated protein(s). Vaccination controls Etx-induced disease in the field. However, therapeutic measures are currently lacking. This review initially introduces *C. perfringens* toxins, subsequently focusing upon the Etx and its biochemistry, disease characteristics in various animals that include laboratory models (*in vitro* and *in vivo*), and finally control mechanisms (vaccines and therapeutics).

Reprinted from *Toxins*. Cite as: Stiles, B.G.; Barth, G.; Barth, H.; Popoff, M.R. *Clostridium perfringens* Epsilon Toxin: A Malevolent Molecule for Animals and Man? *Toxins* **2013**, *5*, 2138-2160.

1. Introduction

Clostridium perfringens is a Gram-positive, spore-forming anaerobe residing in soil, water as well as the gastrointestinal tracts of various mammals, including humans. This ubiquitous bacillus is one of the most "toxic" of all known bacteria, collectively producing more than 15 different protein toxins/enzymes with diverse modes of action [1,2]. Pathogenic *Clostridium* species synthesize some of the most potent toxins that include tetanus and botulinum neurotoxins, respectively produced by *Clostridium tetani* and *Clostridium botulinum*. These bacteria are found in similar environments (*i.e.*, soil) as long-lasting, quiescent spores that await a mammalian host and infection/intoxication opportunity.

C. perfringens was first isolated by William Welch and George Nuttall (1892) at Johns Hopkins Hospital (Baltimore, MD, USA) following autopsy of a cancer/tuberculosis patient, eight hours post-death. In particular, they noted flammable gas bubbles perfused throughout the cadaver and especially within blood vessels. Gas (carbon dioxide plus hydrogen) and organic acids (acetic, butyric, lactic, *etc.*) are common byproducts of anaerobic metabolism by *C. perfringens*. *C. perfringens* has also been successively known in the literature as *Bacillus aerogenes capsulatus*, *Bacillus welchii* or *Clostridium welchii* [1–3]. Various diseases of animals and humans caused by *C. perfringens* are linked to protein toxins, and the next section succinctly describes the "major" and "minor" toxins produced by this bacterium. One of those toxins, epsilon, is this review's focus as this protein impacts in many ways the veterinary and biodefense fields throughout the world.

2. *Clostridium perfringens* Toxins: Major and Minor (A Brief Overview)

Protein toxins are important virulence factors of *C. perfringens* and have been a research focus of various laboratories around the world. For bacterial pathogens, toxins possessing diverse modes of action often play critical roles during disease, including food gathering and suppressing the host's immune system. The four major toxins produced by *C. perfringens* either affect cell membranes directly by increasing permeability and causing ion imbalances (alpha, beta and epsilon toxins), or destroy the actin cytoskeleton (iota toxin) [1,2]. Intoxication by any of these clostridial proteins ultimately leads to cell dysfunction and death, as well as host suffering that can become fatal. Like the spores formed by other *Clostridium* and *Bacillus* species that enable survival in soil, protein toxins can play a pivotal role in bacteria surviving and subsequently thriving in an animal or human host.

There are five toxin types (A, B, C, D and E) of *C. perfringens* based upon the production of one or more major protein toxins [1,2] (Table 1). These toxins are linked to diverse diseases/intoxications of humans and/or animals (Table 2).

Table 1. Major toxins for *C. perfringens* typing.

Toxin	*C. perfringens* Type					Cellular Target (mode of action)
	A	B	C	D	E	
Alpha	+	+	+	+	+	Membrane (phospholipid destruction)
Beta		+	+			Membrane (pore formation)
Epsilon		+		+		Membrane (pore formation)
Iota					+	Actin (cytoskeleton destruction)

Table 2. *C. perfringens* toxin types and associated diseases.

Toxin Type	Disease
A	Myonecrosis (gas gangrene in humans and animals); Necrotic enteritis of fowl plus piglets; Human food poisoning and antibiotic-associated diarrhea
B	Hemorrhagic enteritis in calves, foals and sheep; Dysentery in lambs
C	Necrotizing enteritis in humans (also popularly called pigbel, darmbrand or fire-belly), as well as in pigs, calves, goats and foals; Enterotoxemia in sheep (alias struck)
D	Enterotoxemia in lambs (known as pulpy kidney disease), goats and cattle
E	Enterotoxemia in calves and lambs. Similar enteric disease induced by iota-like toxin in rabbits, caused by *Clostridium spiroforme*

For diagnostic purposes, major toxins of *C. perfringens* found in field samples or cultured isolates *in vitro* were historically neutralized in the laboratory by type-specific antisera in mouse-lethal and guinea-pig dermonecrotic assays [3]. Rapid genetic methods employing multiplex polymerase chain reaction (PCR) are now much more common for typing *C. perfringens* [4,5]. This technique is accurate and rapid; however, PCR merely suggests a gene's presence and indicates neither expression levels nor quantities of an effector molecule (biologically-active toxin) that are ultimately responsible for causing physiological changes to a cell.

Detection of all major *C. perfringens* toxins has also been reported by various laboratories using ELISA technology [6–9]. Quantitation of epsilon toxin protein is also possible using a novel, mass spectrometry technique [10]. In contrast to ELISA or mass spectrometry, animal assays and toxin-susceptible cell cultures can effectively determine if biologically-active toxin (in conjunction with toxin-specific antibody use) exists in a sample. For any biological protein in a suspect sample, structural integrity is linked to many factors that include how quickly the sample is collected post-mortem, room temperature *vs.* refrigerated/frozen storage, how long the sample is stored before testing, *etc*. In particular, enterotoxins found in digesta can be altered by many "factors" such as proteases and non-specific protein binding naturally found in the intestinal tract.

2.1. Alpha Toxin

Type A strains of *C. perfringens* are most commonly found throughout the environment and linked to gas gangrene of animals and humans [1,11–15]. Alpha toxin facilitates gas gangrene due to *C. perfringens* infection, a life-threatening myonecrotic disease historically common with battlefield wounds [11,12]. Deep, penetrating wounds contaminated by soil harboring various clostridial species, including *C. perfringens*, are often to blame for this quickly advancing disease of thick-muscled body regions that include the buttocks, shoulder, arm and leg [11,14,15]. Prior to the 20th century and a true understanding of antisepsis, with a knowledge of specific etiological agents that cause infectious diseases like gangrene, an extremity wound could quickly result in an amputation that simply would not occur with today's modern medicine [15]. Rapid treatment involving extensive surgical debridement, various antibiotics that include beta-lactams, clindamycin and/or metronidazole, as well as hyperbaric oxygen prove effective for most cases of *C. perfringens*-induced gangrene. Anti-toxin (historically, serum antibodies of equine origin) is another possible therapy for mitigating alpha-toxin induced myonecrosis [11,14,16]. Recombinant-based technology generating human monoclonal antibodies could replace equine polyclonal, circumventing potentially life-threatening serum sickness. Regarding a prophylaxis perspective, vaccine studies from various groups using the carboxy-terminal (cell binding) domain of alpha toxin show protection in mice against either toxin-induced lethality or bacterial challenge in a gangrene model [17,18].

Alpha toxin is a zinc-containing phospholipase C (43 kDa), composed of two structural domains, that destroys eukaryotic cell membranes [19,20]. In 1941, *C. perfringens* alpha was the first bacterial toxin ascribed enzymatic activity [21]. Like beta, but unlike epsilon and iota, the alpha toxin is relatively susceptible to proteolysis by serine-type proteases such as trypsin and chymotrypsin. The amino-terminal domain contains a catalytic site and ganglioside (GM1a) binding motif, the latter being curiously similar to that found on *C. botulinum* neurotoxin [22]. Interaction of GM1a with alpha toxin promotes clustering/activation of tyrosine kinase A involved in signal transduction. The carboxy-terminal domain of alpha toxin binds to membrane phospholipids.

2.2. Beta Toxin

Originally purified in 1977, beta toxin is a 35 kDa protein that shares sequence similarity with the alpha and gamma hemolysins of *Staphylococcus aureus* [23,24]. The toxin is responsible for fatal necrotic enteritis in animals and humans involving intestinal necrosis and bloody stools. In humans, diseases such as pigbel (Papua, New Guinea) or Darmbrand (post-World War II Germany) follow consumption of meat by individuals on a minimal protein diet with a low-basal level of pancreatic trypsin [25]. For some pigbel cases, individuals may have consumed trypsin inhibitor via sweet potatoes (a staple component of the normal diet) and/or be infected by round worms (*Ascaris lubricoides*) that release trypsin inhibitor into the intestinal lumen. Unusually high concentrations of protein in the intestinal tract facilitate *C. perfringens* types B (animal) or C (human) overgrowth, leading to lethal levels of beta toxin, which forms cation-selective channels in lipid membranes [26]. Tachykinin (neuropeptide) receptors play a role in beta-toxin induced fluid release from the circulatory system into tissue, suggesting involvement of the sensory nervous system [27]. This particular study in murine dermis reveals that beta intoxication is inhibited by tachykinin NK_1 antagonists, capsaicin and an omega conotoxin (*Conus magus* MVIIA) that specifically blocks N-type calcium-channels.

2.3. Iota Toxin

Iota toxin, discovered in 1943 by Bosworth [28], is unique among the major toxins of *C. perfringens* by consisting of two non-linked proteins that form a multimeric complex on susceptible cells, akin to the edema and lethal binary toxins from *Bacillus anthracis* [29]. Like *C. perfringens*, *B. anthracis* is a Gram-positive sporulating bacillus found in soil that uniquely causes various forms of anthrax (dermal, enteric and pneumonic) in animals as well as humans. For iota toxin, the cell-binding component (Ib or iota b) is an 81 kDa monomer that forms a heptamer following proteolytic activation of the 94 kDa protoxin. The complementary enzymatic component is an ADP-ribosyl transferase (Ia or iota a) of 45 kDa, which in complex with the Ib heptamer, travels from the cell surface via lipid rafts into endosomes. The Ia component translocates from an acidified endosome into the cytosol, through a transmembrane pore formed by the Ib heptamers. In the cytosol, Ia covalently transfers an ADP-ribose moiety from nicotinamide adenine dinucleotide (NAD) to Arg177 on globular (G) actin, which then prevents filamentous (F) actin formation. Destruction of the cytoskeleton ensues, ultimately killing an intoxicated cell unable to maintain intracellular trafficking necessary for homeostasis.

In addition to the major toxins listed above, there is a sporulation-linked *C. perfringens* enterotoxin that causes a major form of food poisoning linked to meat consumption [30]. Finally, there are other *C. perfringens* proteins designated as minor toxins that can play prominent roles in pathogenesis. A list of these latter toxins/enzymes with putative activities is presented in Table 3.

Table 3. Minor toxins/enzymes of *C. perfringens*.

Toxin/Enzyme	Activity
Beta 2	?
Delta	Cytolysin
Eta	?
Gamma	?
Kappa	Collagenase
Lambda	Protease
Mu	Hyaluronidase
Nu	Deoxyribonuclease
NanI, NanJ, NanH	Neuraminidase
NetB	Hemolysin
Theta (perfringolysin O)	Oxygen-labile Hemolysin
TpeL	Glucosylation of Ras

3. Epsilon Toxin (Etx): *C. perfringens* Most Toxic Toxin

3.1. Natural Occurrence and a Potential Biological Warfare/Terrorism Agent?

Etx produced by *C. perfringens* types B and D is involved in animal (goats, sheep and less frequently cattle) enterotoxemias that can be rapidly fatal and economically devastating [1,2,31]. Etx is the most potent of all *C. perfringens* toxins as determined by a 50% lethal dose (LD$_{50}$ of ~70 ng/kg body weight, ranking behind only the *C. botulinum* and *C. tetani* neurotoxins in classic mouse-lethal assays commonly used for clostridial toxins. *C. perfringens* is considered normal intestinal flora in ruminants. Resident types B and D can cause life-threatening illness in a "naïve" digestive system shortly after birth or following a diet change involving more carbohydrates. Given proper conditions, *C. perfringens* can rapidly proliferate in the intestines and concomitantly produce life-threatening levels of toxins, including Etx. The disease is strictly a toxemia that spills into the circulatory system, since bacterial invasion of intestinal tissue is not common. Experts naturally associate the Etx with veterinary maladies. Unlike various studies from multiple groups involving different animal species, the existing literature simply does not suggest that human intoxication by Etx naturally occurs even infrequently.

Due to national and international concerns involving biological warfare/terrorism, *C. perfringens* Etx has thus received much attention from various governments [32]. The potential nefarious use of Etx against humans by rogues provides chilling insight into human psychology and fear mongering. The United States Department of Agriculture and Centers for Disease Control and Prevention have classified Etx as a select agent, like some bacterial diseases (brucellosis, glanders and typhus) plus other protein toxins such as ricin and staphylococcal enterotoxin B (SEB). However, modification in December 2012 of the select agents list has now removed *C. perfringens* Etx [33]. In France, Etx is still classified as a potential biological weapon requiring special authorization for laboratory work from the Agence Nationale de Securite du Medicament (ANSM). Varying opinions clearly exist around the world regarding *C. perfringens* Etx, its potential nefarious use to promote societal fear, and subsequently imposed oversight to protect the governed masses.

3.2. Chemical and Physical Properties

C. perfringens Etx is coded on a plasmid and secreted as a protoxin (32.9 kDa), subsequently activated by extracellular serine-type proteases (trypsin/chymotrypsin) that remove 10–13 amino- and 22 or 29 carboxy-terminal residues, depending on the protease [31]. The protoxin contains a typical leader sequence (32 residues) that facilitates secretion from the bacterium's cytosol into the environment. Activated toxin (29 kDa) is relatively resistant to proteases in the gastrointestinal tracts of mammals.

Etx is an elongated, beta-sheet (100 Å × 20 Å × 20 Å) composed of three domains sharing conformation with other bacterial pore-forming toxins (PFTs) of the aerolysin-like family, such as *Aeromonas hydrophila* aerolysin and *C. perfringens* enterotoxin (Figure 1) [30,34,35].

Figure 1. Crystal structures of *C. perfringens* epsilon toxin, *C. perfringens* enterotoxin, *L. sulphureus* lectin and *A. hydrophila* aerolysin. A basic, three-domain structure is evident for each protein [30,34–37]. Amino acids involved in pore-formation are in red.

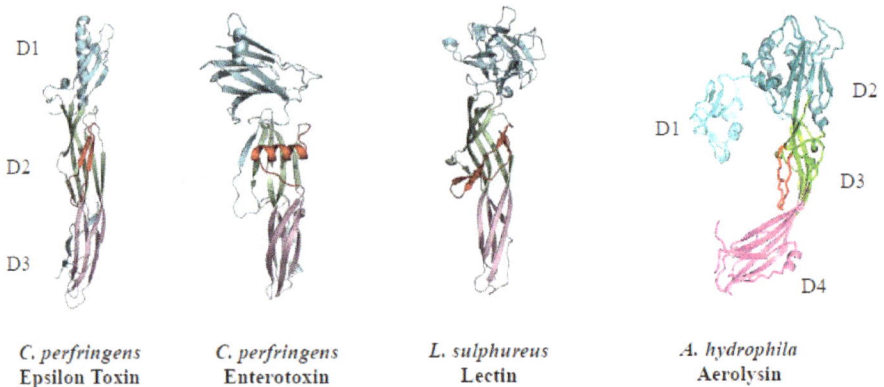

| *C. perfringens* Epsilon Toxin | *C. perfringens* Enterotoxin | *L. sulphureus* Lectin | *A. hydrophila* Aerolysin |

Furthermore, this protein family is produced by other diverse life forms that include a Brazilian tree (*Enterolobium contortisiliquum*), mushroom (*Laetiporus sulphureus*) and freshwater hydra (*Chlorohydra viridis*) [35]. Although primary sequence homology among the aerolysin-like family is relatively low (< 20% identity), conformations are strikingly similar. A hallmark of these convergently-evolving proteins involves synthesis as monomers that generate homo-oligomeric (hexameric or heptameric) complexes on a cell-surface membrane, leading to transmembrane pore formation.

Two groups of PFTs exist, based upon using either amphipathic alpha-helix (alpha-PFT) or beta-hairpin (beta-PFT) structures for membrane insertion. Etx is a beta-PFT. Putative roles for Etx domains include: receptor binding (domain I–amino terminus); membrane insertion plus channel formation (domain II–central region); and proteolysis activation plus monomer-monomer interaction sites (domain III–carboxy terminus) [34,35]. Following loss of a carboxy-terminal peptide from epsilon protoxin, there are subsequent monomer-monomer interactions that lead to homo-heptamer formation [38].

Proteolysis is a common process that activates many bacterial toxins. For epsilon, this process induces conformational changes that facilitate homo-oligomerization of activated toxin on the external surface of a targeted, eukaryotic cell. This "protein priming" enables Etx to quickly act after binding to diverse target cells that include those of neuronal, renal and endothelial origins (described in detail below). Loss of the amino- and carboxy-termini from the epsilon protoxin generates a more acidic protein (isoelectric point of 5.4 vs. 8.0), perhaps favoring more productive receptor interactions [31,39]. For bacterial toxins produced intestinally and requiring proteolysis, proteases synthesized by resident bacteria (including C. perfringens lambda toxin, a 35 kDa thermolysin-like metalloenzyme) and the host are abundant [38,40]. Recent evidence, contrary to the existing paradigm, reveals that Etx can be activated intracellularly (in a select type D strain lacking lambda toxin), remains in C. perfringens until stationary/death phase and is then released into the environment after autolysis [41]. Identity of this protease, and a better understanding of this novel activation mode perhaps found in other protoxin-producing pathogens, remains a mystery.

3.3. Cellular Target (Mode of Action)

Etx induces pore formation in eukaryotic cell membranes via detergent-resistant, cholesterol-rich membrane domains (lipid rafts) that promote aggregation of toxin monomers into homo-heptamers [42,43]. The pro- and activated-toxin forms bind to lipid rafts, but the former do not generate oligomers. Lipid rafts play important roles in many diseases caused by bacteria (as well as associated toxins) and viruses [44]. In particular, caveolins 1 and 2 found in these extracellular domains on ACHN (human adenocarcinoma kidney) cells bind to Etx and facilitate toxin oligomerization [45]. Caveolins (1–3) are a family of integral-membrane proteins (~20 kDa each) important in caveolae-mediated endocytosis and cell signaling. An Etx oligomer (155 kDa), critical for biological activity: (1) rapidly (within 15 minutes) forms at 37 °C on the surface of Madin-Darby canine kidney (MDCK) cells; (2) is more stable towards sodium dodecyl sulfate and heat (100 °C) when formed at 37 °C vs. 4 °C; and (3) becomes internalized from the cell surface, promoting vacuole formation in the late endosomes and lysosomes [46,47]. Epsilon-treated MDCK cells readily swell, form multiple blebs, and ultimately lyse; however, blocking of acid-mediated endocytosis via chloroquine, monensin, or bafilomycin A1 has no effect upon the intoxication process [46]. Pretreatment of cells with proteinase K results in a smaller (95 kDa) Etx complex, suggesting involvement of a surface protein(s) during toxin oligomerization [46].

Sialidases (neuraminidases) produced by C. perfringens enhance binding of the bacterium and Etx to cultured cells [48]. Interestingly, certain mammalian cell lines upregulate their sialidase genes upon contact with C. perfringens. Increased toxin binding following sialidase activity has been described previously for another enteric pathogen, *Vibrio cholerae*, perhaps suggesting a conserved pathogenic mechanism between diverse, enterotoxin-producing bacteria [49].

When applied to cells, Etx forms transmembrane pores (≥2 nm diameter) that rapidly facilitate: (1) free passage of 1 kDa-sized molecules; (2) decreased intracellular potassium levels; and (3) increased intracellular levels of chloride and sodium [46,50]. In addition to altering the membrane, secondary effects of epsilon intoxication involve cytoskeletal dysfunction [51] affecting integrity of cell monolayers [42]. Disruption of cell monolayers provides further understanding of the subsequent dysfunction of vascular endothelium, edema and crossing of the blood-brain barrier by the toxin as well as albumin-sized (~65 kDa) molecules [52,53].

4. A Veterinary Perspective on Etx: Field and Laboratory Findings

Sheep and goats are the most frequent, natural hosts for *C. perfringens* Etx, cattle less so [54,55]. *C. perfringens* type D strains are more common than type B for epsilon-induced disease, although both toxin types produce Etx. Various factors can play a role in disease, including: (1) consumption of feed rich in fermentable carbohydrates; (2) poor nutritional status; (3) parasite infestation; (4) pregnancy toxemia; (5) tranquilizer (phenothiazine) use; and (6) overdose of a broad spectrum anthelmintic (netobimin) [54,55]. Although Etx can be found in the heart, lungs, liver and stomach following intoxication, it noticeably accumulates in the kidneys, causing what veterinarians classically refer to as pulpy kidney, or overeating, disease [1,2,31,56–58]; however, evidence for pulpy kidney disease is inconsistent, probably a post-mortem effect and thus not considered diagnostic. A combination of toxin detected in the small intestines and brain lesions is a strong diagnostic. Another indicator that kidneys are a primary target of Etx is that the very few, susceptible cell lines discovered to date are mostly of kidney descent from dog, mouse and human [31].

Post-mortem findings in kidneys from either lambs or mice given Etx show similar results, including congestion, interstitial hemorrhage and degenerated epithelium in the distal tubules. Toxin accumulating in the kidney may be a host defense attempting to prevent lethal toxin concentrations in the brain [58]. During epsilon enterotoxemia of lambs, glucose excretion occurs via the urine and is perhaps a result of liver-released glycogen [59].

C. perfringens Etx rapidly disrupts the blood-brain barrier, binds neuronal cells and causes lethality [31,52,56,60]. Among neuronal cell populations, the neurons are most susceptible followed by oligodendrocytes and astrocytes [61]. There can be swelling, vacuolation and necrosis in the brain. Edema in the brain (rat) post-Etx exposure increases aquaporin-4 levels in astrocytes, which may be a defensive attempt to reduce osmotic pressure surrounding sensitive neurons [62]. Aquaporins (aquaglyceroporins) are a family (n = 13 members in mammals) of 30 kDa-sized proteins that form tetrameric, membrane-channel complexes regulating water flow in various cell types, perhaps representing an exploitable therapeutic target against Etx. There is already a keen medical interest in aquaporins as: (1) a diagnostic indicator for certain autoimmune diseases (*i.e.*, Sjogren's syndrome and nephrogenic diabetes insipidus); and (2) a therapy for glaucoma, meningitis, stroke, epilepsy, cancer, pain as well as weight regulation [63]. Cerebral swelling and necrosis of the brain following Etx exposure can be due to multiple factors like reduced blood flow, hypoxia and/or direct toxicity on various cell types within the brain.

Clinical signs linked to Etx given intravenously to calves, lambs and kid goats are dose dependent and occur within minutes for calves and up to three hours in lambs [64,65]. For kids, either deprived or reared with colostrum by non-vaccinated dams, and given a 120, 185 or 250 mouse (M) LD$_{50}$/kg body weight dose of Etx, clinical signs are evident in all within 95 minutes [64]. In this same study, weaned lambs (colostrum-reared from non-vaccinated dams) given a 120 or

250 MLD$_{50}$/kg body weight dose of Etx show clinical signs within 170 minutes. These animals experience a myriad of symptoms involving labored breathing (*i.e.*, foam-filled airways, alveolar edema, *etc.*), excited/exaggerated movements, intermittent convulsions, loss of consciousness and death. Additional signs of epsilon intoxication include elevated blood pressure, fluid in the lungs and brain congestion with edema [31,65]. There are remarkable histological differences between brain lesions in kids, calves and lambs. Perivascular leakage of protein is evident in lamb and calf, but not kid, brains; however, all of these animal types suffer from central nervous system distress post-Etx exposure.

More natural than an intravenous injection of purified toxin, duodenal inoculation of Angora goats (12–14 kg kids) with either whole culture, culture supernatant or washed cells of *C. perfringens* type D leads to: (1) diarrhea (dark green and foul smelling containing bowel mucosa, fibrin and sometimes blood); (2) respiratory distress (lung edema and froth in trachea/bronchi); (3) glycosuria (although not a uniform response by all animals); and (4) central nervous system dysfunction (*i.e.*, recumbency, bleating, convulsions and opisthotonos) [66]. Similar symptoms are also evident in lambs, minus diarrhea, pseudomembrane formation or any overt histological changes in the intestines [54,67]. In lambs there can be sudden death or neurological manifestations during acute disease that include struggling, opisthotonos, convulsions, lateral recumbency and violent paddling.

Etx differentially affects sheep and goats, as the former have more overt brain effects (*i.e.*, lesions) while the latter are more affected in the gut (*i.e.*, diarrhea with structural damage to the colon, but not small intestine) [68]. Regarding the intestinal tract, transit time of digesta is approximately the same for goats and sheep; therefore, differences in susceptibility within this organ might include other factors such as protease types/concentrations within the intestinal lumen and/or cell-surface receptor densities on the mucosal epithelia [66]. Gross pathology of dose- and time-based experiments upon ligated ileal and colonic loops in goats and sheep show that the ileum is not affected by Etx; however, in the colon of either species the goblet cells are missing and there is an excess of mucus, leukocytes and sloughed epithelial cells into the lumen [69]. Colonic lesions are more severe in goats and the colonic loops of either species retain more water at two and four hours, *vs.* controls, post-Etx administration. There is also fluid volume change in toxin-treated ileal loops of goats, but minimally so in sheep (fluid evidenced only at four, but not two, hours following a maximum dose). Sodium efflux into the lumen may play a role in fluid accumulation within the ileum and colon, post-Etx exposure.

4.1. Small Animal Models

The mode of action for Etx *in vivo* involves ion imbalance, endothelial disruption and edema. A vicious cycle is established by the toxin within the digestive tract and includes increased intestinal permeability that promotes higher, circulating levels of toxin throughout the body [61]. To more deeply understand Etx, or any other toxin, it is necessary to develop economically feasible, readily reproducible laboratory models *in vivo* that often include small animals such as mice. Furthermore, results from animal models critically advance vaccine and therapeutic developments.

Various laboratories have published efforts involving different routes of murine intoxication by Etx. One laboratory reveals that an intravenous injection of Etx into mice (2–4 LD$_{50}$) yields seizures within 60 minutes [70]. Excessive glutamate release is proposed to cause neuronal damage (cell death) targeting the hippocampus, particularly pyramidal cells.

In mice injected intraperitoneally with sub-lethal/lethal doses of Etx, electron microscopy of brain reveals that within just 30 minutes the end-feet of astrocytes (cerebellum) begin swelling with evident damage to the capillary endothelium [71]. Cytoplasmic boundaries of Etx-affected astrocytes are not well defined and often ruptured. Neurons, relative to astrocytes, are not affected by the toxin in this model. Other noticeable changes elicited by Etx in the brain, over time, include: 1 h post-toxin–vacuole formation in endothelial cells; 2 h post-toxin–more prominent blebbing from endothelial cells and rupturing of astrocytes; 3 h post-toxin–morphological changes in the capillary endothelium that include multiple vacuoles with organelle disintegration, as well as a shrunken nucleus in granule cells; 5 h post-toxin–a few neurons become mildly shrunk with some vacuole formation; and 24 h post-toxin (following multiple sub-lethal injections)–platelet aggregation, swelling of astrocyte end-feet corresponding to clear, perivascular spaces. Perhaps hypoxia linked to capillary damage affects more drastically cell types with relatively low levels of oxidative enzymes, such as astrocytes and granule cells. This extensive study by Finnie also points out that varying results from similar studies involving Etx-induced damage to the brain could be linked to fixation techniques: immersion *vs.* perfusion [71].

One study with *C. perfringens* type D culture supernatants (late-log phase, $n = 39$ disease-causing strains collected from the 1940s–present), injected intravenously into mice, shows that Etx is required for lethality [72]. Trypsin added to these Etx-containing supernatants further increases lethal potency, while an Etx-specific monoclonal antibody prevents lethality. There is no correlation between alpha toxin and perfringolysin O concentrations with lethality in this model. For unknown reasons, not all supernatants of type D strains (approximately five percent) tested positive for Etx *in vitro*, yet by definition each of these bacteria contain the toxin gene. Such findings of silent Etx genes clearly have implications upon diagnosis and taxonomy. These studies nicely mimic those subsequently done in sheep, goats and mice, revealing that a genetic knockout of Etx derived from a wild-type D strain (sheep isolate, CN1020) does not cause disease following intraduodenal injection [73]. Complementation of this mutant with the wild-type toxin gene generates a phenotype that elicits Etx-based disease. In this study, clinical signs of Etx-intoxication in mice include depression, ataxia, circling and dyspnea. Overall, the above results strongly suggest that clostridial vaccines meant to prevent enterotoxemia should target Etx as an antigen [72,73].

An extensive study by Goldstein *et al.* shows that orogastric-administered Etx causes fluid accumulation in the murine ileum, with toxin binding preferentially throughout the villus, *vs.* crypt, surface [74]. Such a model is more cost-effective than large animal models (sheep and goats) for studying the intestinal/systemic effects of *C. perfringens* Etx. Interestingly, this same study reveals that Ussing chamber experiments of either mouse or rat ileum, incubated with Etx (8000 LD$_{50}$ on mucosal surface for 60 minutes), does not disrupt transepithelial resistance. However, a similar experiment with basolateral application of toxin (30 LD$_{50}$) shows damage within 40 minutes, likely disrupting tight junction gaps. Entry of toxin into the blood stream could facilitate this latter event.

In this same study, horseradish peroxidase (44 kDa) minimally passes from the intestinal lumen into the zonula occludens of murine loops injected with Etx, suggesting minimal intestinal damage from the mucosal surface [74]. Intravenously-injected Evans blue dye, which binds to plasma proteins, leaks into the intestinal lumen within three hours after Etx administration in intestinal loops. Microscopy of rat and mouse intestinal loops reveals minimal effects upon the epithelium, with some shrunken cells and debris. Apoptotic cells are evident and contain condensed, fragmented nuclei plus degenerated organelles. Mild edema exists in the lamina propia of some, but not all, intestinal loops from animals treated with Etx.

Another study by Fernandez-Miyakawa *et al.* uses oral administration of Etx-producing *C. perfringens* type D ($n = 10$ strains given at 8×10^9 colony forming units (CFU)/mouse), leading to strain-dependent degrees of lethality (0%–100%) [75]. Seizures, hyper-excitability, depression, tubular necrosis of kidney and lung edema are signs of intoxication in these animals. Mice are passively protected by a monoclonal antibody against Etx, further suggesting a critical role of the latter in this model. However, there is no correlation between *in vitro* production (protein amounts) of Etx and lethality.

4.2. Detection

The classic mouse assay involving toxin neutralization with *C. perfringens* type-specific antisera is used less these days, given non-animal alternatives. Relatively large numbers of *C. perfringens* type D (10^4–10^7 CFU/mL) can be isolated from Etx-affected sheep, but such data for goats are lacking to our knowledge [1]. Use of ELISA technology for specifically detecting Etx in intestinal contents is evidently, to date, one of the best ways to confirm intoxication [6,76]. As an example, ELISA testing (commercial kit) of various body fluids in Merino lambs ($n = 15$), each inoculated intraduodenally with *C. perfringens* type D culture (300 mL; 200–800 MLD_{50} of Etx), reveals respective 92%, 64% and 57% detection rates of Etx (0.075 MLD_{50}/mL detection limit) in the ileum, duodenum and colon [6]. Animals were euthanized upon showing severe clinical signs (two to 26 hours post-toxin administration), and all fluids immediately frozen at -20 °C until processed within two months. Etx is also detected in 7% of the pericardial and aqueous humor fluids, but not in abdominal fluid or urine.

An extensive, ovine-based study by Uzal *et al.* assesses four techniques for detecting Etx (200–0.0075 MLD_{50}/mL) spiked in intestinal contents, pericardial fluid and aqueous humor [76]. A capture ELISA format with intestinal contents, using polyclonal antibody (0.075 MLD_{50}/mL detection limit) adsorbed to the plate well, is more sensitive than a monoclonal antibody capture ELISA (25 MLD_{50}/mL), counter-immunoelectrophoresis (50 MLD_{50}/mL) or toxin neutralization in mice (6 MLD_{50}/mL). One major caveat with the polyclonal capture ELISA, counter-immunoelectrophoresis and mouse-lethal neutralization assays is that all use polyclonal antibodies against *C. perfringens* type D that could recognize antigens other than Etx. However, specificity of each assay is confirmed by using intestinal contents from sheep ($n = 12$) that died of causes other than Etx. Further studies using intestinal contents from goats with experimental or natural epsilon intoxication essentially confirm results from the spiked toxin experiments. Altogether, these results and those from other laboratories reveal that a properly crafted ELISA can be an effective method for detecting various toxins of *C. perfringens* [6–9].

A novel liquid chromatography–mass spectrometry technique uses immunoaffinity beads to initially concentrate Etx or protoxin from a complex matrix, such as serum or milk [10]. Detection limits of the assay are 5 ng/mL for either toxin or protoxin, and results are possible within four hours of sample processing. Mass spectrometry avoids cross-reactivity issues inherent in any antibody-based assay; however, ELISA and mass spectrometry do not determine whether the detected protein toxin is biologically active.

Payne *et al.* discovered that a cell culture assay employing MDCK cells, purified Etx diluted in culture medium and specific neutralizing antibodies (monoclonal) against Etx correlates well with the mouse lethal assay [77]. Of the twelve different cell lines tested, the MDCK is the only one susceptible (~15 ng Etx/mL detection limit). The toxin-resistant cell lines are quite varied and include: (1) kidney (African green monkey, Vero and MA104; bovine, MDBK; rat, NRK-59F; porcine, LLC-PK1; feline, CRFK); (2) lung (rat, JTC-19); (3) neuronal (rat, B65); (4) B lymphocyte (human, B12); (5) intestinal epithelium (rat, IEC-6); and (6) monocyte/macrophage (murine, P388.D1). Readout involves inhibited metabolism of a viability stain (tetrazolium salt) by Etx-intoxicated cells, quantitated by spectrophotometry. Experiments are not reported using "dirty" field samples, such as intestinal contents, that can contain potential confounding factors. Various lines used in cell culture assays with various clostridial toxins, and neutralizing antibodies, have proven useful over time.

In addition to established cell lines, freshly isolated human kidney cells (tubular epithelial) behave like MDCK cells following Etx, which includes blebbing and large complex formation [46,78]. Perhaps future *in vitro* studies that support discovery of therapeutics and vaccines for biodefense should use human, *vs.* canine, kidney cells?

5. Management of Etx Intoxication: Therapy and Prophylaxis

Effective vaccines against Etx (described below) are readily available for animal use, thus obviating the need for a therapeutic in susceptible populations given this prophylaxis. In fact, vaccines that target *C. perfringens* type D (in particular Etx) are common reagents used for managing sheep and goat herds around the world [54]. There is certainly nothing (therapeutic or vaccine) against Etx approved for human use at this time. Findings from different laboratories and various *in vivo/in vitro* studies suggest that therapy is possible. Perhaps a proteomics-based approach following Etx exposure can reveal even more interesting, and unique, host-based targets for therapeutic intervention? This has recently been done using mice given Etx intravenously, with subsequent analysis of select organs (*i.e.*, brain and kidney), plasma and urine for differentially expressed proteins [79]. The study reveals 136 different proteins with altered expression, post-toxin exposure. Similar to staphylococcal enterotoxins and expansion of distinct Vβ-bearing T cells [80], unique patterns of up- and/or down-regulated proteins might be useful in the future for diagnosing epsilon intoxication.

A murine-based study using brain slices incubated with Etx and Etx-specific antibody shows that the toxin binds preferentially to the cerebellum, particularly oligodendrocytes and granule cells [81]. Current-clamp experiments with granule cells reveal that Etx decreases resistance, induces membrane depolarization and ultimately increases glutamate release. In addition to release of lactate dehydrogenase into the medium, incubation of primary-cultured granule cells with Etx causes membrane blebbing, rapid increase of intracellular calcium levels (within five minutes of toxin exposure) and glutamate release. Overall, these studies importantly provide a brain cell-specific target/assay *in vitro* for testing novel therapeutics.

Related murine endeavors *in vivo* by Miyamoto *et al.* reveal that riluzole, a benzothiazole (234 Da) therapeutic for human amyotrophic lateral sclerosis that prevents presynaptic glutamate release, minimizes murine seizures as well as glutamate release induced by Etx [70,82]. These results occurred after an intraperitoneal dose of riluzole (16 mg/kg body weight) given 30 minutes before an intravenous dose (2 or 4 LD_{50}) of Etx. However, the drug was evidently not used as a therapeutic (*i.e.*, administered after toxin exposure). Experiments with rats also show that riluzole (8 mg/kg body weight), as well as glutamate antagonists MK-801 or CNQX (3 mg/kg body weight and 100 nmole, respectively), decrease hippocampus damage following intravenous injection of Etx (100 ng/kg body weight, a minimum lethal dose in rats) [70,82]. Etx-induced damage to the brain differs between rats and mice, as in rats the cortex and hippocampus (pyramidal cells) are most affected *vs.* the mouse cerebral cortex and granular layer of the cerebellum [82]. Overt effects of toxin in rats include upper body tremor, limb rigidity and muscular incoordination followed by hypotonus and paralysis. Perhaps varying susceptibility of brain regions between species may be linked to receptor densities, an aspect not clarified in subsequent literature to our knowledge.

A small-molecule library (151,616 compounds) and high-throughput screening have also been used with a MDCK cell-based assay (384-well plates) for discovering novel therapeutics against Etx [83]. In this study, Lewis *et al.* ultimately found three, structurally-unique inhibitors that afford protection against Etx but do not prevent toxin binding or oligomerization. These inhibitors seemingly affect pore function and/or an unidentified co-factor important in Etx intoxication. Two of these compounds (*N*-cycloalkylbenzamide and furo[2,3-*b*]quinoline) protect cells using various criteria. One experiment used a constant concentration of inhibitor added concomitantly with increasing concentrations of Etx needed to kill 50 percent of the cells (CT_{50}). A post-exposure experiment shows that addition of either compound (50 μM) up to 10 minutes after toxin (25 nM) exposure is protective *in vitro*. Such results logically lead to efficacy studies in animals, yet to be done or at least published to our knowledge. Furthermore, these inhibitors are specific for Etx as the related *A. hydrophila* aerolysin is not inhibited in similar assays [83].

Another therapeutic approach against Etx includes dominant-negative inhibitors. These protein-based therapeutics have been used for other oligomer-forming bacterial toxins produced by Gram-positive and Gram-negative pathogens such as *Bacillus anthracis*, *Escherichia coli* and *Helicobacter pylori* [84–86]. Dominant-negative proteins are a recombinantly-attenuated version of a toxin generated by peptide deletion or amino acid substitution(s). Integration of a dominant-negative protein(s) into a wild-type toxin complex in solution or on a cell surface generates a non-functional toxin oligomer. Two dominant-negative inhibitors for Etx have been created via cysteine substitutions of isoleucine 51/alanine 114 or valine 56/phenylalanine 118 [87]. These particular paired mutations facilitate an intramolecular disulfide bond, restrict toxin insertion into the membrane and cause oligomer dysfunction (decreased heat/detergent stability plus poor pre-pore to pore transition) that ultimately inactivates Etx *in vitro*. Although used only as an Etx inhibitor *in vitro* (MDCK cells), dose-dependently effective at a 1, 2, 4 or 8 (wild-type toxin):1 (dominant negative) mole mixture, these unique constructs should perhaps be tested *in vivo* for therapeutic and vaccine potential.

Additional studies in mice show that the epsilon protoxin delays time to death when given intravenously before activated toxin. Protection presumably occurs by competitive occupation of cell-surface receptors, namely in the brain, by the protoxin [56]. These data further suggest the feasibility of a receptor-targeted approach for prophylaxis/therapy. In fact, Buxton had discovered many years prior to these studies that a formalin toxoid of the protoxin protects mice up to 100 minutes after Etx exposure [88]. These collective results indeed make sense as the protoxin and toxin share the same binding site and dissociation constant ($K_d \sim 4$–6 nM) on MDCK cells [89]. Plasma membrane integrity plus an unidentified O-linked glycoprotein are important for toxin binding, as determined by detergent solubilization, pronase or lectin pretreatment of cells. The detergent type (Triton X-100, sodium deoxycholate or sodium cholate), concentration and temperature also affect toxin binding. The latter results suggest cryptic receptors naturally hidden by membrane protein(s), lipid(s) and/or carbohydrate(s). This same study shows that Etx targets the distal and collecting tubules, but not those proximal, in cryosectioned mouse kidneys [89]. To date, work with receptor antagonists other than the epsilon protoxin have evidently not been pursued (at least published) by various laboratories.

Identifying a toxin's cell-surface receptor and understanding how toxin molecularly interacts with it can be very useful in formulating receptor-based therapies. The latter include toxin-binding antagonists and therapeutic targeting of susceptible cell populations throughout the body. Early receptor-binding studies employing radio-iodinated Etx reveal a heat-labile sialoglycoprotein, as pretreatment of rat synaptosome membranes with heat (70–80 °C for 10 minutes), neuraminidase, lipase or pronase effectively reduces saturable binding of the toxin [90]. The K_d values of Etx binding to rat brain homogenates and synaptosomal membrane fractions are 2.5 and 3.3 nM, respectively, thus very similar to K_d values in kidney cells [89,90]. Furthermore, a snake-venom presynaptic neurotoxin (beta-bungarotoxin from the many-banded krait, *Bungarus multicinctus*) that blocks acetylcholine release dose-dependently decreases binding of Etx, suggesting a common receptor for these protein toxins from diverse organisms [90]. However, the presynaptic neurotoxin produced by *C. botulinum* type A had no effect upon binding of Etx. Furthermore, sialidase pre-treatment of kidney cells and synaptosomes suggests different receptors for Etx [48,90].

Unique studies involving gene-trap mutagenesis show that hepatitis A virus cellular receptor 1 (HAVCR1) is a receptor, or co-receptor, for Etx in MDCK cells [91,92]. It is also possible that HAVCR1 promotes Etx-induced: (1) intracellular signaling due to toxin binding or increased ion flow; and/or (2) protein-protein interactions (*i.e.*, homo-oligomer formation). There are eight other proteins, including sphingomyelin synthase 2, that when not expressed lead to varying cell resistance towards Etx [91]. These results suggest that multiple cell factors play a role during epsilon intoxication; however, HAVCR1 is the only protein consistently linked to increased expression and Etx susceptibility in other cell lines.

The natural role of HAVCR1 (also known as KIM-1, Kidney Injury Molecule-1) involves T-regulatory cells and maintaining immunological balance throughout the body. HAVCR1 is a class I, integral-membrane, O-linked glycoprotein containing multiple isoforms varying within a mucin-like domain containing multiple glycosylation sites. The 100, but not 90, kDa variant of HAVCR1 binds Etx, perhaps due to increased length of the extracellular, mucin-like domain containing approximately 57 glycosylation sites [91]. Domain I tyrosines (residues 29, 30, 36, 196) on Etx are surface-accessible and contribute to toxin binding to HAVCR1 [34,92]. Replacement of any of these tyrosines with glutamic acid yields a non-toxic protein unable to bind cells (MDCK), yet these molecules possess a similar circular dichroism (CD) spectrum and resistance to trypsin digestion as the wild-type toxin. In contrast, mutagenesis of tyrosines 16 and 20, or phenylalanine 37 does not decrease epsilon cytotoxicity. Replacement of phenylalanine 199 with glutamic acid yields a non-toxic protein with a unique CD spectrum, suggesting conformational differences *vs.* wild-type toxin. Although recent studies with Etx and HAVCR1 are intriguing, further studies must be done to more clearly understand the intimate interactions between these molecules.

A subsequent, and contrasting, study by Bokori-Brown *et al.* interestingly suggests that aforementioned tyrosines 29, 30, 36 and 196 (when individually changed to alanine) do not play a role in binding/cytotoxicity of ACHN cells, in contrast to MDCK cells [92,93]. Such findings might suggest an alternative receptor(s) exploited by Etx, perhaps mediated by a beta-octyl-glucoside binding site within domain III. It is also quite possible that replacement of these tyrosines with an electroneutral alanine, *vs.* an electronegative glutamic acid, is too subtle of a change. Cumulative data from different laboratories indeed make this aspect of understanding epsilon intoxication complex and clearly unresolved to date.

As stated earlier, Etx is primarily of veterinary concern and vaccines are used in the field [94–96]. Before the use of commercial vaccines against *C. perfringens* type D (Etx), more lambs died from enterotoxemia than all other diseases combined [94]. Lambs from ewes vaccinated three to four weeks before parturition have higher passive antibody titers against Etx *vs.* lambs from ewes vaccinated six weeks before parturition. Ewe-derived antibodies can evidently protect lambs up to twelve weeks of age. Recommendations that lambs be vaccinated twice before six weeks of age, regardless of ewe vaccination status, are evidently not warranted [94]. From a cost/benefit perspective, lambs marketed by five months of age may not need vaccinations if ewes are appropriately vaccinated against clostridial enterotoxemia [94].

Parenteral hyperimmune sera can also generate passive protection for three to four weeks in weaned lambs [97]. However, animals showing clinical signs of epsilon intoxication cannot be saved by anti-toxin. Colostrum (bovine) containing anti-Etx antibodies, following multiple immunizations with a clostridial multicomponent vaccine, can also provide protection (circulating antibody) against Etx when fed to lambs within 48 hours of birth [98]. This antibody–rich product can be particularly useful for weak and/or orphaned lambs, as well as stored in relatively large quantities for long periods at −20 °C. It is also possible that a monoclonal antibody targeting a critical epitope, like the membrane insertion region of Etx, could be a better characterized, purified therapeutic *vs.* a heterogeneous population of antibodies in serum or colostrum [99,100].

Antibodies generated by either active or passive immunization can effectively thwart *C. perfringens* Etx, when timely present.

Human and animal vaccines have been quite effective over time against various diseases. However, many veterinary vaccines, such as those against *C. perfringens* (or other bacterial pathogens) and associated toxins, are frequently formaldehyde toxoids of culture filtrates that may also contain whole cells [101]. Although these vaccines can be efficacious and relatively inexpensive, they are typically too crude for human use [102]. Current veterinary vaccines containing epsilon toxoid can vary in protective efficacy [95]. This latter study tested seven commercially available, epsilon-toxoids available in Brazil and remarkably found only two of sufficient potency. Clearly, quality control can vary greatly which ultimately affects the bottom line: protection against disease.

The vaccination schedule to counter epsilon enterotoxemia varies too, depending upon the animal species [96]. Two doses, containing aluminum hydroxide adjuvant, are usually given two to six weeks apart and then followed by an annual (sheep) or quarterly (goat) boost. Evidently immunoreactivity (antibody titers and duration) towards Etx-based vaccine is less in goats *vs.* sheep. An attempt to remedy this problem involved a specific study in goats, comparing liposome-*vs.* aluminum hydroxide- based vaccines (three doses at three week intervals) of epsilon toxoid [96]. The latter proved much more superior. In fact, the liposome-based vaccine did not elicit a significant antibody response towards epsilon toxoid. In animals, and likely humans too, an efficacious vaccine against Etx will probably not be a "one and done" that generates lasting protection.

In contrast to relatively crude (multiple antigen) vaccines for Etx, recombinant technology can be helpful in many ways towards a better vaccine. The gene for Etx was first successfully cloned, sequenced and expressed in 1992, making subsequent work possible [103]. As one example, recombinant *E. coli* expressing Etx that is then inactivated by 0.5% formaldehyde at room temperature for 10 days can evidently be used as a cost-effective vaccine [95]. This vaccine can be administered at a much lower protein concentration (0.2 mg/dose) than current, chemically-detoxified epsilon toxoids derived from crude cultures of *C. perfringens* [95]. The standard recommendation by the National Institute for Biological Standards and Control (United Kingdom) is 9.2 mg of native Etx per dose. This latter study used a 1:1 emulsion of toxoid (0.2 mg/dose):aluminum hydroxide (2.5%–3.5%) for vaccination of goats, sheep and cattle [95]. Use of a more purified vaccine also affords easier quality control. A human vaccine against Etx will likely involve chemically (*i.e.*, formaldehyde) or recombinantly (*i.e.*, mutation of critical residues needed for receptor binding and/or oligomerization) detoxified versions of purified protein.

In regards to recombinant-based attenuation, replacement of just one amino acid (histidine 106) with proline results in a non-toxic form of Etx (H106P) when tested by MDCK cytotoxicity and mouse lethality [104]. In contrast, a serine or alanine replacement of histidine 106 does not eliminate toxicity. The H106P protein (0.27 nmole/mouse given intraperitoneally with Freund's incomplete adjuvant) affords vaccine-based protection against 1000 LD_{50} of intravenous, wild-type toxin.

Further vaccine refinement, with minimal risk of aforementioned toxicity, involves use of a linear, B-cell epitope (amino acids 40–62 of Etx) linked to a carrier molecule (*E. coli* heat-labile toxin B subunit) [105]. In theory this could be useful, but a small protein fragment may adopt an unnatural conformation (*vs.* that found on native protein) and subsequently elicit antibodies that don't readily bind native protein. This particular study did not determine if this construct elicits protective antibodies in an animal model for epsilon intoxication. In our opinion, replacement of select amino acids involved in a critical step during the intoxication process (*i.e.*, cell binding, oligomerization, *etc.*) typically does not grossly alter native conformation and thus represents a better vaccine strategy offering multiple epitopes to the immune system.

6. Conclusions

C. perfringens is one of the most "toxic" bacteria known, using different protein toxins in different ways to perpetuate itself. Many of these toxins are commonly linked to various diseases in many mammalian species. In particular, the Etx has been studied by various groups and is primarily a veterinary concern for some large animals. Vaccines of varying quality are available to combat epsilon enterotoxemia, for veterinary use only. A human equivalent for biodefense is not commercially available. As a therapeutic or prophylactic for humans, toxin-specific immunoglobulins might be helpful following nefarious application of Etx. Clearly, there is much more to learn regarding Etx, how it works and how to protect against it. Such knowledge goes a long way towards bettering the lives of animals and humans.

Acknowledgments

G.B. and B.G.S. thank Wilson College for generous use of facilities that include computers, copiers and library services.

Conflicts of Interest

The authors declare no conflict of interest.

References

1. Songer, J.G. Clostridial enteric diseases of domestic animals. *Clin. Microbiol. Rev.* **1996**, *9*, 216–234.
2. McDonel, J.L. *Clostridium perfringens* toxins (Type A, B, C, D, E). *Pharmacol. Ther.* **1980**, *10*, 617–655.
3. Oakley, C.L.; Warrack, G.H. Routine typing of *Clostridium welchii*. *J. Hyg. Camb.* **1953**, *51*, 102–107.
4. Goldstein, M.R.; Kruth, S.A.; Bersenas, A.M.; Holowaychuk, M.K.; Weese, J.S. Detection and characterization of *Clostridium perfringens* in the feces of healthy and diarrheic dogs. *Can. J. Vet. Res.* **2012**, *76*, 161–165.
5. Albini, S.; Brodard, I.; Jaussi, A.; Wollschlaeger, N.; Frey, J.; Miserez, R.; Abril, C. Real-time multiplex PCR assays for reliable detection of *Clostridium perfringens* toxin genes in animal isolates. *Vet. Microbiol.* **2008**, *127*, 179–185.

6. Layana, J.E.; Fernandez-Miyakawa, M.E.; Uzal, F.A. Evaluation of different fluids for detection of *Clostridium perfringens* type D epsilon toxin in sheep with experimental enterotoxemia. *Anaerobe* **2006**, *12*, 204–206.

7. Carman, R.J.; Stevens, A.L.; Lyerly, M.W.; Hiltonsmith, M.F.; Stiles, B.G.; Wilkins, T.D. *Clostridium difficile* binary toxin (CDT) and diarrhea. *Anaerobe* **2011**, *17*, 161–165.

8. Macias Rioseco, M.; Beingesser, J.; Uzal, F.A. Freezing or adding trypsin inhibitor to equine intestinal contents extends the lifespan of *Clostridium perfringens* beta toxin for diagnostic purposes. *Anaerobe* **2012**, *18*, 357–360.

9. Hale, M.L.; Stiles, B.G. Detection of *Clostridium perfringens* alpha toxin using a capture antibody ELISA. *Toxicon* **1999**, *37*, 471–484.

10. Seyer, A.; Fenaille, F.; Feraudet-Tarisse, C.; Volland, H.; Popoff, M.R.; Tabet, J.C.; Junot, C.; Becher, F. Rapid quantification of clostridial epsilon toxin in complex food and biological matrixes by immunopurification and ultraperformance liquid chromatography-tandem mass spectrometry. *Anal. Chem.* **2012**, *84*, 5103–5109.

11. Langley, F.H.; Winkelstein, L.B. Gas gangrene: A study of 96 cases treated in an evacuation hospital. *JAMA* **1945**, *128*, 783–792.

12. Bryant, A.E.; Stevens, D.L. The Pathogenesis of Gas Gangrene. In *The Clostridia: Molecular Biology and Pathogenesis*; Rood, J.I., McClane, B.A., Songer, J.G., Titball, R.W., Eds.; Academic Press: San Diego, CA, USA, 1997; Chapter 11, pp. 185–196.

13. Smith, L.D.; Gardner, M.V. The occurrence of vegetative cells of *Clostridium perfringens* in soil. *J. Bacteriol.* **1949**, *58*, 407–408.

14. MacLennan, J.D.; MacFarlane, M.G. The treatment of gas gangrene. *Br. Med. J.* **1944**, *1*, 683–685.

15. Smith, L.D. Clostridia in gas gangrene. *Bact. Rev.* **1949**, *13*, 233–254.

16. Evans, D.G.; Perkins, F.T. Fifth international standard for gas-gangrene antitoxin (perfringens) (*Clostridium welchii* type A antitoxin). *Bull. World Health Organ.* **1963**, *29*, 729–735.

17. Williamson, E.D.; Titball, R.W. A genetically engineered vaccine against the alpha-toxin of *Clostridium perfringens* protects mice against experimental gas gangrene. *Vaccine* **1993**, *11*, 1253–1258.

18. Stevens, D.L.; Titball, R.W.; Jepson, M.; Bayer, C.R.; Hayes-Schroer, S.M.; Bryant, A.E. Immunization with the C-domain of alpha-toxin prevents lethal infection, localizes tissue injury, and promotes host response to challenge with *Clostridium perfringens*. *J. Infect. Dis.* **2004**, *190*, 767–773.

19. Naylor, C.E.; Eaton, J.T.; Howells, A.; Justin, N.; Moss, D.S.; Titball, R.W.; Basak, A.K. Structure of the key toxin in gas gangrene. *Nat. Struct. Biol.* **1998**, *5*, 738–746.

20. Sakurai, J.; Nagahama, M.; Oda, M. *Clostridium perfringens* alpha-toxin: Characterization and mode of action. *J. Biochem.* **2004**, *136*, 569–574.

21. MacFarlane, M.G.; Knight, B.C. The biochemistry of bacterial toxins. The lecithinase activity of *Cl. welchii* toxins. *Biochem. J.* **1941**, *35*, 884–902.

22. Oda, M.; Kabura, M.; Takagishi, T.; Suzue, A.; Tominaga, K.; Urano, S.; Nagahama, M.; Kobayashi, K.; Furukawa, K.; Furukawa, K.; *et al. Clostridium perfringens* alpha-toxin recognizes the GM1a/TrkA complex. *J. Biol. Chem.* **2012**, *287*, 33070–33079.

23. Hunter, S.E.C.; Brown, J.E.; Oyston, P.C.F.; Sakurai, J.; Titball, R.W. Molecular genetic analysis of beta-toxin of *Clostridium perfringens* reveals sequence homology with alpha-toxin, gamma-toxin, and leukocidin of *Staphylococcus aureus. Infect. Immun.* **1993**, *61*, 3958–3965.

24. Sakurai, J.; Duncan, C.L. Purification of beta-toxin from *Clostridium perfringens* type C. *Infect. Immun.* **1977**, *18*, 741–745.

25. Walker, P.D.; Batty, I.; Egerton, J.R. The typing of *Cl. perfringens* and the veterinary background. *Papua New Guinea Med. J.* **1979**, *22*, 50–56.

26. Shatursky, O.; Bayles, R.; Rogers, M.; Jost, B.H.; Songer, J.G.; Tweten, R.K. *Clostridium perfringens* beta-toxin forms potential-dependent, cation-selective channels in lipid bilayers. *Infect. Immun.* **2000**, *68*, 5546–5551.

27. Nagahama, M.; Morimitsu, S.; Kihara, A.; Akita, M.; Setsu, K.; Sakurai, J. Involvement of tachykinin receptors in *Clostridium perfringens* beta-toxin-induced plasma extravasation. *Br. J. Pharmacol.* **2003**, *138*, 23–30.

28. Bosworth, T. On a new type of toxin produced by *Clostridium welchii. J. Comp. Path.* **1943**, *53*, 245–255.

29. Stiles, B.G.; Wigelsworth, D.J.; Popoff, M.R.; Barth, H. Clostridial binary toxins: Iota and C2 family portraits. *Front. Cell. Infect. Microbiol.* **2011**, doi:10.3389/fcimb.2011.00011.

30. Briggs, D.C.; Naylor, C.E.; Smedley, J.G.; Lukoyanova, N.; Robertson, S.; Moss, D.S.; McClane, B.A.; Basak, A.K. Structure of the food-poisoning *Clostridium perfringens* enterotoxin reveals similarity to the aerolysin-like pore-forming toxins. *J. Mol. Biol.* **2011**, *413*, 138–149.

31. Popoff, M.R. Epsilon toxin: A fascinating pore-forming toxin. *FEBS J.* **2011**, *278*, 4602–4615.

32. Huebner, K.D.; Wannemacher, R.W.; Stiles, B.G.; Popoff, M.R.; Poli, M.A. *Textbook of Military Medicine: Medical Aspects of Biological Warfare*; Dembek, Z.F., Ed.; Office of The Surgeon General, Borden Institute: Washington, DC, USA, 2007; Chapter 17, p. 355.

33. HHS and USDA Select Agents and Toxins. 7 CFR Part 331, 9 CFR Part 121, and 42 CFR Part 73. Availavle online: http://www.selectagents.gov/resources/List_of_Select_Agents_and_Toxins_2012-12-4-English.pdf (accessed on 1 October 2013).

34. Cole, A.R.; Gibert, M.; Popoff, M.R.; Moss, D.S.; Titball, R.W.; Basak, A.K. *Clostridium perfringens* epsilon-toxin shows structural similarity to the pore-forming toxin aerolysin. *Nat. Struct. Mol. Biol.* **2004**, *11*, 797–798.

35. Knapp, O.; Stiles, B.G.; Popoff, M.R. The aerolysin-like toxin family of cytolytic, pore-forming toxins. *Open Toxinol. J.* **2010**, *3*, 53–68.

36. Mancheno, J.M.; Tateno, H.; Goldstein, I.J.; Martinez-Ripoll, M.; Hermoso, J.A. Structural analysis of the *Laetiporus sulphureus* hemolytic pore-forming lectin in complex with sugars. *J. Biol. Chem.* **2005**, *280*, 17251–17259.

37. Parker, M.W.; Buckley, J.T.; Postma, J.P.; Tucker, A.D.; Leonard, K.; Pattus, F.; Tsernoglou, D. Structure of the *Aeromonas* toxin proaerolysin in its water-soluble and membrane-channel states. *Nature* **1994**, *367*, 292–295.

38. Miyata, S.; Matsushita, O.; Minami, J.; Katayama, S.; Shimamoto, S.; Okabe, A. Cleavage of a C-terminal peptide is essential for heptamerization of *Clostridium perfringens* epsilon-toxin in the synaptosomal membrane. *J. Biol. Chem.* **2001**, *276*, 13778–13783.

39. Petit, L.; Gibert, M.; Henri, C.; Lorin, V.; Baraige, F.; Carlier, J.P.; Popoff, M.R. Molecular basis of the activity of *Clostridium perfringens* toxins. *Curr. Topics Biochem. Res.* **1999**, *1*, 19–35.

40. Jin, F.; Matsushita, O.; Katayama, S.; Jin, S.; Matsushita, C.; Minami, J.; Okabe, A. Purification, characterization, and primary structure of *Clostridium perfringens* lambda-toxin, a thermolysin-like metalloprotease. *Infect. Immun.* **1996**, *64*, 230–237.

41. Harkness, J.M.; Li, J.; McClane, B.A. Identification of a lambda toxin-negative *Clostridium perfringens* strain that processes and activates epsilon prototoxin intracellularly. *Anaerobe* **2012**, *18*, 546–552.

42. Petit, L.; Gibert, M.; Gourch, A.; Bens, M.; Vandewalle, A.; Popoff, M.R. *Clostridium perfringens* epsilon toxin rapidly decreases membrane barrier permeability of polarized MDCK cells. *Cell. Microbiol.* **2003**, *5*, 155–164.

43. Miyata, S.; Minami, J.; Tamai, E.; Matsushita, O.; Shimamoto, S.; Okabe, A. *Clostridium perfringens* epsilon-toxin forms a heptameric pore within the detergent-insoluble microdomains of Madin-Darby canine kidney cells and rat synaptosomes. *J. Biol. Chem.* **2002**, *277*, 39463–39468.

44. Lafont, F.; Abrami, L.; van der Goot, F.G. Bacterial subversion of lipid rafts. *Curr. Opin. Microbiol.* **2004**, *7*, 4–10.

45. Fennessey, C.M.; Sheng, J.; Rubin, D.H.; McClain, M.S. Oligomerization of *Clostridium perfringens* epsilon toxin is dependent upon caveolins 1 and 2. *PLoS One* **2012**, *7*, e46866.

46. Petit, L.; Gibert, M.; Gillet, D.; Laurent-Winter, C.; Boquet, P.; Popoff, M.R. *Clostridium perfringens* epsilon-toxin acts on MDCK cells by forming a large membrane complex. *J. Bacteriol.* **1997**, *179*, 6480–6487.

47. Nagahama, M.; Itohayashi, Y.; Hara, H.; Higashihara, M.; Fukatani, Y.; Takagishi, T.; Oda, M.; Kobayashi, K.; Nakagawa, I.; Sakurai, J. Cellular vacuolation induced by *Clostridium perfringens* epsilon-toxin. *FEBS J.* **2011**, *278*, 3395–3407.

48. Li, J.; Sayeed, S.; Robertson, S.; Chen, J.; McClane, B.A. Sialidases affect the host cell adherence and epsilon toxin-induced cytotoxicity of *Clostridium perfringens* type D strain CD3718. *PLoS Pathog.* **2011**, *7*, e1002429.

49. Galen, J.E.; Ketley, J.M.; Fasano, A.; Richardson, S.H.; Wasserman, S.S.; Kaper, J.B. Role of *Vibrio cholerae* neuraminidase in the function of cholera toxin. *Infect. Immun.* **1992**, *60*, 406–415.

50. Petit, L.; Maier, E.; Gibert, M.; Popoff, M.R.; Benz, R. *Clostridium perfringens* epsilon toxin induces a rapid change of cell membrane permeability to ions and forms channels in artificial lipid bilayers. *J. Biol. Chem.* **2001**, *276*, 15736–15740.

51. Donelli, G.; Fiorentini, C.; Matarrese, P.; Falzano, L.; Cardines, R.; Mastrantonio, P.; Payne, D.W.; Titball, R.W. Evidence for cytoskeletal changes secondary to plasma membrane functional alterations in the *in vitro* cell response to *Clostridium perfringens* epsilon-toxin. *Comp. Immunol. Microbiol. Infect. Dis.* **2003**, *26*, 145–156.

52. Zhu, C.; Ghabriel, M.N.; Blumbergs, P.C.; Reilly, P.L.; Manavis, J.; Youssef, J.; Hatami, S.; Finnie, J.W. *Clostridium perfringens* prototoxin-induced alteration of endothelial barrier antigen (EBA) immunoreactivity at the blood-brain barrier (BBB). *Exp. Neurol.* **2001**, *169*, 72–82.

53. Finnie, J.W.; Hajduk, P. An immunohistochemical study of plasma albumin extravasation in the brain of mice after the administration of *Clostridium perfringens* type D epsilon toxin. *Aust. Vet. J.* **1992**, *69*, 261–262.

54. Uzal, F.A.; Songer, J.G. Diagnosis of *Clostridium perfringens* intestinal infections in sheep and goats. *J. Vet. Diagn. Invest.* **2008**, *20*, 253–265.

55. Uzal, F.A.; Pasini, I.; Olaechea, F.V.; Robles, C.A.; Elizondo, A. An outbreak of enterotoxaemia caused by *Clostridium perfringens* type D in goats in Patagonia. *Vet. Rec.* **1994**, *135*, 279–280.

56. Nagahama, M.; Sakurai, J. Distribution of labeled *Clostridium perfringens* epsilon toxin in mice. *Toxicon* **1991**, *29*, 211–217.

57. Soler-Jover, A.; Blasi, J.; Gomez de Aranda, I.; Navarro, P.; Gibert, M.; Popoff, M.R.; Martin-Satue, M. Effect of epsilon toxin-GFP on MDCK cells and renal tubules *in vivo*. *J. Histochem. Cytochem.* **2004**, *52*, 931–942.

58. Tamai, E.; Ishida, T.; Miyata, S.; Matsushita, O.; Suda, H.; Kobayashi, S.; Sonobe, H.; Okabe, A. Accumulation of *Clostridium perfringens* epsilon-toxin in the mouse kidney and its possible biological significance. *Infect. Immun.* **2003**, *71*, 5371–5375.

59. Gardner, D.E. Pathology of *Clostridium welchii* type D enterotoxaemia. 3. Basis of the hyperglycaemic response. *J. Comp. Pathol.* **1973**, *83*, 525–529.

60. Wioland, L.; Dupont, J.L.; Bossu, J.L.; Popoff, M.R.; Poulain, B. Attack of the nervous system by *Clostridium perfringens* epsilon toxin: From disease to mode of action on neural cells. *Toxicon* **2013**, *75*, 122–135.

61. Finnie, J.W. Pathogenesis of brain damage produced in sheep by *Clostridium perfringens* type D epsilon toxin: A review. *Aust. Vet. J.* **2003**, *81*, 219–221.

62. Finnie, J.W.; Manavis, J.; Blumbergs, P.C. Aquaporin-4 in acute cerebral edema produced by *Clostridium perfringens* type D epsilon toxin. *Vet. Pathol.* **2008**, *45*, 307–309.

63. Verkman, A.S. Aquaporins in clinical medicine. *Annu. Rev. Med.* **2012**, *63*, 303–316.

64. Uzal, F.A.; Kelly, W.R. Effects of the intravenous administration of *Clostridium perfringens* type D epsilon toxin on young goats and lambs. *J. Comp. Pathol.* **1997**, *116*, 63–71.

65. Uzal, F.A.; Kelly, W.R.; Morris, W.E.; Assis, R.A. Effects of intravenous injection of *Clostridium perfringens* type D epsilon toxin in calves. *J. Comp. Pathol.* **2002**, *126*, 71–75.

66. Uzal, F.A.; Kelly, W.R. Experimental *Clostridium perfringens* type D enterotoxemia in goats. *Vet. Pathol.* **1998**, *35*, 132–140.

67. Uzal, F.A.; Kelly, W.R.; Morris, W.E.; Bermudez, J.; Baison, M. The pathology of peracute experimental *Clostridium perfringens* type D enterotoxemia in sheep. *J. Vet. Diagn. Invest* **2004**, *16*, 403–411.

68. Finnie, J.W. Neurological disorders produced by *Clostridium perfringens* type D epsilon toxin. *Anaerobe* **2004**, *10*, 145–150.

69. Fernandez-Miyakawa, M.E.; Uzal, F.A. The early effects of *Clostridium perfringens* type D epsilon toxin in ligated intestinal loops of goats and sheep. *Vet. Res. Commum.* **2003**, *27*, 231–241.

70. Miyamoto, O.; Sumitani, K.; Nakamura, T.; Yamagami, S.; Miyata, S.; Itano, T.; Negi, T.; Okabe, A. *Clostridium perfringens* epsilon-toxin causes excessive release of glutamate in the mouse hippocampus. *FEMS Microbiol. Lett.* **2000**, *189*, 109–113.

71. Finnie, J.W. Ultrastructural changes in the brain of mice given *Clostridium perfringens* type D epsilon toxin. *J. Comp. Path.* **1984**, *94*, 445–452.

72. Sayeed, S.; Fernandez-Miyakawa, M.E.; Fisher, D.J.; Adams, V.; Poon, R.; Rood, J.I.; Uzal, F.A.; McClane, B.A. Epsilon-toxin is required for most *Clostridium perfringens* type D vegetative culture supernatants to cause lethality in the mouse intravenous injection model. *Infect. Immun.* **2005**, *73*, 7413–7421.

73. Garcia, J.P.; Adams, V.; Beingesser, J.; Hughes, M.L.; Poon, R.; Lyras, D.; Hill, A.; McClane, B.A.; Rood, J.I.; Uzal, F.A. Epsilon toxin is essential for the virulence of *Clostridium perfringens* type D infection in sheep, goats and mice. *Infect. Immun.* **2013**, *81*, 2405–2414.

74. Goldstein, J.; Morris, W.E.; Loidl, C.F.; Tironi-Farinatti, C.; McClane, B.A.; Uzal, F.A.; Fernandez-Miyakawa, M.E. *Clostridium perfringens* epsilon toxin increases the small intestinal permeability in mice and rats. *PLoS One* **2009**, *4*, e7065.

75. Fernandez-Miyakawa, M.E.; Sayeed, S.; Fisher, D.J.; Poon, R.; Adams, V.; Rood, J.I.; McClane, B.A.; Saputo, J.; Uzal, F.A. Development and application of an oral challenge mouse model for studying *Clostridium perfringens* type D infection. *Infect. Immun.* **2007**, *75*, 4282–4288.

76. Uzal, F.A.; Kelly, W.R.; Thomas, R.; Hornitzky, M.; Galea, F. Comparison of four techniques for the detection of *Clostridium perfringens* type D epsilon toxin in intestinal contents and other body fluids of sheep and goats. *J. Vet. Diagn. Invest.* **2003**, *15*, 94–99.

77. Payne, D.W.; Williamson, E.D.; Havard, H.; Modi, N.; Brown, J. Evaluation of a new cytotoxicity assay for *Clostridium perfringens* type D epsilon toxin. *FEMS Microbiol. Lett.* **1994**, *116*, 161–168.

78. Fernandez-Miyakawa, M.E.; Zabal, O.; Silberstein, C. *Clostridium perfringens* epsilon toxin is cytotoxic for human renal tubular epithelial cells. *Hum. Exp. Toxicol.* **2010**, *30*, 275–282.

79. Kumar, B.; Alam, S.I.; Kumar, O. Host response to intravenous injection of epsilon toxin in mouse model: A proteomic view. *Proteomics* **2012**, *13*, 89–107.

80. Ferry, T.; Thomas, D.; Perpoint, T.; Lina, G.; Monneret, G.; Mohammedi, I.; Chidiac, C.; Peyramond, D.; Vandenesch, F.; Etienne, J. Analysis of superantigenic toxin Vbeta T-cell signatures produced during cases of staphylococcal toxic shock syndrome and septic shock. *Clin. Microbiol. Infect.* **2008**, *14*, 546–554.

81. Lonchamp, E.; Dupont, J.L.; Wioland, L.; Courjaret, R.; Mbebi-Liegois, C.; Jover, E.; Doussau, F.; Popoff, M.R.; Bossu, J-L.; Barry, J.; Poulain, B. *Clostridium perfringens* epsilon toxin targets granule cells in the mouse cerebellum and stimulates glutamate release. *PLoS One* **2011**, *5*, e13046.

82. Miyamoto, O.; Minami, J.; Toyoshima, T.; Nakamura, T.; Masada, T.; Nagao, S.; Negi, T.; Itano, T.; Okabe, A. Neurotoxicity of *Clostridium perfringens* epsilon-toxin for the rat hippocampus via the glutamatergic system. *Infect. Immun.* **1998**, *66*, 2501–2508.

83. Lewis, M.; Weaver, C.D.; McClain, M.S. Identification of small molecule inhibitors of *Clostridium perfringens* epsilon-toxin cytotoxicity using a cell-based high-throughput screen. *Toxins* **2010**, *2*, 1825–1847.

84. Aulinger, B.A.; Roehrl, M.H.; Mekalanos, J.J.; Collier, R.J.; Wang, J.Y. Combining anthrax vaccine and therapy: A dominant-negative inhibitor of anthrax toxin is also a potent and safe immunogen for vaccines. *Infect. Immun.* **2005**, *73*, 3408–3414.

85. Wai, S.N.; Westermark, M.; Oscarsson, J.; Jass, J.; Maier, E.; Benz, R.; Uhlin, B.E. Characterization of dominantly negative mutant ClyA cytotoxin proteins in *Escherichia coli. J. Bacteriol.* **2003**, *185*, 5491–5499.

86. Genisset, C.; Galeotti, C.L.; Lupetti, P.; Mercati, D.; Skibinski, D.A.; Barone, S.; Battistutta, R.; de Bernard, M.; Telford, J.L. A *Helicobacter pylori* vacuolating toxin mutant that fails to oligomerize has a dominant negative phenotype. *Infect. Immun.* **2006**, *74*, 1786–1794.

87. Pelish, T.M.; McClain, M.S. Dominant-negative inhibitors of the *Clostridium perfringens* ε-toxin. *J. Biol. Chem.* **2009**, *284*, 29446–29453.

88. Buxton, D. Use of horseradish peroxidase to study the antagonism of *Clostridium welchii* (*Cl. perfringens*) type D epsilon toxin in mice by the formalinized epsilon protoxin. *J. Comp. Pathol.* **1976**, *86*, 67–72.

89. Dorca-Arevalo, J.; Martin-Satue, M.; Blasi, J. Characterization of the high affinity binding of epsilon toxin from *Clostridium perfringens* to the renal system. *Vet. Microbiol.* **2012**, *157*, 179–189.

90. Nagahama, M.; Sakurai, J. High-affinity binding of *Clostridium perfringens* epsilon-toxin to rat brain. *Infect. Immun.* **1992**, *60*, 1237–1240.

91. Ivie, S.E.; Fennessey, C.M.; Sheng, J.; Rubin, D.H.; McClain, M.S. Gene-trap mutagenesis identifies mammalian genes contributing to intoxication by *Clostridium perfringens* epsilon-toxin. *PLoS One* **2011**, *6*, e17787.

92. Ivie, S.E.; McClain, M.S. Identification of amino acids important for binding of *Clostridium perfringens* epsilon toxin to host cells and to HAVCR1. *Biochemistry* **2012**, *51*, 7588–7595.

93. Bokori-Brown, M.; Kokkinidou, M.D.; Savva, C.G.; Fernandes da Costa, S.P.; Naylor, C.E.; Cole, A.R.; Moss, D.S.; Basak, A.K.; Titball, R.W. *Clostridium perfringens* epsilon toxin H149A mutant as a platform for receptor binding studies. *Protein Sci.* **2013**, *22*, 650–659.

94. De la Rosa, C.; Hogue, D.E.; Thonney, M.L. Vaccination schedules to raise antibody concentrations against epsilon-toxin of *Clostridium perfringens* in ewes and their triplet lambs. *J. Anim. Sci.* **1997**, *75*, 2328–2334.

95. Lobato, F.C.; Lima, C.G.; Assis, R.A.; Pires, P.S.; Silva, R.O.; Salvarani, F.M.; Carmo, A.O.; Contigli, C.; Kalapothakis, E. Potency against enterotoxemia of a recombinant *Clostridium perfringens* type D epsilon toxoid in ruminants. *Vaccine* **2010**, *28*, 6125–6127.

96. Uzal, F.A.; Wong, J.P.; Kelly, W.R.; Priest, J. Antibody response in goats vaccinated with liposome-adjuvanted *Clostridium perfringens* type D epsilon toxoid. *Vet. Res. Commun.* **1999**, *23*, 143–150.

97. Odendaal, M.W.; Visser, J.J.; Botha, W.J.; Prinsloo, H. The passive protection of lambs against *Clostridium perfringens* type D with semi-purified hyperimmune serum. *Onderstepoort J. Vet. Res.* **1988**, *55*, 47–50.

98. Clarkson, M.J.; Faull, W.B.; Kerry, J.B. Vaccination of cows with clostridial antigens and passive transfer of clostridial antibodies from bovine colostrum to lambs. *Vet. Rec.* **1985**, *116*, 467–469.

99. El-Enbaawy, M.I.; Abdalla, Y.A.; Hussein, A.Z.; Osman, R.M.; Selim, S.A. Production and evaluation of a monoclonal antibody to *Clostridium perfringens* type D epsilon toxin. *Egypt. J. Immunol.* **2003**, *10*, 77–81.

100. McClain, M.S.; Cover, T.L. Functional analysis of neutralizing antibodies against *Clostridium perfringens* epsilon-toxin. *Infect. Immun.* **2007**, *75*, 1785–1793.

101. Walker, P.D. Bacterial vaccines: Old and new, veterinary and medical. *Vaccine* **1992**, *10*, 977–990.

102. Titball, R.W. *Clostridium perfringens* vaccines. *Vaccine* **2009**, *27*, D44–D47.

103. Hunter, S.E.; Clarke, I.N.; Kelly, D.C.; Titball, R.W. Cloning and nucleotide sequencing of the *Clostridium perfringens* epsilon-toxin gene and its expression in *Escherichia coli*. *Infect. Immun.* **1992**, *60*, 102–110.

104. Oyston, P.C.; Payne, D.W.; Havard, H.L.; Williamson, E.D.; Titball, R.W. Production of a non-toxic site-directed mutant of *Clostridium perfringens* epsilon toxin which induces protective immunity in mice. *Microbiology* **1998**, *144*, 333–341.

105. Kaushik, H.; Deshmukh, S.; Mathur, D.D.; Tiwar, A.; Garg, L.C. Recombinant expression of *in silico* identified B-cell epitope of epsilon toxin of *Clostridium perfringens* in translational fusion with a carrier protein. *Bioinformation* **2013**, *9*, 617–621.

Recent Insights into *Clostridium perfringens* Beta-Toxin

Masahiro Nagahama, Sadayuki Ochi, Masataka Oda, Kazuaki Miyamoto, Masaya Takehara and Keiko Kobayashi

Abstract: *Clostridium perfringens* beta-toxin is a key mediator of necrotizing enterocolitis and enterotoxemia. It is a pore-forming toxin (PFT) that exerts cytotoxic effect. Experimental investigation using piglet and rabbit intestinal loop models and a mouse infection model apparently showed that beta-toxin is the important pathogenic factor of the organisms. The toxin caused the swelling and disruption of HL-60 cells and formed a functional pore in the lipid raft microdomains of sensitive cells. These findings represent significant progress in the characterization of the toxin with knowledge on its biological features, mechanism of action and structure-function having been accumulated. Our aims here are to review the current progresses in our comprehension of the virulence of *C. perfringens* type C and the character, biological feature and structure-function of beta-toxin.

Reprinted from *Toxins*. Cite as: Nagahama, M.; Ochi, S.; Oda, M.; Miyamoto, K.; Takehara, M.; Kobayashi, K. Recent Insights into *Clostridium perfringens* Beta-Toxin. *Toxins* **2015**, *7*, 396–406.

1. Introduction

Clostridium perfringens is a gram-positive, rod-shaped bacterium. This organism is an anaerobic microorganism, but not a strictly anaerobic bacteria. *C. perfringens* strains elaborate four major toxins, named alpha-, beta-, epsilon- and iota-toxins, which have lethal, necrotic and cytotoxic activities, among others, and are categorized into five groups (types A to E) [1–5]. *C. perfringens* strains are known to be correlates with a variety of infectious disease: myonecrosis in humans and animals is due to type A strains, type B strains cause lamb dysentery, type C strains are associated with necrotizing enterocolitis (e.g., Darmbrand and Pig-bel), type D strains are correlated with enterotoxemia of sheep, and type E strains are the cause of enterotoxemia in calves and lambs. Individual major toxins have been considered to be an essential pathogeic agent in these diseases. *C. perfringens* type C, which produces alpha- and beta-toxin, causes hemorrhagic serious ulceration or mucous necrosis of the small bowel in humans, swine, and cattle [3–5]. Beta-toxin is appreciated to be the aetiological factor in necrotizing enterocolitis caused by type C strains [3–5].

Beta-toxin belongs to a β-pore-forming toxin family, which includes *Staphylococcus aureus* alpha-toxin, leukocidin, and gamma-toxin [5–7]. Cell lines that are susceptible to this toxin have been found. In addition, knowledge on its biological features, pathogenic role and action mechanism of beta-toxin has also been pooled. This review outlines recent knowledge on this issue and deals with the mechanism of beta-toxin.

2. Pathogenesis of *C. perfringens* Type C

Beta-toxin is elaborated by *C. perfringens* type B and C strain isolates and is the essential pathogenic agent of necro-hemorrhagic enteritis induced by *C. perfringnes* type C [4–7]. *C. perfringens* type C strain isolates also induce lethal infections ranging from necro-hemorrhagic enterocolitis to enterotoxemia in pigs, cattle, sheep and goats, mainly in neonatal animals of numerous domestic animal species, in which the organism propagates in the small bowel and elaborates toxins [4,8,9]. Even though mature animals can contract such illness, they most often occur in the young animals [10]. Piglets are highly sensitive to type C infectious diseases [11,12], although similar infections occur in newborn calves [13], lambs [14] and goats. During periods of a type C infection, necro-hemorrhagic enteritis can be extensive, following incorporation of beta-toxin from the small bowel into the systemic circulation. Neurological symptoms such as tetanic contraction and opisthotonos have been recognized in those animals prior to death [4], suggesting the related neurological symptoms are attributed to toxins that are elaborated in the bowels but then uptaked into the circulation to influence viscera such as the brain. In naturally occuring necro-hemorrhagic enteritis in piglets, beta-toxin was shown to bind to vessel endothelial cells in the enteric mucosa [15,16]. In unvaccinated herds, the mortalities can reach 100%, causing significant economic losses [4,9].

In humans, strains of type C induce food-borne necrotizing enterocolitis (also named as Darmbrand or Pig-bel), which is an endemic disease in the Highland of Papua New Guinea [17,18]. The human-type infection is historically most strongly related to the Highland of Papua New Guinea, where it is recognized as Pig-bel and occurs in individuals after the ingestion of insufficiently cooked pork during certain ritualistic ceremonies [17,19]. Affected individuals with Pig-bel in Papua New Guinea present with serious bloody diarrhea, abdominal pain, distension and emesis. Surgical excision of necrotic tissues of the intestine is the last way to save patients. As Pig-bel in Papua New Guinea results from an increased consumption of pork, it is proposed that the illness is related to the intake of a high-protein food. More specifically, an essential agent in the severity of the disease is thought to be a change from a low-protein food (based on "kaw kaw", sweet potato containing trypsin inhibitors that contribute to preservation of the protease-labile beta-toxin in Papua New Guinea) to a high-protein food. Low-protein ingestion for long periods generally appears to amount to chronic protein nutritional deficiency in peoples. As such, it is assumed that the circumstance in Germany just after World War II [18] was equal to that in Papua New Guinea. It is suggested that the abovementioned condition may be associated with an enteric canal mainly conditioned to a vegetarian diet unexpectedly being confronted with animal protein-rich diet [20]. Prior to vaccination campaign undertaken during 1970s and 1980s, type C-induced necrotizing enterocolitis was the most common etiology for death in children over the age of one in the Highlands of Papua New Guinea [17–21].

Necrotizing enterocolitis caused by type C isolate is marked by mucosal necrosis influencing several regions of the proximal jejunum, which was validated by the endoscopic examinations observed in several cases. Necrotizing enterocolitis associated with type C isolates has been published in some countries [22–25]. A study of ten sporadic cases (five male, five female) indicated that the mean age was 43.1 years (12 to 66) [25]. Four patients died, showing a high death rate. Six patients had insufficiently controlled or nontreated diabetes, which could be associated with morbidity of infectious diseases, delayed gastric emptying and reduced intestinal mobility. Early manifestations were bloody diarrhea in three cases and abdominal pain in seven cases. Diets that had been taken by the affected individuals were of animal source in three cases, seafood in two cases, turkey or chicken in two cases, food of vegetable source in one case, and none in two cases, indicating that there were no particular associations with particular foodstuffs [25]. The small intestine was influenced in eight cases and the colon in two cases. The infectious disease manifestations varied depending upon the term of illness and its seriousness, but general observations were mucosal necrosis with or without the formation of pseudomembrane, emphysema, and bleeding of the submucosal tissue [25]. Many toxins produced by bacteria were probably related to the mucosal necrosis. The alpha-toxin has phospholipase C and sphingomyelinase activities that destroy cell membranes by cleaving phosphatidylcholine and sphingomyelin. The beta-toxin causes transmural intestinal necrosis [25]. The risk factors for progression to this infectious disease contain decreased production of trypsin responsible for an inadequate protein food or pancreatic disease and the ingestion of diets containing an enrichment of trypsin inhibitor. Individually, these risk factors play a role in persistence of toxin in the alimentary tract during infection of type C. Spontaneous *C. perfringens* type *C*-induced necrotizing enterocolitis in humans showed acute, deep necrosis of the enteric mucosa accompanied with acute necrosis of blood vessel and severe hemorrhage in the submucosa and lamina propria. The disease is also raised in diabetic subjects. Immunohistochemical studies of tissue samples from a diabetic patient who died of necrotizing enterocolitis showed endothelial binding of the toxin in enteric lesion sites [25].

The administration of the toxoid type C filtrates containing of beta-toxin to Papua New Guinea tribes people resulted in a dramatic decrease in the frequency of necrotizing enterocolitis [26,27]. Springer and Selbitz [28] also reported that vaccination of sows with *C. perfringens* type C toxoid-containing vaccine and simultaneous treatment of a penicillin antibiotic preparation in piglets results in a marked decrease in piglet losses. Furthermore, we showed that immunization with the oligomer of the toxin, the inactive conformation, rescued mice from an intraperitoneal admistration with a culture filtrate of type C strains derived from Pig-bel patients [5]. Previous vaccination with an inactivated crude toxin preparation gained from culture filtrates of type C strains also protected against the development of the disease [1]. On the occasion of agricultural domestic livestock, immunization against type C enterocolitis is generally proposed in order to avoid destructive losses [4,20]. Beta-toxin is largely responsible for the mortality of type C infections, but we understand very little about its mode of action or the sensitive cell types. Several groups recently reported recombinant attenuated beta-toxin vaccines. Recombinant beta-toxoid produced by *E. coli* showed the elicitation of protective antitoxin antibody in rabbits [29]. Salvarani *et al.* [30] also reported that *C. perfringens* recombinant alpha-toxoid and beta-toxoid can be regarded as candidates for the establishment of a vaccine against *C. perfringens* type C. Recently, Bhatia *et al.* [31] reported that recombinant B-subunit of *E. coli* heat labile enterotoxin-putative B-cell epitope of beta-toxin (140–156 amino acid residues) fusion protein is a potential vaccine candidate against beta-toxin.

3. Evidence for the Involvement of Beta-Toxin

Current experimental investigations employing rabbit intestinal loop model and mice oral or intestinal inoculation models apparently showed that beta-toxin is the important pathogenic agent of type C strains, as described by usage of purified beta-toxin or isogenic null mutants of type C isolates [32–34]. As type C isolates typically produce alpha-toxin, beta-toxin and perfringolysin O, the McClane group evaluated the roles of these toxins to the pathogenesis of type C disease using single and double isogenic toxin null mutants of type C disease isolate CN3685 [32,35–38]. In rabbit intestinal loops, inoculation of wild-type CN3685 induced necrosis of villous tip, which showed that there was early intestinal epithelial injury. On the other hand, the *cpb* null mutant induced neither intestinal necrosis nor accumulation of bloody fluid in rabbit intestinal loops [33]. Additionally, complementing the *cpb* null mutant to recover beta-toxin production notably elevated intestinal pathogenesis. On the other hand, a double mutant of CN3685 that did not produce alpha-toxin or perfringolysin O retained sufficient virulent in rabbit ileal loops. Furthermore, in the presence of trypsin inhibitor, beta-toxin induced bloody fluid accumulation in rabbit ileal loops. In the mouse model, wild-type CN3685 was 100% lethal, yet an isogenic *cpb* null mutant showed largely decreased lethality. In the meantime, an isogenic CN3685 double mutant failed to produce alpha-toxin or perfringolysin O exhibited only a modest reduction in lethality. Gurtner *et al.* [15] showed endothelial localization of beta-toxin in acute lesion sites of spontaneously developing necrotizing enterocolitis in piglets. Furthermore, anti-beta-toxin antibody blocked the intestinal pathogenesis of wild-type CN3685 [32]. Purified beta-toxin produced the intestinal injury of wild-type CN3685, and this damage was protected by anti-beta-toxin antibody [15]. Taking these finding together, the above studies suggest that beta-toxin plays a crucial role in the pathogenesis of type C isolate.

Naturally occurring *C. perfrigens* type C enteritis in piglets showed binding of beta-toxin to vascular endothelial cells in lesion sites of necrotizing enterocolitis, which suggests that beta-toxin could cause vascular necrosis, hemorrhage and then hypoxic necrosis [39]. In a neonatal piglet jejunal loop model, beta-toxin was recognized in microvascular endothelial cells in intestinal villus during initial and development stages of lesion sites caused by *C. perfringens* type C enteritis, indicating that beta-toxin-caused endothelial injury is closely correlated with the initial stage of *C. perfringens* type C infection [40,41]. A direct binding between beta-toxin and endothelial cells might induce vascular necrosis and target destruction of endothelial cells, contributing to the virulence of necrotizing enterocolitis [15,16].

4. Beta-Toxin Gene and Regulation of Expression

C. perfringens type B and C strain isolates carry the *cpb* gene. The beta-toxin gene exists in large plasmid DNAs in *C. perfringens* that also include the insertion sequence IS1151 [42,43]. Type B isolates contain either 65-kb or 90-kb *cpb*-encoding plasmids [44,45]. Type C strains are known to carry the beta toxin-coding gene on some plasmids ranging from ~65 to ~110 kb [45,46].

Vidal *et al.* [38] reported that many type C strains produce alpha-toxin, beta-toxin and perfringolysin O much more rapidly upon close interact with human intestinal Caco-2 cells than during their *in vitro* growth without these cells. Quick Caco-2 cell-caused up-regulation of the production of beta-toxin participates in the VirS/VirR two-component system, since increased *in vivo* transcription of the *cpb* and *pfoA* genes is inhibited by inactivating the *virR* gene. This is regained upon complementation to recover *virR* expression [35]. Moreover, this two-component system was shown to be needed for type C strain CN3685 to induce the production of beta-toxin *in vivo* and cause either enterotoxemia or necrotizing enteritis in experimental animal models [35]. Therefore, the quick Caco-2 cell-caused toxin production denotes the host cell:pathogen crosstalk influencing production of toxin, which controls VirS/VirR [38,45]. *C. perfringens* contains a chromosomal operon with partly homology to the *S. aureus* operon coding components of the accessory gene regulatory (Agr) quorum-sensing system [45]. A recent study showed that the Agr-like quorum sensing system modulates the elaboration of beta-toxin and is needed for CN3685 to induce necrotizing enterocolitis [36]. In particular, by employing agrB null mutants and their complemented mutants, it was shown that the Agr-like quorum sensing system is necessary for CN3685 to induce intestinal damage in experimental animal models [36]. The dependence of CN3685 pathogenesis on the Agr-like quorum sensing system was demonstrated, in part, to be associated with the system's regulation of enteric beta-toxin elaboration [36].

5. Characterization of Beta-Toxin

Beta-toxin is a 34,861 Da protein, with the gene that encodes it being localized on virulence plasmids in *C. perfringens* [46]. The toxin is dermonecrotic and lethal, but non-hemolytic. Beta-toxin is extremely labile and highly sensitive to thiol group reagents and proteases [5,47,48]. The toxin (monomer) easily changes into the non-toxic oligomer in buffer [5]. Thereby, the beta-toxin-evoked pathogenesis can only occur under specific conditions.

Beta-toxin is remarkably related at the amino acid levels to pore-forming toxins produced by *Staphylococcus aureus*: the A and B components of gamma toxin (22% and 28% similarity), alpha-toxin (28% similarity), and the S and F components of leukocidin (17% and 28% similarity) [5,49]. *S. aureus* alpha-toxin forms heptameric oligomer and inserts itself into cell membranes, leading to formation of functional pore [49]. The alpha-toxin is the archetype of a family member of those toxins with membrane-injuring activity. The above mentioned sequence similarity, thus, suggests that beta-toxin is also a member of pore-forming toxins [49–51]. Sakurai *et al.* [47,52] showed that beta-toxin activity is inhibited by sulfhydryl group-reactive reagents. This toxin has one cysteine residue at position 265, but the substitution of Cys-265 with Ser or Ala did not influence its activity [53]. However, the replacement of Cys-265 with a bulky side chain reduced its activity. Furthermore, replacement of residues (Y266, L268 and W275) in the vicinity of Cys-265 resulted in a complete loss of lethal activity [53]. These results suggest that the area near the Cys-265 in the toxin is needed for binding to this toxin's receptor or formation of oligomer. Although the structure-activity relationship of beta-toxin has remained unclear, an early site-directed mutagenic study reported that the receptor-binding site of the toxin may be present in its *C*-terminal region [51].

6. Mechanism of Action of Beta-Toxin

After beta-toxin was administered intravenously into rats, an elevation in blood pressure and a reduction in heart rate were concomitantly recognized [54,55]. Beta-toxin attacks on the autonomic nervous system control and then causes arterial contraction by the liberation of catecholamines, so it elevates the blood pressure. When injected into the skin of mice, the toxin induces dermonecrosis and edema. We previously described that beta-toxin causes the liberation of substance P, an agonist of tachykinin NK_1 receptor, which is involved in the subsequent neurogenic plasma extravasation [56]. Furthermore, substance P liberated by the beta-toxin from sensory neurons causes the liberation of TNF-α, and these agents are responsible for plasma extravasation [57]. These results indicate that beta-toxin directly or indirectly attacks the central and the peripheral nerves.

The Pothaus group revealed the binding of the toxin to intestinal endothelial cells in naturally infected pigs, a human, and in experimental pig models [39,41]. Furthermore, beta-toxin was shown to be extremely virulent to primary porcine and human endothelial cells and caused acute cell death [15,16]. Hence, it appears that a direct action of beta-toxin on endothelial cells in the bowel plays a role in the virulence of intestinal damage caused by *C. perfringens* type C strains. Beta-toxin was also found to cause necrosis in porcine endothelial cells [58]. Incubation of these cells with beta-toxin resulted in the typical biochemical and morphological behaviors of cells that had died due to necrotic cell death. Beta-toxin-caused necrosis included stimulation of the cell signaling events participating in calpain activation [58].

Steinthorsdottir *et al.* [59] found that beta-toxin formed an oligomeric complex on human umbilical vein endothelial cells. This toxin is known to shift readily to a multimeric form *in vitro*. The Tweten group [60,61] reported that beta-toxin could form potential-dependent, cation-selective channels in planar lipid bilayers composed of phosphatidylcholine and cholesterol (1:1). The pore sizes were determined to be approximately 12 Å in diameter, indicating that the toxin is an oligomerizing, pore-forming protein toxin. Beta-toxin causes swelling and disruption of the human leukemic HL-60 cell line [62]. Given that the incubation of cells with the beta-toxin caused the concurrent incorporation of Ca^{2+}, Na^+ and Cl^- into them, and that the toxin-caused Ca^{2+} incorporation and swelling were strikingly inhibited by polyethylene glycols 600 and 1000, it seems likely that the morphological alterations in the cells caused by the beta-toxin is caused via functional pores formed by the toxin in the cell membrane [62]. We reported that beta-toxin formed an oligomeric pore of 228 kDa, which is related to its cytotoxic effect, in lipid raft microdomains of HL-60 cells [62]. Methyl-beta-cyclodextrin, an agent involved in the selective encapsulation of membrane cholesterol, inhibited beta-toxin binding to lipid raft microdomains and cytotoxicity caused by the toxin. Additionally, the treatment of liposomes with beta-toxin induced the oligomer formation of the 228 kDa of the toxin in liposomal membranes and the release of carboxyfluorescein from them. It seems likely that the complex of 228 kDa is a pore, showing that beta-toxin easily forms an oligomer as a functional pore in plasma membranes [62]. From these observations, it is clear that the toxin binds to specific receptors located principally in the lipid raft microdomains of HL-60 cells, forms functional pores in these rafts and causes cytotoxicity. We also reported that beta-toxin induce cell death of five human hematopoietic tumor cell lines (HL-60, U937, THP-1, MOLT-4 and BALL-1) [63]. U937 and THP-1 were highly susceptible to beta-toxin compared with HL-60. We further indicated that the toxin bound preferentially to susceptible cells. We showed that the toxin acts on various immune cells. Potassium efflux by the toxin in THP-1 cells caused the phosphorylation of p38 MAP and JNK kinases. This stimulation was not necessary for the toxin-caused cell death; partly, it was a stress reaction to beta-toxin for survival of cells, indicating that the MAP kinase-signaling cascade plays an important role in the protection from beta-toxin-induced cytotoxicity on THP-1 cells [63].

7. Conclusions

Beta-toxin produced by *C. perfringens* types B and C is known as the major virulence agent of necrotizing enterocolitis and enterotoxemia in humans and animals. Beta-toxin formed a functional pore of 228 kDa, which is associated with its cytotoxicity, in the lipid raft microdomains of sensitive cells. Additional study of the mode of action of beta-toxin has provided great insights into the implementation of new preventive strategies and the discovery of novel treatments. Our comprehension of the effect of beta-toxin in virulence of type C infectious disease is restricted, but the identification of its action mechanism should confer a basis for further investigation that may clarify its effect at the molecular and cellular levels and its predominant contribution to virulence. Elucidation of the accurate association of toxins with specific receptor would permit to design specific pharmacological inhibitors or to generate toxin molecules with modified specificity. Some progress has been made in developing vaccines against beta-toxin. Vaccines of varying quality are available to combat type C infection, for animals and humans use. Clearly, there is much more to learn regarding beta-toxin, how it works and how to protect against it.

Conflicts of Interest

The authors declare no conflict of interest.

References

1. McDonel, J.L. *Pharmacology of Bacterial Toxins*; Pergamon Press: New York, NY, USA, 1986; pp. 477–517.
2. Sakurai, J. Toxins of *Clostridium perfringens. Rev. Med. Microbiol.* **1995**, *6*, 175–185.
3. Sakurai, J.; Nagahama, M.; Ochi, S. Major toxins of *Clostridium perfringens. J. Toxicol. Toxin Rev.* **1997**, *16*, 195–214.
4. Songer, J.G. *Clostridial enteric* diseases of domestic animals. *Clin. Microbiol. Rev.* **1996**, *9*, 216–234.
5. Sakurai, J.; Nagahama, M. *Clostridium perfringens* beta-toxin: Characterization and action. *Toxin Rev.* **2006**, *25*, 89–108.
6. Popoff, M.R.; Bouvet, P. *Clostridial* toxins. *Future Microbiol.* **2009**, *4*, 1021–1064.
7. Popoff, M.R. *Clostridial* pore-forming toxins: Powerful virulence factors. *Anaerobe* **2014**, *30*, 220–238.
8. Uzal F.A.; McClane, B.A. Recent progress in understanding the pathogenesis of *Clostridium perfringens* type C infections. *Vet. Microbiol.* **2011**, *153*, 37–43.
9. Ma, M.; Gurjar, A.; Theoret, J.R.; Garcia, J.P.; Beingesser, J.; Freedman, J.C.; Fisher, D.J.; McClane, B.A.; Uzal, F.A. Synergistic effects of *Clostridium perfringens* enterotoxin and beta toxin in rabbit small intestinal loops. *Infect. Immun.* **2014**, *82*, 2958–2970.
10. Timoney, J.F.; Gillespie, J.H.; Scott, F.W.; Barlough, J.E. *Hagan and Bruner's Microbiology and Infectious Diseases of Domestic Animals*; Comstock Publishing Associates: New York, NY, USA, 1988.

11. Fitzgerald, G.R.; Barker, T.; Welter, M.W.; Welter, C.J. Diarrhea in young pigs: Comparing the incidence of the five most common infectious agents. *Vet. Med.* **1988**, *83*, 80–86.

12. Johnson, M.W.; Fitzgerald, G.R.; Welter, M.W.; Welter, C.J. The six most common pathogens responsible for diarrhea in newborn pigs. *Vet. Med.* **1992**, *87*, 382–386.

13. Griner, L.A.; Bracken, K.F. *Clostridium perfringens* (type C) in acute hemorrhagic enteritis in calves. *J. Am. Vet. Med. Assoc.* **1953**, *122*, 99–102.

14. Griner, L.A.; Johnson, H.W. *Clostridium perfringens* type C in hemorrhagic enterotoxemia of lambs. *J. Am. Vet. Med. Assoc.* **1954**, *125*, 125–127.

15. Gurtner, C.; Popescu, F.; Wyder, M.; Sutter, E.; Zeeh, F.; Frey, J.; von Schubert, C.; Posthaus, H. Rapid cytopathic effects of *Clostridium perfringens* beta-toxin on porcine endothelial cells. *Infect. Immun.* **2010**, *78*, 2966–2973.

16. Popescu, F.; Wyder, M.; Gurtner, C.; Frey, J.; Cooke, R.A.; Greenhill, A.R.; Posthaus, H. Susceptibility of primary human endothelial cells to *C. perfringens* beta-toxin suggesting similar pathogenesis in human and porcine necrotizing enteritis. *Vet. Microbiol.* **2011**, *153*, 173–177.

17. Johnson, S.; Gerding, D.N. Enterotoxemic infections. In *The Clostridia: Molecular Biology and Pathogenesis*; Rood, J.I., McClane, B.A., Songer, J.G., Titball, R.W., Eds.; Academic Press: London, UK, 1997; pp. 117–140.

18. Ma, M.; Li, J.; McClane, B.A. Genotypic and phenotypic characterization of *Clostridium perfringens* isolates from Darmbrand cases in post-World War II Germany. *Infect. Immun.* **2012**, *80*, 4354–4363.

19. Lawrence, G.W.; Lehmann, D.; Anian, G.; Coakley, C.A.; Saleu, G.; Barker, M.J.; Davis, M.W. Impact of active immunization against enteritis necroticans in Papua New Guinea. *Lancet* **1990**, *336*, 1165–1167.

20. Murrell, T.G.; Roth, L.; Egerton, J.; Samels, J.; Walker, P.D. Pig-bel: Enteritis necroticans. A study in diagnosis and management. *Lancet* **1966**, *1*, 217–222.

21. McClane, B.A.; Uzal, F.A.; Miyakawa, M.F.; Lyerly, D.; Wilkins, T. The Enterotoxic clostridia. In *The Prokaryotes*; Dworkin, M., Falkow, S., Rosenberg, E., Schleifer, K.H., Stackebrandt, E., Eds.; Springer: New York, NY, USA, 2004; pp. 698–752.

22. Lee, H.; Bueschel, D.M.; Nesheim, S.R. Enteritis necroticans (Pigbel) in a diabetic child. *N. Eng. J. Med.* **2000**, *342*, 1250–1253.

23. Gui, L.; Subramony, C.; Fratkin, J.; Hughson, M.D. Fatal enteritis necroticans (Pigbel) in a diabetic adult. *Mod. Pathol.* **2002**, *15*, 66–70.

24. Sobel, J.; Mixter, C.G.; Kolhe, P.; Gupta, A.; Guarner, J.; Zaki, S.; Hoffman, N.A.; Songer, J.G.; Fremont-Smith, M.; Fischer, M.; *et al.* Necrotizing enterocolitis associated with *Clostridium perfringens* type A in previously healthy North American adults. *J. Am. Coll. Surg.* **2005**, *201*, 48–56.

25. Matsuda, T.; Okada, Y.; Tanabe, Y.; Shimizu, Y.; Nagashima, K.; Sakurai, J.; Nagahama, M.; Tanaka, S. Enteritis necroticans "pigbel" in a Japanese diabetic adult. *Pathol. Int.* **2007**, *57*, 622–626.

26. Lawrence, G.; Shann, F.; Freestone, D.S.; Walker, P.D. Prevention of necrotizing enteritis in Papua New Guinea by active immunization. *Lancet* **1979**, *1*, 227–230.

27. Davis, M.; Lawrence, G.; Shann, F.; Walker, P.D. Longevity of protection by active immunization against necrotising enteritis in Papua New Guinea. *Lancet* **1982**, *2*, 389–390.

28. Springer, S.; Selbitz, H.J. The control of necrotic enteritis in sucking piglets by means of a *Clostridium perfringens* toxoid vaccine. *FEMS. Immunol. Med. Microbiol.* **1999**, *24*, 333–336.

29. Milach, A.; de los Santos, J.R.; Turnes, C.G.; Moreira, A.N.; de Assis, R.A.; Salvarani, F.M.; Lobato, F.C.; Conceição, F.R. Production and characterization of *Clostridium perfringens* recombinant β toxoid. *Anaerobe* **2012**, *18*, 363–365.

30. Salvarani, F.M.; Conceição, F.R.; Cunha, C.E.; Moreira, G.M.; Pires, P.S.; Silva, R.O.; Alves, G.G.; Lobato, F.C. Vaccination with recombinant *Clostridium perfringens* toxoids α and β promotes elevated antepartum and passive humoral immunity in swine. *Vaccine* **2013**, *31*, 4152–4155.

31. Bhatia, B.; Solanki, A.K.; Kaushik, H.; Dixit, A.; Garg, L.C. B-cell epitope of beta toxin of *Clostridium perfringens* genetically conjugated to a carrier protein: expression, purification and characterization of the chimeric protein. *Protein Expr. Purif.* **2014**, *102*, 38–44.

32. Sayeed, S.; Uzal, F.A.; Fisher, D.J.; Saputo, J.; Vidal, J.E.; Chen, Y.; Gupta, P.; Rood, J.I.; McClane, B.A. Beta toxin is essential for the intestinal virulence of *Clostridium perfringens* type C disease isolate CN3685 in a rabbit ileal loop model. *Mol. Microbiol.* **2008**, *67*, 15–30.

33. Uzal, F.A.; Saputo, J.; Sayeed, S.; Vidal, J.E.; Fisher, D.J.; Poon, R.; Adams, V.; Fernandez-Miyakawa, M.E.; Rood, J.I.; McClane, B.A. Development and application of new mouse models to study the pathogenesis of *Clostridium perfringens* type C enterotoxemias. *Infect. Immun.* **2009**, *77*, 5291–5299.

34. Vidal, J.E.; McClane, B.A.; Saputo, J.; Parker, J.; Uzal, F.A. Effects of *Clostridium perfringens* beta-toxin on the rabbit small intestine and colon. *Infect. Immun.* **2008**, *76*, 4396–4404.

35. Ma, M.; Vidal, J.; Saputo, J.; McClane, B.A.; Uzal, F. The VirS/VirR two-component system regulates the anaerobic cytotoxicity, intestinal pathogenicity, and enterotoxemic lethality of *Clostridium perfringens* type C isolate CN3685. *MBio* **2011**, *2*, doi:10.1128/mBio.00338-10.

36. Vidal, J.E.; Ma, M.; Saputo, J.; Garcia, J.; Uzal, F.A.; McClane, B.A. Evidence that the Agr-like quorum sensing system regulates the toxin production, cytotoxicity and pathogenicity of *Clostridium perfringens* type C isolate CN3685. *Mol. Microbiol.* **2012**, *83*, 179–194.

37. Garcia, J.P.; Beingesser, J.; Fisher, D.J.; Sayeed, S.; McClane, B.A.; Posthaus, H.; Uzal, F.A. The effect of *Clostridium perfringens* type C strain CN3685 and its isogenic beta toxin null mutant in goats. *Vet. Microbiol.* **2012**, *157*, 412–419.

38. Vidal, J.E.; Ohtani, K.; Shimizu, T.; McClane, B.A. Contact with enterocyte-like Caco-2 cells induces rapid upregulation of toxin production by *Clostridium perfringens* type C isolates. *Cell. Microbiol.* **2009**, *11*, 1306–1328.

39. Miclard, J.; Jäggi, M.; Sutter, E.; Wyder, M.; Grabscheid, B.; Posthaus, H. *Clostridium perfringens* beta-toxin targets endothelial cells in necrotizing enteritis in piglets. *Vet. Microbiol.* **2009**, *137*, 320–325.

40. Schumacher, V.L.; Martel, A.; Pasmans, F.; van Immerseel, F.; Posthaus, H. Endothelial binding of beta toxin to small intestinal mucosal endothelial cells in early stages of experimentally induced *Clostridium perfringens* type C enteritis in pigs. *Vet. Pathol.* **2013**, *50*, 626–629.

41. Miclard, J.; van Baarlen, J.; Wyder, M.; Grabscheid, B.; Posthaus, H. *Clostridium perfringens* beta-toxin binding to vascular endothelial cells in a human case of enteritis necroticans. *J. Med. Microbiol.* **2009**, *58*, 826–828.

42. Gibert, M.; Perelle, S.; Daube, G.; Popoff, M.R. *Clostridium spiroforme* toxin genes are related to *C. perfringens* iota toxin genes but have a different genomic localization. *Syst. Appl. Microbiol.* **1997**, *20*, 337–347.

43. Katayama, S.; Dupuy, B.; Daube, G.; China, B.; Cole, S.T. Genome mapping of *Clostridium perfringens* strains with I-CeuI shows many virulence genes to be plasmid-borne. *Mol. Gen. Genet.* **1996**, *251*, 720–726.

44. Sayeed, S.; Li, J.; McClane, B.A. Characterization of virulence plasmid diversity among *Clostridium perfringens* type B isolates. *Infect. Immun.* **2010**, *78*, 495–504.

45. Li, J.; Adams, V.; Bannam, T.L.; Miyamoto, K.; Garcia, J.P.; Uzal, F.A.; Rood, J.I.; McClane, B.A. Toxin plasmids of *Clostridium perfringens*. *Microbiol. Mol. Biol. Rev.* **2013**, *77*, 208–233.

46. Gurjar, A.; Li, J.; McClane, B.A. Characterization of toxin plasmids in *Clostridium perfringens* type C isolates. *Infect. Immun.* **2010**, *78*, 4860–4869.

47. Sakurai, J.; Fujii, Y.; Matsuura, M. Effect of oxidizing agents and sulfhydryl group reagents on beta toxin from *Clostridium perfringens* type C. *Microbiol. Immunol.* **1980**, *24*, 595–601.

48. Sakurai, J.; Fujii, Y. Purification and characterization of *Clostridium perfringens* beta toxin. *Toxicon* **1987**, *25*, 1301–1310.

49. Hunter, S.E.; Brown, J.E.; Oyston, P.C.; Sakurai, J.; Titball, R.W. Molecular genetic analysis of beta-toxin of *Clostridium perfringens* reveals sequence homology with alpha-toxin, gamma-toxin, and leukocidin of *Staphylococcus aureus*. *Infect. Immun.* **1993**, *61*, 3958–3965.

50. Geny, B.; Popoff, M.R. Bacterial protein toxins and lipids: Pore formation or toxin entryinto cells. *Biol. Cell.* **2006**, *98*, 667–678.

51. Steinthorsdottir, V.; Fridriksdottir, V.; Gunnarsson, E.; Andrésson, O.S. Site-directed mutagenesis of *Clostridium perfringens* beta-toxin: expression of wild-type and mutant toxins in *Bacillus subtilis*. *FEMS Microbiol. Lett.* **1998**, *158*, 17–23.

52. Sakurai, J.; Fujii, Y.; Nagahama, M. Effect of p-chloromercuribenzoate on *Clostridium perfringens* beta-toxin. *Toxicon.* **1992**, *30*, 323–330.

53. Nagahama, M.; Kihara, A.; Miyawaki, T.; Mukai, M.; Sakaguchi, Y.; Ochi, S.; Sakurai, J. *Clostridium perfringens* beta-toxin is sensitive to thiol-group modification but does not require a thiol group for lethal activity. *Biochim. Biophys. Acta* **1999**, *1454*, 97–105.

54. Sakurai, J.; Fujii, Y.; Matsuura, M.; Endo, K. Pharmacological effect of beta toxin of *Clostridium perfringens* type C on rats. *Microbiol. Immunol.* **1981**, *25*, 423–432.

55. Sakurai, J.; Fujii, Y.; Dezaki, K.; Endo, K. Effect of *Clostridium perfringens* beta toxin on blood pressure of rats. *Microbiol. Immunol.* **1984**, *28*, 23–31.

56. Nagahama, M.; Morimitsu, S.; Kihara, A.; Akita, M.; Setsu, K.; Sakurai, J. Involvement of tachykinin receptors in *Clostridium perfringens* beta-toxin-induced plasma extravasation. *Br. J. Pharmacol.* **2003**, *138*, 23–30.

57. Nagahama, M.; Kihara, A.; Kintoh, H.; Oda, M.; Sakurai, J. Involvement of tumor necrosis factor-alpha in *Clostridium perfringens* beta-toxin-induced plasma extravasation in mice. *Br. J. Pharmacol.* **2008**, *153*, 1296–1302.

58. Autheman, D.; Wyder, M.; Popoff, M.; D'Herde, K.; Christen, S.; Posthaus, H. *Clostridium perfringens* beta-toxin induces necrostatin-inhibitable, calpain-dependent necrosis in primary porcine endothelial cells. *PLoS One* **2013**, *8*, doi:10.1371/journal.pone.0064644.

59. Steinthorsdottir, V.; Halldórsson, H.; Andrésson, O.S. *Clostridium perfringens* beta-toxin forms multimeric transmembrane pores in human endothelial cells. *Microb. Pathog.* **2000**, *28*, 45–50.

60. Shatursky, O.; Bayles, R.; Rogers, M.; Jost, B.H.; Songer, J.G.; Tweten, R.K. *Clostridium perfringens* beta-toxin forms potential-dependent, cation-selective channels in lipid bilayers. *Infect. Immun.* **2000**, *68*, 5546–5551.

61. Tweten, R.K. *Clostridium perfringens* beta-toxin and *Clostridium septicum* alpha-toxin: Their mechanisms and possible role in pathogenesis. *Vet. Microbiol.* **2001**, *82*, 1–9.

62. Nagahama, M.; Hayashi, S.; Morimitsu, S.; Sakurai, J. Biological activities and pore formation of *Clostridium perfringens* beta-toxin in HL 60 cells. *J. Biol. Chem.* **2003**, *278*, 36934–36941.

63. Nagahama, M.; Shibutani, M.; Seike, S.; Yonezaki, M.; Takagishi, T.; Oda, M.; Kobayashi, K.; Sakurai, J. The p38 MAPK and JNK pathways protect host cells against *Clostridium perfringens* beta-toxin. *Infect. Immun.* **2013**, *81*, 3703–3708.

Binding Studies on Isolated Porcine Small Intestinal Mucosa and *in vitro* Toxicity Studies Reveal Lack of Effect of *C. perfringens* Beta-Toxin on the Porcine Intestinal Epithelium

Simone Roos, Marianne Wyder, Ahmet Candi, Nadine Regenscheit, Christina Nathues, Filip van Immerseel and Horst Posthaus

Abstract: Beta-toxin (CPB) is the essential virulence factor of *C. perfringens* type C causing necrotizing enteritis (NE) in different hosts. Using a pig infection model, we showed that CPB targets small intestinal endothelial cells. Its effect on the porcine intestinal epithelium, however, could not be adequately investigated by this approach. Using porcine neonatal jejunal explants and cryosections, we performed *in situ* binding studies with CPB. We confirmed binding of CPB to endothelial but could not detect binding to epithelial cells. In contrast, the intact epithelial layer inhibited CPB penetration into deeper intestinal layers. CPB failed to induce cytopathic effects in cultured polarized porcine intestinal epithelial cells (IPEC-J2) and primary jejunal epithelial cells. *C. perfringens* type C culture supernatants were toxic for cell cultures. This, however, was not inhibited by CPB neutralization. Our results show that, in the porcine small intestine, CPB primarily targets endothelial cells and does not bind to epithelial cells. An intact intestinal epithelial layer prevents CPB diffusion into underlying tissue and CPB alone does not cause direct damage to intestinal epithelial cells. Additional factors might be involved in the early epithelial damage which is needed for CPB diffusion towards its endothelial targets in the small intestine.

Reprinted from *Toxins*. Cite as: Roos, S.; Wyder, M.; Candi, A.; Regenscheit, N.; Nathues, C.; van Immerseel, F.; Posthaus, H. Binding Studies on Isolated Porcine Small Intestinal Mucosa and *in vitro* Toxicity Studies Reveal Lack of Effect of *C. perfringens* Beta-Toxin on the Porcine Intestinal Epithelium. *Toxins* **2015**, *7*, 1235-1252.

1. Introduction

The anaerobic, Gram-positive, spore-forming bacterium *C. perfringens* causes different diseases in humans and animals, such as septicemia, myonecrosis, enterotoxemia, food poisoning and enteritis [1]. Classification into five types is based on the production of four major toxins: Alpha- (CPA), beta- (CPB), epsilon (ETX)- and iota-toxin (ITX) [2]. *C. perfringens* type C produces CPA and CPB; however, additional toxins, such as beta-2 toxin, enterotoxin, perfringolysin and TpeL can be secreted [3,4]. *C. perfringens* type C causes necrotizing enteritis (NE) in newborn animals and in humans [1]. Piglets are most commonly affected and, as in all affected hosts, the hallmark lesion of NE is a severe, segmental, necro-hemorrhagic jejunitis [5]. The exact role of different toxins in the pathogenesis of the disease is not known yet. Experimental studies using genetic approaches and animal models of disease clearly demonstrated that CPB is the essential virulence factor of type C strains [6].

CPB is a soluble 35 kDa monomer protein that is thermo-labile and highly sensitive to degradation by trypsin [4]. It is a member of the beta-barrel pore-forming toxin family and forms oligomeric pores in several susceptible immune cell lines [7,8]. Other studies showed that CPB is required for *C. perfringens*-induced necrotic enteritis in rabbit ileal loops and that purified CPB can produce similar lesions [6]. In this model, lesions were reminiscent of early epithelial damage and thus CPB might also directly act on intestinal epithelial cells. So far, it is however unknown whether epithelial necrosis is caused by a direct effect of the pore-forming toxin on intestinal epithelial cells or indirectly [9,10]. Our recent work showed that endothelial cells are highly susceptible to CPB [11,12] and that pore-formation in endothelial cells rapidly leads to necrotic cell death [13]. In spontaneous disease CPB can be demonstrated in endothelial cells [14,15] and this also correlated to early vascular lesions in an experimental infection model in pigs [16]. Based on these results, we hypothesized that endothelial cells in the small intestine are the primary target for CPB. This however would require that CPB can pass the intestinal epithelial barrier to penetrate into deeper mucosal tissue. A direct toxic effect of CPB on porcine small intestinal epithelial cells is possible and this would facilitate toxicity on endothelial cells. Our hypothesis was however based on analyses of spontaneous end stage lesions [14] and early lesions in experimentally infected piglets [16]. Although we could not demonstrate CPB binding to epithelial cells in these studies, they were not suitable to investigate early epithelial toxicity as they represent stages where tissue necrosis has already begun. Additionally, the piglet intestinal loop model would require a relatively high number of lethal infectious experiments to closely monitor initial interactions of CPB with the small intestinal epithelium. The aim of our study was therefore to evaluate binding of CPB to cells in the porcine small intestinal mucosa and the toxic effect of CPB on cultured porcine small intestinal epithelial cells. To achieve this, we chose two experimental approaches: (I) CPB binding studies using neonatal porcine jejunal cryosections and tissue explants; (II) cellular toxicity studies using a porcine jejunal epithelial cell line (IPEC-J2) and primary porcine jejunal epithelial cells.

2. Results

2.1. Cellular Binding of CPB in Porcine Jejunal Cryosections

Immunohistochemical stainings of cryosections incubated with supernatants of *C. perfringens* type C NCTC 3180 and type C JF 3721 localized CPB at the endothelium in all layers of the jejunal wall (Figure 1A,B). In contrast, no CPB signal could be demonstrated at the epithelium.

2.2. Cellular Binding of CPB in Porcine Jejunal Explants

Histological sections of the porcine neonatal jejunal explants revealed no morphological differences between explants incubated with *C. perfringens* type C or type A supernatants or toxin free control medium. All explants, regardless of the incubation protocol, remained well preserved for the first 4 h of the experiment and afterwards showed early signs of autolysis, such as detachment of epithelial cells. The maximum length of the experiments was therefore set at 6 h of incubation.

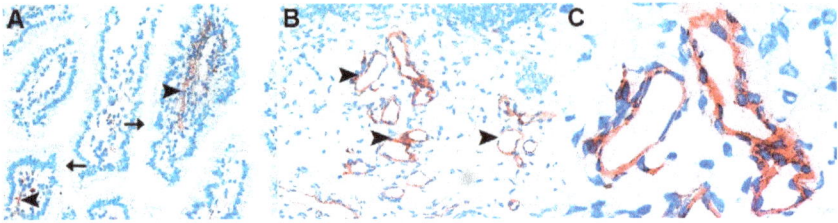

Figure 1. Immunohistochemical localization of CPB (*Clostridium perfringens* beta-toxin) in porcine jejunal cryosections. Cryosections of porcine neonatal jejunum were incubated with *C. perfringens* type C NCTC 3180 supernatant (diluted 1:10 in PBS). CPB was subsequently detected by immunohistochemistry on endothelial cells (arrowheads) in the lamina propria (**A**) and submucosa (**B**), but not on epithelial cells (arrows). (**C**) Magnification of B depicting endothelial CPB signal. Representative pictures from one out of two independent experiments, magnification 400×.

Immunohistochemical staining of all explants incubated with *C. perfringens* type C supernatants showed binding of CPB to the endothelium (Figure 2A). No CPB signal at the epithelium was detectable. In all explants, a weak endothelial signal in vessels and capillaries located directly at the cut margins of the explants was visible from 1 h onwards. These signals were not present in vessels located more than 50 µm from the cut border and were due to diffusion of CPB into the tissue from the cut edges and therefore neglected for our analyses. In unmodified explants, weak endothelial signals in the superficial lamina propria of the villi became evident after 2 h of incubation (Figure 2B); a weak signal in the submucosa of unmodified explants was detectable after 3 h of incubation (Figure 2C). Mechanical damage to the epithelium resulted in the rapid appearance of moderate to strong signals at the endothelium in the villi (Figure 2A,B). Binding to the endothelium in the submucosa was moderate in these preparations (Figures 2C and S1). Detachment of the tunica serosa and tunica muscularis led to the rapid appearance of a stronger CPB signal at the endothelium, especially within the submucosa compared to unmodified explants (Figures 2A,C and S1). Detachment of the tunica serosa and tunica muscularis, however, did not lead to an increased signal intensity at endothelial cells in the villi compared to unmodified explants up to 3 h of incubation (Figure 2A,B). Signal intensities were only increased compared to unmodified explants at 4–6 h of incubation. Pre-incubation of the *C. perfringens* type C supernatants with neutralizing anti-CPB antibodies (mAb-CPB) resulted in complete inhibition of CPB signals in all preparations (Figure S2). Multi-way ANOVA, corrected for animal and time, showed significant differences in scores between the three preparations in the binding of CPB to villous and submucosal endothelium (p-values <0.0001). In the case of villous endothelium, all preparation groups showed a significant difference from each other in Tukey-Kramer multiple comparison tests. Explants with a damaged epithelium had a significantly higher chance of exhibiting a moderate or strong signal at the endothelium in villi compared to unmodified explants (OR: 55.3; 95% CI: 15.3–199.0) and explants with a detached mucosa (OR: 4.5; 95% CI: 1.5–13.5) (Figure 2B). For the binding site endothelium-submucosa, detached mucosa explants differed significantly from explants with damaged epithelium or unmodified explants. The chance of detecting a moderate or high signal at the endothelium in the submucosa in explants with a detached mucosa was significantly higher than in unmodified explants (OR: 94.0; 95% CI: 24.9–355.3) and explants with a mechanically damaged epithelium (OR: 59.5; 95% CI: 17.5–202.1) (Figure 2C). There was no significant difference in signal intensities at submucosal endothelia between explants with mechanically damaged epithelium and unmodified explants (OR: 1.6; 95% CI: 0.49–5.0) (Figure 2C).

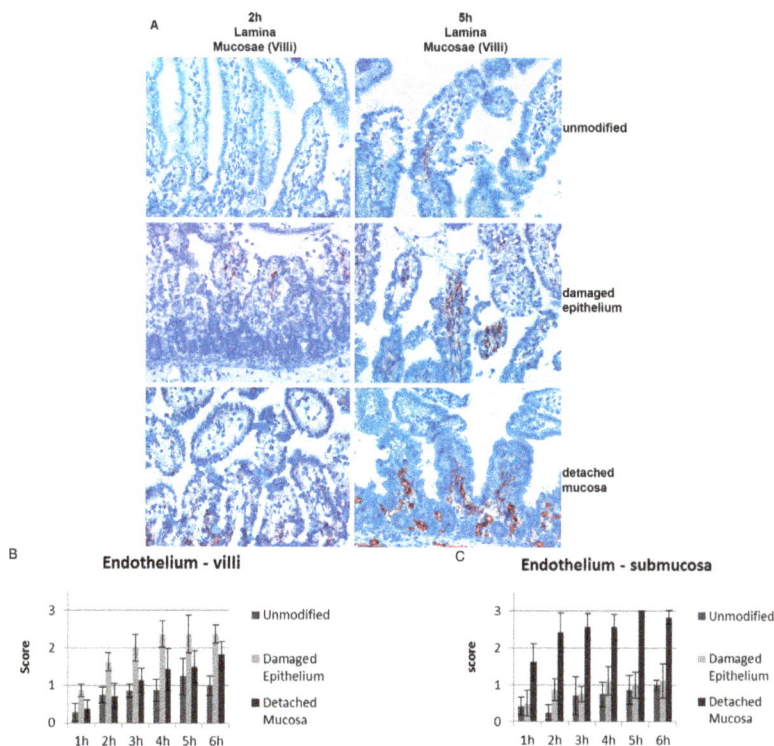

Figure 2. Immunohistochemical localization of CPB in the subepithelial lamina propria of porcine jejunal explants. (**A**) Jejunal explants were either unmodified (unmodified), the epithelium was mechanically damaged (damaged epithelium), or the tunica serosa and tunica muscularis were mechanically detached (detached mucosa). Explants were incubated with *C. perfringens* type C supernatant (JF 3721 diluted 1:10 in RPMI), fixed, and CPB was detected by immunohistochemistry. After 2 h of incubation, mild to moderate CPB signals were detected at endothelial cells close to the damaged epithelium whereas undamaged explants showed no or only very mild signals at 2 h. After 5 h of incubation CPB signals were overall stronger in tissue with a mechanically damaged epithelium. Binding of CPB to the epithelium was not detectable in any explant. Representative pictures from one out of three independent experiments, magnification 400×. (**B**) Mean CPB signal scores and standard deviation at endothelium-villi of eight explants in three independent time course experiments. (**C**) Mean CPB signal scores and standard deviation at endothelium-submucosa (fotographs depicted in Figure S1) of eight explants in three independent time course experiments.

These results showed that CPB has a high affinity to endothelial cells in porcine jejunal explants. Despite the lack of epithelial signals in immunohistochemistry, binding of small amounts of CPB to epithelial cells, which could be missed using this technique, cannot be excluded. However, intact small intestinal epithelium seems to act as a diffusion barrier for CPB and prevent tissue penetration of the toxin to reach endothelial cells in the lamina propria and deeper intestinal layers.

2.3. Trans Epithelial Electrical Resistance Measurements (TEER)

2.3.1. Apical Exposure of Polarized Porcine Small Intestinal Cells Layers to C. perfringens Type C Culture Supernatants and rCPB

To evaluate the effect of C. perfringens type C culture supernatants and purified recombinant CPB (rCPB) on the porcine intestinal epithelium, a porcine jejunal epithelial cell line (IPEC-J2) forming a polarized continuous and tight epithelial layer on Transwells® [17] was apically incubated with late log phase culture supernatants of two C. perfringens type C strains diluted in cell culture medium. Non-inoculated, sterile TGY and supernatants of a C. perfringens type A strain (JF 3693) were used as negative controls. Apical incubation with undiluted sterile TGY reduced the TEER below 1 kΩ cm^2 but did not do so at a 1:2 dilution. Therefore, all culture supernatants were used at a 1:2 dilution in cell culture medium to achieve maximum CPB concentrations. Similar to TGY, the C. perfringens type A supernatant did not have a significant effect on TEER. As a positive control, we used the pore-forming toxin aerolysin (100 ng/mL) which is known to damage epithelial cells [18]. Apical incubation of IPEC-J2 with aerolysin resulted in a significant drop of TEER values below 1 kΩ cm^2 within 1 h and a subsequent further decline (Figure 3A). The NCTC 3180 supernatant used at a 1:2 dilution (7.5 µg CPB/mL) induced a drop of TEER values below 2 kΩ cm^2 at 2 h which subsequently further declined (Figure 3A). The supernatant of JF 3721 (2.2 µg CPB/mL) induced a drop in TEER below 2 kΩ cm^2 starting at 12 h (Figure 3A). Pre-incubation of both type C supernatants with neutralizing mAb-CPB did not reduce this effect. Additionally, both C. perfringens type C culture supernatants used at a 1:10 dilution, which contained CPB at a concentration of 1.5 µg/mL (NCTC 3180) and 440 ng/mL (JF 3721), did not cause a drop of TEER below 2 kΩ cm^2 over 48 h (Figure S3).

Dunn's Z multiple comparison test revealed a significant difference between TEER values of Transwells® of the TGY control group and Transwells® which had been incubated with aerolysin, neutralized or non-neutralized NCTC 3180 supernatants for 1 h–48 h. TEER values of Transwells® which had been incubated with neutralized JF 3721 supernatant differed significantly from the TGY control group at 12 h–48 h, whereas TEER values of Transwells® incubated with JF 3721 differed significantly at 24 h–48 h. No significant differences were present between Transwells® incubated with type C supernatants (NCTC 3180, JF 3721) and the corresponding supernatants pre-incubated with neutralizing anti-CPB antibodies. No significant differences were detected between TEER values of Transwells® that had been incubated with C. perfringens type A and the TGY control group.

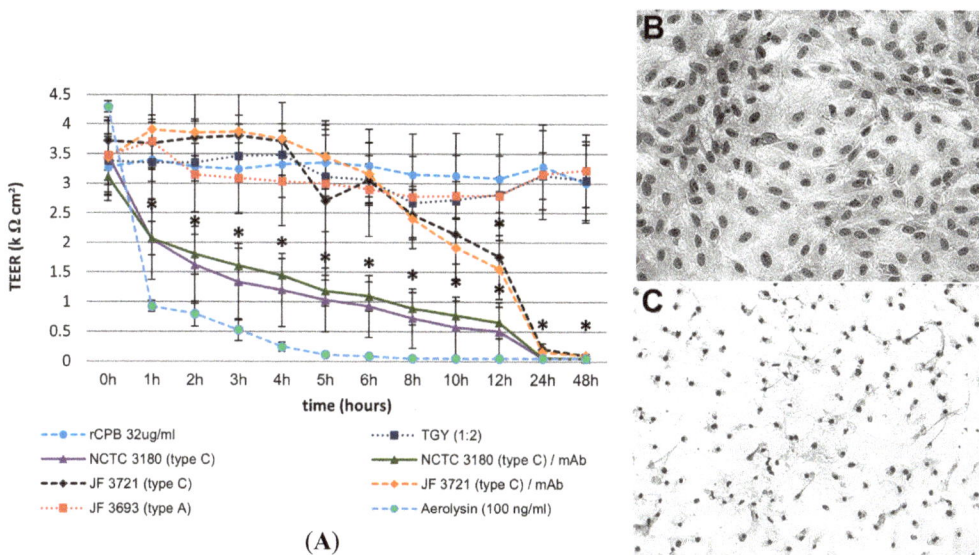

Figure 3. (A) Apical exposure of IPEC-J2 cell layers to rCPB does not affect the TEER. IPEC-J2 cells grown on collagen coated Transwells® were incubated with *C. perfringens* supernatants (diluted 1:2 in cell culture medium), TGY (1:2), or rCPB (32 µg/mL) in the apical compartment. Aerolysin used as positive control caused a rapid and sustained drop in TEER. Apical incubation of type C supernatants caused a drop of TEER values within 2 h (NCTC 3180) or 12 h (JF 3721) below 2 kΩ cm^2. rCPB did not cause a drop of TEER below 2 kΩ cm^2, similar to type A supernatants and TGY used as controls. Values represent means of three independent experiments with a total of eight separate Transwells®. Error bars represent two-fold standard deviation. Asterisks indicate values which differ significantly from the TGY control group. **(B)** Endothelial cells (PAEC) co-cultivated in the basolateral compartment of Transwells® of IPEC-J2 cells apically exposed to rCPB (32 µg/mL) did not exhibit a cytopathic effect. **(C)** Mechanical damage of the IPEC-J2 layer resulted in cytopathic effects in co-cultivated PAEC when the same supernatant as in B was added to the apical compartment. Co-cultivation was performed for 48 h.

Our results using *C. perfringens* type C culture supernatants indicated that CPB was not likely to be the epithelium damaging factor in our experiments. To further evaluate whether CPB was directly toxic to the intestinal epithelium, we incubated polarized IPEC-J2 cells with purified rCPB at a concentration of 32 µg/mL. Recombinant CPB was previously shown to be highly toxic to primary endothelial cells at concentrations of 13 ng/mL, which was in the same range of toxicity as *C. perfringens* type C derived CPB [11]. Even at these approximately 2000-fold higher concentrations, rCPB did not induce a drop of TEER in polarized IPEC-J2 cells (Figure 3A). A lack or the complete loss of cytotoxic activity of rCPB was excluded by incubating PAEC monolayers grown in 96-well plates with apical medium/toxin preparations at the end of each experiment. After 48 h of apical incubation of IPEC-J2, rCPB preparations still induced cell death in PAEC. This effect was inhibited by antibody mediated neutralization of CPB. To evaluate whether rCPB was able to pass the polarized IPEC-J2 layer in the Transwells®, PAEC were co-cultivated in the basolateral chamber. Using this approach, no cytopathic effect was observed after 48 h of apical incubation of the IPEC-J2 with rCPB (Figure 3B). Mechanical destruction of the IPEC-J2 layer before apical incubation with rCPB resulted in marked cytotoxicity in co-cultivated PAEC (Figure 3C).

2.3.2. Basolateral Exposure of Polarized Porcine Small Intestinal Cells Layers to *C. perfringens* Type C Culture Supernatants and rCPB

Basolateral exposure of IPEC-J2 cell layers to aerolysin as a positive control resulted in a rapid and complete loss of the TEER of polarized IPEC-J2 layers (Figure 4). TGY did not have an effect on the TEER up to a dilution of 1:5 and the control *C. perfringens* type A supernatant also did not induce a drop in TEER below 2 kΩ cm^2 at this concentration. Basolateral incubation of polarized IPEC-J2 cells with the supernatant of NCTC 3180 diluted 1:10 in cell culture medium resulted in a rapid decline of TEER. After 1 h, TEER values dropped below 2 kΩ cm^2 and subsequently declined (Figure 4). Pre-incubation of the supernatants with neutralizing mAb-CPB did not reduce this effect. Basolateral incubation with supernatant of JF 3721 diluted 1:10 in cell culture medium did not induce a drop of TEER values below 2 kΩ cm^2 (Figure S4), however increasing the concentration to a 1:5 dilution resulted in a significant drop at 24 h with a subsequent further decline (Figure 4). Pre-incubation with mAb-CPB had no effect on this drop in TEER. Recombinant CPB at a concentration of 32 µg/mL did not induce a significant drop in TEER when applied basolaterally. Dunn's Z multiple comparison tests revealed a significant difference between TEER values of the TGY control group and TEER values of IPEC-J2 grown on Transwells® which had been incubated with aerolysin and the supernatant of NCTC 3180 (1:10) neutralized with mAb-CPB from 1 h–48 h. TEER values of non-neutralized NCTC 3180 differed significantly from 2 h to 48 h from the TGY control group. TEER values of Transwells® incubated with supernatant of JF 3721 neutralized or non-neutralized differed significantly from 24 h to 48 h from the TGY control group. Similar to apical exposure no significant differences were present between Transwells® incubated with type C supernatants (NCTC 3180, JF 3721) and the corresponding neutralized supernatants. Similar to apical exposure experiments, 32 µg/mL CPB in the basolateral chamber of the Transwells® did not reduce the TEER below 2 kΩ cm^2.

Culture supernatants were considerably more toxic when IPEC-J2 cell layers where exposed basolaterally compared to apical exposure. However, lack of inhibition of the toxic effect by neutralization of CPB and the lack of effect of high rCPB concentrations again indicated that CPB was not essential for this effect.

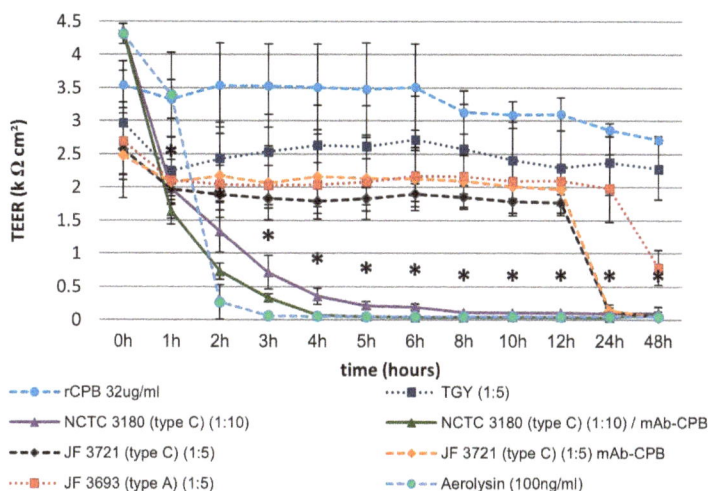

Figure 4. Reduction of TEER upon basolateral exposure of IPEC-J2 cells to *C. perfringens* type C culture supernatants but not rCPB. IPEC-J2 cells were incubated in the basolateral compartment of the Transwell® with the same toxins and supernatants (diluted 1:10 and 1:5) as in Figure 3. Aerolysin and the *C. perfringens* type C supernatant NCTC 3180 (1:10) caused a rapid and sustained drop in TEER. This effect could not be inhibited by neutralization of CPB using mAb-CPB. Supernatants of JF 3721 induced a drop in TEER within 12 h and supernatants of JF 3693 within 48 h. rCPB (32 µg/mL), or TGY (1:5) resulted in a small drop of TEER values, which however never dropped below 2 kΩ cm² and subsequently increased towards the end of the experiment. Values represent means of three independent experiments with a total of eight separate Transwells®. Error bars represent two-fold standard deviation. Asterisks indicate values which differ significantly from the TGY control group.

2.4. Exposure of Primary Porcine Small Intestinal Epithelial Cells to C. perfringens Type C Culture Supernatants and rCPB

To exclude that the lack of effect of CPB on polarized IPEC-J2 cell layers was cell line specific, we additionally cultured primary porcine neonatal jejunal epithelial cells. Exposure of primary jejunal epithelial cell monolayers to NCTC 3180 supernatants caused a cytopathic effect at a dilution of 1:20–1:40 (Figure 5). JF 3721 supernatants caused a cytopathic effect at a dilution of 1:2–1:20 (Figure 5). For both type C supernatants, this cytopathic effect was still detectable after pre-incubation with neutralizing antibody mAb-CPB (Figure 5). Recombinant CPB at a concentration of 32 µg/mL (Figure 5) or supernatant of *C. perfringens* type A JF 3693 (data not shown) did not cause a cytopathic effect in primary jejunal epithelial cells.

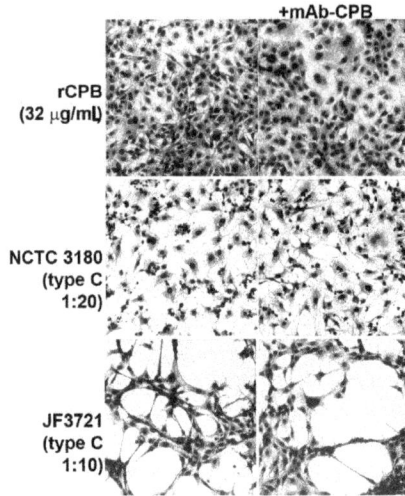

Figure 5. Lack of cytopathic effect of CPB to primary porcine jejunal epithelial cells. rCPB added to the cell culture medium did not cause a cytopathic effect on primary porcine intestinal epithelial cells. Exposure of these cells to *C. perfringens* type C NCTC 3180 and JF 3721 supernatants caused a cytopathic effect, which was not inhibited by pre-incubation with mAb-CPB. Incubation time 48 h.

3. Discussion

C. perfringens type C causes lethal necrotizing enteritis in animals and humans. We previously showed that endothelial cells are targets for CPB in the porcine small intestine. However, a direct toxic effect of CPB on intestinal epithelial cells was also possible. Using porcine small intestinal mucosal explants and cryosections incubated with *C. perfringens* type C culture supernatants, we now confirmed that CPB preferentially binds to endothelial cells in the porcine small intestinal mucosa but not to epithelial cells. Mechanical damage to the epithelium of intestinal explants led to a more rapid binding of CPB to endothelial cells and overall stronger CPB signals at the endothelial lining of vessels. Our results suggest that undamaged intestinal epithelium acts as a barrier against tissue penetration of CPB. To further evaluate the potential toxic effect of CPB on porcine small intestinal epithelial cells, we investigated the effect of *C. perfringens* type C culture supernatants and purified rCPB on polarized IPEC-J2 cells. These cells form continuous and tight polarized epithelial layers when grown on permeable filters [17] and have frequently been used as a model to investigate the interaction between the porcine intestinal epithelium and bacterial pathogens [19,20]. Apical and basolateral exposure of these cells to medium containing 32 μg/mL rCPB, a concentration which is approximately 2000-fold higher than the reported toxicity to primary porcine endothelial cells [11,12], did not affect the TEER of IPEC-J2 cell layers. This indicates that CPB alone was not toxic to the polarized epithelium. Additionally, we could not detect any toxicity on co-cultivated endothelial cells in the basolateral compartment when IPEC-J2 cells were exposed apically to rCPB, despite the fact that the toxin in the apical chamber was still active on endothelial cells at the end of the experiment. Additionally, rCPB did not affect primary porcine intestinal epithelial cells.

In contrast to purified rCPB, the supernatants of two *C. perfringens* type C strains, one NCTC reference strain and a pathogenic porcine isolate, disrupted the epithelial IPEC-J2 layer when incubated at the apical and basolateral side. The supernatants were generated in TGY bacterial culture medium under conditions optimized for CPB production and were used at the maximum concentration which allowed exclusion of adverse effect by the bacterial culture medium itself. They contained 7.5/1.4 (apical/basolateral exposure to NCTC 3180) or 2.2/0.88 (apical/basolateral exposure to JF 3721) µg/mL CPB, concentrations that were 150–500 fold higher than the toxic concentration for endothelial cells [11]. The same supernatants, albeit at lower concentrations, were also toxic to primary porcine small intestinal epithelial cells. Importantly, the toxic effect of type C supernatants was not reduced when they were pre-incubated with neutralizing anti-CPB antibodies. This indicates that CPB was not responsible for the cytopathic effects.

From our results, we conclude that CPB is non-toxic to polarized IPEC-J2 and primary porcine jejunal epithelial cells at concentrations which are regularly secreted by *C. perfringens* type C isolates under normal growth conditions during late log phase.

In addition, our results show that CPB by itself cannot pass the intact IPEC-J2 polarized epithelial layer under the culture conditions used in our experiments. Together with our data showing decreased tissue penetration in unaltered mucosal explants, this suggests that CPB by itself does not readily pass the intact small intestinal epithelial layer. The most likely explanation for these observations is that intestinal epithelial cells are inert to CPB and that the toxin cannot pass the epithelium either by the paracellular route, which would require damage to tight junctions, or transcellularly. It should be mentioned that the IPEC-J2 cell line grown on Transwells® are unlikely to fully match all physiological properties of the porcine neonatal jejunal epithelium. For example, the small intestinal epithelium of newborn piglets also contains vacuolated fetal-type enterocytes (VFE), which enable the transfer of intestinal contents across the epithelium [21]. Therefore, for example trans- or para-cellular transport of CPB occurring *in vivo* at the jejunal epithelial barrier in newborn piglets cannot be excluded by our results.

Our results further indicate that other secreted factors can induce cytopathic effects in intestinal epithelial cell cultures. In contrast to the effects in cell culture systems, lytic effects of culture supernatants on mucosal explants were not detected. However, as this approach is limited by the onset of post-mortem autolytic effects, it is not well suited to detect such effects. The effect observed in cell culture could be due either to a direct toxicity of secreted factors to epithelial cells or the result of damage to the underlying collagen layers which were used in both cell culture systems. Given the complex pathogenesis of clostridial enteric infections, it is likely that several factors contribute to an initial epithelial damage that, in the case of *C. perfringens* type C infections, enables CPB to penetrate the epithelial barrier and act on endothelial cells. Such effects do not necessarily have to be related to direct toxicity by a particular toxin or enzyme. Recently several publications suggested synergistic effects of secreted *C. perfringens* toxins in gas gangrene [22] but also enteric infections. Verherstraeten *et al.* [23] reported on synergistic effects of PFO and CPA in bovine *C. perfringens* type A induced necrohemorrhagic enteritis and Ma *et al.* [24] demonstrated a synergistic effect of enterotoxin (CPE) and CPB from pathogenic human *C. perfringens* type C isolates in rabbit ileal loops.

Our strains did not carry the *cpe* gene; thus, CPE as a contributing factor can be excluded in our experiments.

In limited evaluations we were not able to correlate the toxic effects to the level of alpha-toxin (CPA), perfringolysin (PFO) and collagenase. We did not investigate the contribution of further toxins known to be produced by *C. perfringens* type C strains such as Tpel or beta-2-toxin. Further studies including exposure to purified toxins or enzymes and toxin gene knockout mutants of *C. perfringens* would be needed to identify particular virulence factors which are involved in intestinal epithelial damage. In addition, genome and plasmid sequencing of *C. perfringens* type C strains and analyses of culture supernatants might reveal additional virulence factors in the future.

Interestingly, purified CPB alone injected into rabbit ileal loops rapidly causes necro-hemorrhagic lesions [25] which raises the question whether CPB alone could act on epithelial cells. However, it should be taken into account that ileal loop models in any species represent an unphysiological condition in the intestine. Disruption of normal intestinal motility and passage will unequivocally lead to changes in the microflora and could have rapid functional effects on the small intestinal mucosal barrier, which might not be depicted by morphological studies. In addition, the ligations themselves will induce pressure and ischemic damage to the intestinal wall and also epithelium which, at least locally, will disrupt the epithelial barrier of the intestine. In this respect, it is noteworthy that control loops in the experiments of Ma *et al.* [24] inoculated with culture medium alone histologically also showed very minor lesions. These might be sufficient to allow CPB to pass the intestinal epithelium and target susceptible cells in the lamina propria and deeper layers of the mucosa. Another reason for the discrepancy between our results and the effects of purified CPB in rabbit intestinal loops and a recently developed mouse oral and duodenal inoculation model [26] could be different susceptibilities of cells in rabbits, mice and pigs to CPB. CPB is a beta-barrel pore forming toxin [27]. These toxins are secreted as monomers and bind to susceptible target cells via specific receptors, where they oligomerize and finally form a membrane spanning pore [27]. Cellular receptors of CPB and therefore their distribution on different cells and in different species are still unknown. Beta-toxin has been shown to form oligomeric pores in several human immune cell lines [7,8] and endothelial cells [28]. In immune cell lines and endothelial cells [13] this leads to rapid cell death. Many other cells, such as Hela-, Vero-, CHO-, MDCK-, Cos-7-, P-815, PC12 and fibroblasts cells, were reported to be insensitive to CPB [7,11,12,29,30]; however, literature comparing cellular susceptibilities to CPB is rare. The most likely explanation for this effect is the differential expression of CPB receptors on these cells. Similarly, receptors could be differentially expressed between different species and therefore the different models used to study the mode of action of CPB might not directly be comparable.

4. Conclusions

In conclusion, we confirmed that in the porcine small intestine, CPB preferentially binds to endothelial cells and that it does not appear to bind to small intestinal epithelial cells nor does it have a direct toxic effect on the porcine small intestinal epithelium. This further supports the hypothesis that local damage to the intestinal vasculature mediated by CPB is a key trigger for the development of the disease. Additional secreted factors but also intestinal environmental changes occurring during the onset of *C. perfringens* type C enteritis are potentially involved in primary epithelial damage, which is required for the diffusion of CPB through the intestinal epithelial barrier to reach its main target: endothelial cells. Together with results from recent studies by other groups, we can hypothesize on several key events in the pathogenesis of *C. perfringens* type C enteritis (Figure 6). Further research will be needed on the complex interaction of *C. perfringens* with the intestinal microenvironment, the mucosa and associated immune system, epithelium damaging factors and receptor identification of different toxins in order to gradually complete our picture of the complex pathogenesis of clostridial enteric diseases.

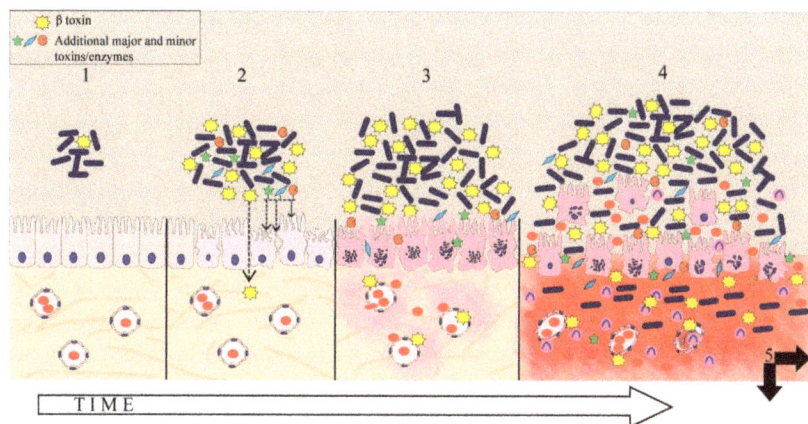

Figure 6. Hypothesis on key events of *C. perfringens* type C induced necrotizing enteritis in pigs triggered by CPB: Enteric disease most likely results from a coordinated interplay between different events and factors. (1) The disease starts with colonization and rapid proliferation of *C. perfringens* type C. (2) Initial epithelial damage/irritation could be caused by secreted toxins and/or be due to the altered luminal environment. This enables CPB to pass the epithelial barrier and diffuse into the lamina propria. 3. Due to the high susceptibility of endothelial cells to CPB, even small amounts of CPB reaching the endothelium result in local vascular leakage. (3) This increases epithelial damage and leakage of blood and tissue components into the intestinal lumen which favors clostridial growth and toxin production [31]. (4) Due to the high proliferative capacity of *C. perfringens* under these local environmental conditions, a vicious cycle of CPB induced hemorrhage, accelerated clostridial growth and toxin secretion develops. (5) This would explain the rapidly progressing hemorrhage and necrosis of the small intestine, the hallmark lesion of *C. perfringens* type C enteritis. In addition, toxins from the intestinal lumen can be absorbed into the systemic circulation and cause enterotoxemia. Toxicity of CPB to immune cells [8] could additionally contribute to the disease at various stages. Not depicted in our hypothesis model is the mucous layer on the epithelium. Currently, we have no knowledge about interactions of *C. perfringens* type C with this important protective layer in the small intestine.

5. Experimental Section

5.1. Production of C. perfringens Culture Supernatants, Recombinant CPB and Aerolysin

C. perfringens strains used were: The porcine *C. perfringens* type C strain JF 3721 [11] (genotype: *cpa*, *cpb*, *cpb2*, *pfo*, *tpel*) isolated from an outbreak of NE in 2006, the type C reference strain NCTC 3180 (sheep peritonitis isolate; genotype: *cpa*, *cpb*, *cpb2*, *pfo*, *tpel*) and the porcine *C. perfringens* type A strain JF 3693 [11] (porcine isolate; genotype *cpa*, *cpb2*, *pfo*) were grown anaerobically on blood agar plates. Overnight anaerobic cultures were produced using liquid TGY broth (trypsin glucose yeast extract broth, 3% tryptic soy broth (BD 286220), 2% D(+) glucose (Merck Millipore, Darmstadt, Germany), 1% yeast extract (BD 212750), 0.1% L-cysteine hydrochloride (Sigma-Aldrich, St. Louis, MO, USA) as described previously [11]. Aliquots of these cultures were transferred to 50 mL TGY and cultivated until they reached late-log phase. The cultures were then chilled on ice and centrifuged (Hettich, Rotanta 460 R, 2602 g, 4 °C, 20 min). Supernatants were sterile filtered using 0.22 μm Millex GV filter units (Merck Millipore, Darmstadt, Germany) and stored on ice until use in the experiments within the next three hours. The amount of CPA in the supernatant was quantified using an Alpha Toxin Elisa Kit (Bio-X Diagnostics, Jemelle, Belgium) with *C. perfringens* phospholipase C (Sigma-Aldrich, St. Louis, MO, USA) as standard. Collagenase activity was quantified using the EnzChek® Gelatinase/Collagenase Assay Kit (Molecular Probes®, Lifetechnologies™, Carlsbad, CA, USA) according to the manufacturer's recommendation. Western-blots to quantify CPB in culture supernatants were carried out as described [11]. Perfringolysin activity was measured using a dilution horse erythrocyte hemolysis assay as described previously [32,33]. Toxin concentrations of the *C. perfringens* strains were: <u>JF 3721</u>: 4.4 μg/mL CPB, 24 μU/mL CPA, 3343.56 U/mL collagenase activity, 4 PFO activity Log_2 (titer); <u>NCTC 3180</u>: 15 μg/mL CPB, 110 μU/mL CPA, 5179.04 U/mL collagenase activity, 5.75 PFO activity Log_2 (titer); <u>JF 3693</u>: 33 μU/mL CPA, 4041.69 U/mL collagenase activity, 4.25 PFO activity Log_2 (titer). Expression and purification of rCPB was performed as described [11]. Cytotoxic activity of rCPB and *C. perfringens* type C culture supernatants to primary porcine endothelial cells (PAEC) was verified before every experiment and was similar to our previous study, where diluted *C. perfringens* type C supernatants containing 15 ng/mL CPB and rCPB concentrations of 13 ng/mL were still toxic to PAEC [11]. Proaerolysin was kindly provided by F. G. van der Goot (Global Health Institute, Ecole Polytechnique Federale de Lausanne, Lausanne, Switzerland) and activated as described [34].

5.2. Jejunal Explants, Histology, Immunohistochemistry

All animal procedures were approved by the Bernese Cantonal Veterinary Office and the ethical committee of the Faculty of Veterinary Medicine, Ghent University. Four neonatal, colostrum deprived piglets were deeply anaesthetized (Stresnil® 0.05 mL/kg, Sanochemia Pharmazeutika AG, Neufeld, Austria; Ketanerkon® 0.15 mL/kg, Streuli Pharma AG, Uznach, Switzerland; Morphasol 0.03 mL/kg, Dr. E. Gräub AG, Bern, Switzerland) and euthanized by exsanguination. The jejunum was immediately harvested and transferred to PBS containing 1× antibiotic/antimycotic (Ab/Am,

Gibco®, Lifetechnologies™, Carlsbad, CA, USA). Sections of 10 cm were opened and flushed several times. The samples were transferred to RPMI 1640 (Gibco®, Lifetechnologies™, Carlsbad, CA, USA; Catalog number 21875) containing 1 × Ab/Am and 150 μg/mL trypsin inhibitor (TI, Trypsin inhibitor from Glycine max (soybean), Sigma-Aldrich, St. Louis, MO, USA). Intestinal samples were prepared in three different ways: Explants remained unmodified (unmodified), the epithelial layer was damaged by scraping with a scalpel blade (damaged epithelium), or the tunica serosa and tunica muscularis were mechanically detached by pulling with forceps, with only the submucosa and mucosa remaining (detached mucosa). The explants were then cut in approximately 1 cm sections and transferred to 6-well-plates. Each well contained 6 mL of *C. perfringens* culture supernatant diluted 1:10 in RPMI (1 × Ab/Am, 150 μg/mL TI). TGY diluted 1:10 in the same medium and medium without additives were used as controls. The explants were incubated at a temperature of 37 °C. At different time points (1, 2, 3, 4, 5, 6 h) explants were placed in 10% buffered formalin and fixed for 24 h at room temperature (RT). Tissue sections were then embedded in paraffin and routinely processed for histology, cut into 5 μm sections and stained with hematoxylin and eosin (HE). Immunohistochemical analyses were performed using monoclonal anti-CPB antibodies (mAb-CPB 10 A2, Centre for Veterinary Biologics, Ames, IA, USA) as described previously [15]. Each experiment using tissue explants was performed independently three times on tissue from different piglets. Binding of CPB to endothelial cells within villi or the submucosa and/or epithelial cells was determined by light microscopy and graded according to the following grading scheme: 0 = no signal, 1 = weak signal, only a small number of vessels (up to 30%) were affected, 2 = moderate signal, a moderate number of vessels (30%–70% of vessels) were affected 3 = strong signal, most vessels (more than 70%) were affected.

5.3. Binding Studies in Intestinal Cryosections

Jejunal tissue of a freshly euthanized colostrum deprived neonatal piglet was flushed several times in PBS, embedded in Tissue Tek O.C.T Compound. (Sakura Finetec USA, Torrance, CA, USA) and frozen in liquid nitrogen. Blocks were stored at −80 °C and cut at a temperature of −20 °C into 4.5 μm sections using a cryomicrotome. Tissue sections were mounted on Superfrost® plus glass slides (Thermo Scientific, Waltham, MA, USA) and were stored at −20 °C. Tissue sections were incubated for 15 h at 4 °C with supernatant of *C. perfringens* strains NCTC 3180, JF 3721, JF 3693 diluted 1:10 in PBS. No fixation was performed prior to immunostaining. Sections were washed with PBS and immunohistochemical detection of CPB was performed using the same protocol as for paraffin sections [15]. The experiment was performed independently twice on tissue sections from two different animals.

5.4. Antibody Neutralization Experiments

C. perfringens type C culture supernatants were pre-incubated with neutralizing mAb-CPB for 1 h as previously described [11].

5.5. Cell Cultures

Primary porcine aortic endothelial cells (PAEC) were obtained and grown as described [11]. The intestinal porcine jejunal cell line (IPEC-J2) [19] was grown in cell culture medium (DMEM/Ham's F12, Gibco; 2× Insulin-Transferrin-Selenium (Gibco®, Lifetechnologies™, Carlsbad, CA, USA); 5% fetal calf serum (FCS); 1 × Ab/Am; 1 × L-Glutamine (Gibco®, Lifetechnologies™, Carlsbad, CA, USA). For the isolation of primary porcine intestinal epithelial cells 20 cm segments of jejunum were obtained from a freshly euthanized piglet and transferred to a petri dish containing Hanks' balanced salt solution (HBSS, Gibco; 5 × Ab/Am). Segments were rinsed with HBSS 5 × Ab/Am and opened longitudinally. The mucosa was mechanically detached from the tunica muscularis and tunica serosa and cut into 3 mm pieces. These were washed with HBSS 5 × Ab/Am, transferred to 15 mL HBSS 1 × Ab/Am containing 10 μg/mL Collagenase/Dispase (Roche Diagnostics GmbH, Mannheim, Germany) and incubated for 60 min at 37 °C under gentle agitation. Finally, the tissue was dissociated by incubation in trypsin EDTA, sieved using BD Falcon® cell strainers (70 μm, Fisher Scientific, Lucens, Switzerland, REF 352350) and centrifuged (131 g, RT, 15 min). The cell pellets were resuspended in medium (DMEM; 5% FCS; 1× Insulin-Transferrin-Selenium; 5 ng/mL epidermal growth factor (EGF, Invitrogen™, Lifetechnologies™, Carlsbad, CA, USA); 1 × Ab/Am) and seeded at a density of $3 \times 10^4/cm^2$ onto collagen coated (4 μg collagen (bovine collagen 1, Cultrex®, Trevigen, Gaithersburg, Maryland, USA)/cm^2) 6-well plates. Medium was replaced after 24 h and then every 3 days. Initially isolated cells reached confluency after 7–14 days. Epithelial origin of primary cells was confirmed using immunofluorescence for cytokeratin. These primary porcine intestinal epithelial cells were then seeded on collagen coated 8-well Nun™ Lab Tek™ (Thermo Scientific, Waltham, MA, USA), grown for 5 days until confluency and fixed with acetone for 10 min at −20 °C. Cells were incubated 1 h RT with monoclonal anti-cytokeratin 18 antibody (Sigma-Aldrich, St. Louis, MO, USA) followed by AlexaFluor goat-anti mouse IgG (Lifetechnologies™, Carlsbad, CA, USA). All washing steps were performed with PBS pH 7.5, three times for 5 min.

5.6. Transepithelial Resistance Measurements

IPEC-J2 were seeded onto Transwell® (Sigma-Aldrich, St. Louis, MO, USA) membranes (6.5 mm Transwell®-COL Collagen-Coated 0.4 pore PTFE Membrane Insert) with a density of 10^5/Transwell® and kept in culture for 8–14 days, until trans epithelial electrical resistance (TEER) reached a minimum of 2 kΩ/cm^2. These TEER values are indicative of a continuous epithelial layer with tight junction formation [35]. TEER was measured using the EVOM^2 Voltohmmeter (World Precision Instruments, Sarasota, FL, USA). Diluted rCPB, *C. perfringens* culture supernatants or control medium was added either in the apical (containing 200 μL fluid) or the basolateral chamber (containing 1 mL fluid) of the Transwell®. *C. perfringens* culture supernatants diluted 1:2 for apical incubation and 1:5 or 1:10 for basal incubation in cell culture medium were used. Higher concentrations of supernatants were not used because at these concentrations non-inoculated (sterile) TGY started to affect TEER values. Confluent monolayers of PAEC grown on fibronectin (1 μg fibronectin (Fibronectin, plasma, purified human, Calbiochem)/cm^2) coated glass cover slips were placed in the basolateral compartment (24-well cell culture plates) of Transwell® for co-cultivation experiments. To test for residual CPB activity in medium from the apical chamber at the end of the experiments, twofold dilution steps of the medium in this chamber were added to confluent PAEC grown in 96-well plates. Cytopathic effects in PAEC were evaluated after 24 h of incubation by light microscopy.

5.7. Cytotoxicity in Primary Porcine Jejunal Epithelial Cells

For detection of cytopathic effects in primary intestinal epithelial cells, two-fold dilution steps of rCPB and supernatants of *C. perfringens* strains were added to confluent monolayers grown in collagen coated 96-well plates. Cytopathic effects were evaluated after 24 h of incubation by light microscopy.

5.8. Statistical Analyses

Statistical analyses were carried out using NCSS 9 Data software (NCSS, LLC, Kaysville, UT, USA). Multi-way ANOVA was performed to assess differences in signal intensity between the three different preparation groups (unmodified, damaged epithelium, detached mucosa) of the explant study. Values were corrected for time and animal, at different locations of the endothelium (*i.e.*, in villi and submucosa) and epithelium separately. Identification of different groups was done via Tukey-Kramer multiple comparison tests. Since not all model assumptions were met (e.g., the dependent variable was not truly continuous), a multivariable logistic regression analysis was performed to confirm results of the previous analyses. Therefore, the dependent variable was recoded into two categories (no to weak signal (0,1); moderate to strong signal (2,3)) whereas the same independent variables were used. Differences in signal strength (binary variable) between the binding sites endothelium-villi and endothelium-submucosa for unmodified explants were assessed via chi^2 test. Kruskal-Wallis tests were performed to assess differences in TEER values between the different incubation groups for each point in time separately. Dunn's Z multiple comparison tests were used to identify significant differences between different incubation groups.

Supplementary Materials

Supplementary materials can be accessed at: http://www.mdpi.com/2072-6651/7/4/1235/s1.

Acknowledgments

Anti-CPB antibodies were kindly provided by G. Shrinivas (USDA, Ames, IA, USA), pro-aerolysin was kindly provided by G. van der Goot (Ecole Polytechnique de Lausanne). This study was supported by a Short International Visit Grant (HP) from the Swiss National Science Foundation (IZK0Z3_133953), a visiting foreign researcher grant from Ghent University (FVI, HP), a Swiss government scholarship (SR) and a grant from the Specialization Committee (Spezko) of the Vetsuisse Faculty. We thank Kerry Woods for English language corrections.

188

Author Contributions

S.R., M.W., A.C., N.R., H.P. conducted experiments; H.P. planned study; H.P. and F.V.I. planned explant experiments; C.N. performed statistical analyses.

Conflicts of Interest

The authors declare no conflict of interest.

References

1. Songer, J.G. Clostridial enteric diseases of domestic animals. *Clin. Microbiol. Rev.* **1996**, *9*, 216–234.
2. Petit, L.; Gibert, M.; Popoff, M.R. *Clostridium perfringens*: Toxinotype and genotype. *Trends Microbiol.* **1999**, *7*, 104–110.
3. Amimoto, K.; Noro, T.; Oishi, E.; Shimizu, M. A novel toxin homologous to large clostridial cytotoxins found in culture supernatant of *Clostridium perfringens* type C. *Microbiology* **2007**, *153*, 1198–1206.
4. Popoff, M.R.; Bouvet, P. Clostridial toxins. *Future Microbiol.* **2009**, *4*, 1021–1064.
5. Jäggi, M.; Wollschlager, N.; Abril, C.; Albini, S.; Brachelente, C.; Wyder, M.; Posthaus, H. Retrospective study on necrotizing enteritis in piglets in switzerland. *Schweiz. Arch. Tierheilkd.* **2009**, *151*, 369–375.
6. Sayeed, S.; Uzal, F.A.; Fisher, D.J.; Saputo, J.; Vidal, J.E.; Chen, Y.; Gupta, P.; Rood, J.I.; McClane, B.A. Beta toxin is essential for the intestinal virulence of *Clostridium perfringens* type C disease isolate cn3685 in a rabbit ileal loop model. *Mol. Microbiol.* **2008**, *67*, 15–30.
7. Nagahama, M.; Hayashi, S.; Morimitsu, S.; Sakurai, J. Biological activities and pore formation of *Clostridium perfringens* beta toxin in hl 60 cells. *J. Biol. Chem.* **2003**, *278*, 36934–36941.
8. Nagahama, M.; Shibutani, M.; Seike, S.; Yonezaki, M.; Takagishi, T.; Oda, M.; Kobayashi, K.; Sakurai, J. The p38 mapk and jnk pathways protect host cells against *Clostridium perfringens* beta-toxin. *Inf. Immun.* **2013**, *81*, 3703–3708.
9. Uzal, F.A.; McClane, B.A. Recent progress in understanding the pathogenesis of *Clostridium perfringens* type C infections. *Vet. Microbiol.* **2011**, *153*, 37–43.
10. Los, F.C.; Randis, T.M.; Aroian, R.V.; Ratner, A.J. Role of pore-forming toxins in bacterial infectious diseases. *Microbiol .Mol. Biol. Rev.* **2013**, *77*, 173–207.
11. Gurtner, C.; Popescu, F.; Wyder, M.; Sutter, E.; Zeeh, F.; Frey, J.; von Schubert, C.; Posthaus, H. Rapid cytopathic effects of *Clostridium perfringens* beta-toxin on porcine endothelial cells. *Inf. Immun.* **2010**, *78*, 2966–2973.
12. Popescu, F.; Wyder, M.; Gurtner, C.; Frey, J.; Cooke, R.A.; Greenhill, A.R.; Posthaus, H. Susceptibility of primary human endothelial cells to *C. perfringens* beta-toxin suggesting similar pathogenesis in human and porcine necrotizing enteritis. *Vet. Microbiol.* **2011**, *153*, 173–177.
13. Autheman, D.; Wyder, M.; Popoff, M.; D'Herde, K.; Christen, S.; Posthaus, H. *Clostridium perfringens* beta-toxin induces necrostatin-inhibitable, calpain-dependent necrosis in primary porcine endothelial cells. *PLoS One* **2013**, *8*, e64644.

14. Miclard, J.; Jaggi, M.; Sutter, E.; Wyder, M.; Grabscheid, B.; Posthaus, H. *Clostridium perfringens* β-toxin targets endothelial cells in necrotizing enteritis in piglets. *Vet. Microbiol.* **2009**, *137*, 320–325.

15. Miclard, J.; van Baarlen, J.; Wyder, M.; Grabscheid, B.; Posthaus, H. *Clostridium perfringens* beta-toxin binding to vascular endothelial cells in a human case of enteritis necroticans. *J. Med. Microbiol.* **2009**, *58*, 826–828.

16. Schumacher, V.L.; Martel, A.; Pasmans, F.; van Immerseel, F.; Posthaus, H. Endothelial binding of beta toxin to small intestinal mucosal endothelial cells in early stages of experimentally induced *Clostridium perfringens* type C enteritis in pigs. *Vet. Pathol.* **2013**, *50*, 626–629.

17. Schierack, P.; Nordhoff, M.; Pollmann, M.; Weyrauch, K.D.; Amasheh, S.; Lodemann, U.; Jores, J.; Tachu, B.; Kleta, S.; Blikslager, A.; *et al.* Characterization of a porcine intestinal epithelial cell line for *in vitro* studies of microbial pathogenesis in swine. *Histochem. Cell Biol.* **2006**, *125*, 293–305.

18. Abrami, L.; Fivaz, M.; Glauser, P.E.; Sugimoto, N.; Zurzolo, C.; van der Goot, F.G. Sensitivity of polarized epithelial cells to the pore-forming toxin aerolysin. *Inf. Immun.* **2003**, *71*, 739–746.

19. Boyen, F.; Pasmans, F.; van Immerseel, F.; Donne, E.; Morgan, E.; Ducatelle, R.; Haesebrouck, F. Porcine *in vitro* and *in vivo* models to assess the virulence of salmonella enterica serovar typhimurium for pigs. *Lab. Anim.* **2009**, *43*, 46–52.

20. Brosnahan, A.J.; Brown, D.R. Porcine ipec-j2 intestinal epithelial cells in microbiological investigations. *Vet. Microbiol.* **2012**, *156*, 229–237.

21. Skrzypek, T.; Valverde Piedra, J.L.; Skrzypek, H.; Kazimierczak, W.; Biernat, M.; Zabielski, R. Gradual disappearance of vacuolated enterocytes in the small intestine of neonatal piglets. *J. Physiol. Pharmacol.* **2007**, *58*, 87–95.

22. Awad, M.M.; Ellemor, D.M.; Boyd, R.L.; Emmins, J.J.; Rood, J.I. Synergistic effects of alpha-toxin and Perfringolysin O in *Clostridium perfringens*-mediated gas gangrene. *Inf. Immun.* **2001**, *69*, 7904–7910.

23. Verherstraeten, S.; Goossens, E.; Valgaeren, B.; Pardon, B.; Timbermont, L.; Vermeulen, K.; Schauvliege, S.; Haesebrouck, F.; Ducatelle, R.; Deprez, P.; *et al.* The synergistic necrohemorrhagic action of *Clostridium perfringens* perfringolysin and alpha toxin in the bovine intestine and against bovine endothelial cells. *Vet. Res.* **2013**, *44*, 45.

24. Ma, M.; Gurjar, A.; Theoret, J.R.; Garcia, J.P.; Beingesser, J.; Freedman, J.C.; Fisher, D.J.; McClane, B.A.; Uzal, F.A. Synergistic effects of *Clostridium perfringens* enterotoxin and beta toxin in rabbit small intestinal loops. *Inf. Immun.* **2014**, *82*, 2958–2970.

25. Vidal, J.E.; McClane, B.A.; Saputo, J.; Parker, J.; Uzal, F.A. Effects of *Clostridium perfringens* beta-toxin on the rabbit small intestine and colon. *Inf. Immun.* **2008**, *76*, 4396–4404.

26. Uzal, F.A.; Saputo, J.; Sayeed, S.; Vidal, J.E.; Fisher, D.J.; Poon, R.; Adams, V.; Fernandez-Miyakawa, M.E.; Rood, J.I.; McClane, B.A. Development and application of new mouse models to study the pathogenesis of *Clostridium perfringens* type C enterotoxemias. *Inf. Immun.* **2009**, *77*, 5291–5299.

27. Popoff, M.R. Clostridial pore-forming toxins: Powerful virulence factors. *Anaerobe* **2014**, *30*, 220–238.

28. Steinthorsdottir, V.; Halldorsson, H.; Andresson, O.S. *Clostridium perfringens* beta-toxin forms multimeric transmembrane pores in human endothelial cells. *Microb. Pathog.* **2000**, *28*, 45–50.

29. Manich, M.; Knapp, O.; Gibert, M.; Maier, E.; Jolivet-Reynaud, C.; Geny, B.; Benz, R.; Popoff, M.R. *Clostridium perfringens* delta toxin is sequence related to beta toxin, netb, and staphylococcus pore-forming toxins, but shows functional differences. *PLoS One* **2008**, *3*, e3764.

30. Shatursky, O.; Bayles, R.; Rogers, M.; Jost, B.H.; Songer, J.G.; Tweten, R.K. *Clostridium perfringens* beta-toxin forms potential-dependent, cation-selective channels in lipid bilayers. *Inf. Immun.* **2000**, *68*, 5546–5551.

31. Vidal, J.E.; Ohtani, K.; Shimizu, T.; McClane, B.A. Contact with enterocyte-like caco-2 cells induces rapid upregulation of toxin production by *Clostridium perfringens* type C isolates. *Cell. Microbiol.* **2009**, *11*, 1306–1328.

32. Fisher, D.J.; Fernandez-Miyakawa, M.E.; Sayeed, S.; Poon, R.; Adams, V.; Rood, J.I.; Uzal, F.A.; McClane, B.A. Dissecting the contributions of *Clostridium perfringens* type C toxins to lethality in the mouse intravenous injection model. *Inf. Immun.* **2006**, *74*, 5200–5210.

33. Stevens, D.L.; Mitten, J.; Henry, C. Effects of alpha and theta toxins from *Clostridium perfringens* on human polymorphonuclear leukocytes. *J. Inf. Dis.* **1987**, *156*, 324–333.

34. Krause, K.H.; Fivaz, M.; Monod, A.; van der Goot, F.G. Aerolysin induces g-protein activation and Ca^{2+} release from intracellular stores in human granulocytes. *J. Biol. Chem.* **1998**, *273*, 18122–18129.

35. Fromter, E.; Diamond, J. Route of passive ion permeation in epithelia. *Nat. New Biol.* **1972**, *235*, 9–13.

Clostridium and *Bacillus* Binary Enterotoxins: Bad for the Bowels, and Eukaryotic Being

Bradley G. Stiles, Kisha Pradhan, Jodie M. Fleming, Ramar Perumal Samy, Holger Barth and Michel R. Popoff

Abstract: Some pathogenic spore-forming bacilli employ a binary protein mechanism for intoxicating the intestinal tracts of insects, animals, and humans. These Gram-positive bacteria and their toxins include *Clostridium botulinum* (C2 toxin), *Clostridium difficile* (*C. difficile* toxin or CDT), *Clostridium perfringens* (ι-toxin and binary enterotoxin, or BEC), *Clostridium spiroforme* (*C. spiroforme* toxin or CST), as well as *Bacillus cereus* (vegetative insecticidal protein or VIP). These gut-acting proteins form an AB complex composed of ADP-ribosyl transferase (A) and cell-binding (B) components that intoxicate cells via receptor-mediated endocytosis and endosomal trafficking. Once inside the cytosol, the A components inhibit normal cell functions by mono-ADP-ribosylation of globular actin, which induces cytoskeletal disarray and death. Important aspects of each bacterium and binary enterotoxin will be highlighted in this review, with particular focus upon the disease process involving the biochemistry and modes of action for each toxin.

Reprinted from *Toxins*. Cite as: Stiles, B.G.; Pradhan, K.; Fleming, J.M.; Samy, R.P.; Barth, H.; Popoff, M.R. *Clostridium* and *Bacillus* Binary Enterotoxins: Bad for the Bowels, and Eukaryotic Being. *Toxins* **2014**, *6*, 2626-2656.

1. Introduction

The *Clostridium* and *Bacillus* genera represent ubiquitous bacilli commonly found in soil, water, and gastrointestinal tracts of insects and animals, as well as humans. Both genera grow in low-oxygen environments; however, the clostridia are better adapted for anaerobic life with varying aerotolerance among different species. Pathogenic *Clostridium* and *Bacillus* species have developed unique mechanisms for survival within and outside of numerous host types, as evidenced by the various diseases frequently linked to their protein toxins and spores that include gas gangrene, food poisoning, antibiotic-associated diarrhea, pseudomembranous colitis, and enterotoxemia [1–5]. As subsequently described, a select group of bacterial binary enterotoxins can play pivotal roles in diverse diseases which also further accentuates the differences existing within this toxin family. The similarities, and dissimilarities, among these protein toxins suggest interesting evolutionary routes employed by some pathogenic *Clostridium* and *Bacillus* species. Common themes for these bacterial binary enterotoxins are: (1) the A and B components are secreted from the bacterium as separate proteins (not a holotoxin); and (2) enzymatic modification of globular (G) actin that destroys the filamentous (F) actin-based cytoskeleton and ultimately the intoxicated cell [6].

2. Pathogenic Bacilli and Binary Enterotoxins: Some of the Basics

The protein components of *C. botulinum* C2 toxin [7], *C. difficile* toxin (CDT) [8], *C. perfringens* ι-toxin and binary enterotoxin (BEC) [9,10], *C. spiroforme* toxin (CST) [11] as well as *B. cereus* vegetative insecticidal protein (VIP) [12] are produced as separate A and B molecules not associated in solution. Table 1 lists the gene locations and molecular weights of these toxin components.

Table 1. *Clostridium* and *Bacillus* binary enterotoxins and components.

Toxin Components	Gene Location	Protein M_r (kDa)
C. perfringens type E ι-toxin	140 kb plasmid [13]	
Ia		45 [13]
Ib		94 precursor [13]
		81 activated [13]
C. perfringens type A BEC	54.5 kb plasmid [10]	
BECa		47 [10]
BECb		80 [10]
C. spiroforme CST	chromosome [14]	
Sa		44 [11,14]
Sb		92 precursor [11,14]
		76 activated [11,14]
C. difficile CDT	chromosome [8]	
CDTa		48 [8]
CDTb		99 precursor [8]
		75 activated [8]
C. botulinum types C and D	chromosome [15] or	
C2	107 kb plasmid [16]	
C2I		49 [17]
C2II		81 precursor [15]
		60 activated [18]
B. cereus VIP	chromosome [19]	
VIP2		52 [20]
VIP1		100 precursor [20]
		80 activated [20]

The cell-binding components are enzymatically inert (as ascertained by existing assays) and produced as precursor molecules activated by various serine-type proteases like chymotrypsin or trypsin derived from the bacterium, host, or exogenous addition *in vitro* [21,22]. Loss of an N-terminal peptide (~20 kDa) from B precursor evidently causes conformational changes that facilitate oligomerization and subsequent docking with an A component(s).

The *Clostridium* and *Bacillus* binary enterotoxins are encoded by plasmid or chromosome-based genes with 27%–31% G + C content [23]. As just one specific example, the ι-toxin, there are two open reading frames with 243 non-coding nucleotides that separate the Ia and Ib genes. Mature Ia and Ib respectively consist of 400 and 664 amino acids [13]. The A and B components of *Clostridium* and *Bacillus* binary enterotoxins, except those for C2 or the recently described BEC, are respectively synthesized with a signal peptide of 29–49 and 39–47 residues [23]. The C2 and BEC toxins are uniquely linked to sporulation and released into the environment following sporangium lysis, thus obviating the need for a signal peptide and secretion [10,24]. It remains a curious mystery as to why similar, intestinal-acting toxins like the bacterial binary enterotoxins portrayed in this review are produced under quite different conditions (sporulation *versus* vegetative growth) by the same genus (*Clostridium*). Sequence identities between the sporulation-linked C2 and BEC components are only 29% and 41% for their A and B components, respectively [10].

Further comparisons of amino acid sequences among the *Clostridium* and *Bacillus* binary enterotoxin components reveal common evolutionary paths, as they share: (1) 80%–85% identity within the ι-toxin family; (2) 31%–40% identity between C2 and ι-family (ι, CDT, CST) toxins; and (3) 29%–31% identity between VIP and equivalent clostridial toxin components, which overall suggests that these toxin genes were derived from a common ancestor. Although unproven, it is plausible that the binary enterotoxin genes originated from an ancestral *Clostridium* and were horizontally transferred between *Bacillus* and *Clostridium* species via plasmids capable of inserting them into the bacterial chromosome, as evidenced by the CDT, CST, and C2 toxin genes. In fact, plasmid-borne genes for the ι and C2 toxins are flanked by insertion sequences [13,16,23]. In contrast, BEC appears unique and not simply a variant of these other binary toxins [10].

2.1. Clostridium perfringens: ι-Toxin and Binary Enterotoxin (BEC)

C. perfringens was first discovered in 1891 and consists of five serotypes (A–E), classically based upon four lethal, dermonecrotic toxins (α, β, ε and ι) neutralized by type-specific antiserum in animal assays [2,25–27]. Although not part of the typing scheme, sporulation-linked enterotoxin (*C. perfringens* enterotoxin or CPE) is also mouse lethal, causes erythema in guinea pigs, and linked to a major form of food poisoning found throughout the world [28,29]. Genetic methods involving multiplex PCR are now more commonly used than animal assays by many diagnostic laboratories for toxin typing of *C. perfringens* isolates [30–33].

The ι-toxin was initially described in 1943 by Bosworth [34], and its binary nature elucidated in the mid-1980s by exploiting cross-reaction and neutralization with *C. spiroforme* antiserum [9,35]. ι-toxin consists of iota a (Ia) and iota b (Ib). Individually, Ia or Ib are considered relatively nontoxic but together form a potent cytotoxin lethal to mice and dermonecrotic in guinea pigs [9,35,36]. Ia is an ADP-ribosyltransferase using nicotinamide adenine dinucleotide (NAD) to mono-ADP-ribosylate arginine [37], specifically R177, on muscle and non-muscle types of G-actin [6,38,39]. Ib, which lacks any discernible enzymatic activity, binds to a cell-surface protein(s) and subsequently translocates Ia into the cytosol of a targeted cell via acidified endosomes [40–44].

The ι-toxin is exclusively produced by type E strains and implicated in sporadic diarrheic outbreaks among calves as well as lambs [2,4,34,45–47]. Like the other binary enterotoxins described in this review, ι-toxin requires proteolytic activation as first described by Ross *et al.* in 1949 [21,47]. It was subsequently discovered, after cloning and sequencing of the ι-toxin gene, that proteolytic activation of Ib protomer (Ibp) into Ib occurs at A211 [13] which then facilitates Ia docking [42], formation of temperature- and voltage- dependent, cation-selective channels (K^+ efflux, Na^+ influx) [48,49], as well as SDS-stable heptamers on cell membranes [49,50] and in solution [40,49]. Further studies in artificial membranes and cells focused upon Ib-induced channels show: (1) 6-fold higher permeability for K^+ *versus* Na^+; (2) blockage by Ia only at pH $\leq$ 5.6; (3) decreased pH (to 4.6) does not open Ib channels, and in fact closes them (*i.e.*, pH 3.7 shuts 50% of the channels); (4) that like the C2II component of C2 toxin, Ib-generated channels conduct various quaternary ammonium ions; and (5) that chloroquine does not block ι-induced channels, in contrast to those formed by C2II [48,51]. Ib heptamers generated in solution do not induce K$^+$ release and are readily digested by pronase after binding to Vero (African Green Monkey kidney) cells at 37 °C, unlike Ib heptamers that form on the cell surface [49]. Like C2II, Vero cell-bound Ibp is not subsequently activated over time or after incubation with an excess of trypsin or chymotrypsin [50]. To date, extensive proteolytic activation studies similar to those for C2II and Ib have not been conducted with the other *Clostridium* and *Bacillus* binary enterotoxins.

For unknown reasons, pepsin, proteinase K, subtilisin or thermolysin activate Ibp more efficiently in solution than V8 protease, thrombin or even trypsin [21]. The zinc-dependent, lambda protease produced by some strains of *C. perfringens* also effectively activates ι-toxin, as well as the protoxin form of epsilon [52]. This makes sense as older cultures of *C. perfringens* type E generate proteolytically-activated ι-toxin [9,35,47]. Furthermore, the Ia molecule is also proteolytically activated by these same enzymes with a resultant loss of 9–13 amino acids (*N*-terminus), but it is still uncertain whether proteolysis of Ia increases: (1) docking efficiency to cell-bound Ib; (2) translocation efficiency into the cytosol; and/or (3) ADP-ribosyltransferase activity [21]. Proteolytic activation of Ia is seemingly unique among the *Clostridium* and *Bacillus* binary enterotoxins. It is noteworthy that amongst another family of AB toxins composed of heterologous proteins that form holotoxins in solution, such as *Escherichia coli* heat labile, *Shigella dysenteriae* shiga, and *Vibrio cholerae* cholera enterotoxins, the enzymatic A components are also processed by serine-type proteases. The difference with Ia is that these other A components form A$_1$ and A$_2$ subunits linked by a disulfide bond, subsequently reduced within the target cell's endoplasmic reticulum [53–57].

Intriguingly, a study by Nagahama *et al.* suggests that dose-dependent binding of only Ib ($\geq$100 ng/mL) to A431 (human epithelial) and A549 (human lung) cells (eight different lines tested) can cause rapid loss of ATP and cytotoxicity [58]. This is an interesting twist from the *Clostridium* and *Bacillus* binary enterotoxin paradigm not involving lipid-raft based oligomerization, unlike that previously described for ι-toxin by different groups using different cell lines [59,60]. It is possible that similar analysis with other binary enterotoxins portrayed in this review might yield equivalent results, thus promoting new ways of thinking.

In addition to the ι-toxin, Yonogi *et al.* recently describe a novel, binary enterotoxin of *C. perfringens* (BEC) produced by different type A isolates implicated in two food-borne gastroenteritis outbreaks in Japan [10]. This report reveals two components for BEC [BECa (47 kDa) and BECb (80 kDa)] which share no sequence similarity with the single-chain CPE, yet like CPE, are produced during sporulation. Cultural conditions for BEC production involve Duncan-Strong medium used specifically for sporulation of *C. perfringens*, in which these genetically-distinct outbreak isolates were CPE negative. Crude culture supernatants from either BEC-producing isolate cause fluid accumulation in rabbit intestinal loops and suckling mice.

BEC is coded on a large plasmid containing 55 open reading frames (ORFs), in which 39 ORFs encode for proteins of unknown function. A limited screen of 36 other *C. perfringens* isolates (human intestine) reveals only one that harbors the genes for BEC, suggesting minimal prevalence throughout nature, to date. There is respectively 36%–43% and 28%–44% amino acid sequence identity of BECb and BECa with complimentary components of other *Clostridium* and *Bacillus* binary enterotoxins. The sequence data suggest BEC to be a novel toxin, and not a variant of those previously described in the literature. Although biological activity on Vero cells and in suckling mice is optimal when recombinantly-produced and purified BECa and BECb are combined, BECb alone (≥ 1 µg) can cause fluid accumulation in mice. Furthermore, enterotoxic activity of culture supernatant from a parent strain is knocked out by targeting the *becB* gene. BEC, like all other *Clostridium* and *Bacillus* binary enterotoxins presented in this review, modifies G-actin via ADP-ribosylation using NAD as substrate.

2.2. Clostridium spiroforme Toxin (CST)

Like *C. perfringens*, the distinctly-coiled *C. spiroforme* also causes diarrhea (spontaneous or antibiotic-induced) but only in rabbits [1,61–68]. Although *C. spiroforme* was first isolated from human feces, correlation with human disease has not been definitively proven [69]. Rabbits are most susceptible to *C. spiroforme*-induced diarrhea during stressful periods that include lactation, old age, weaning, and an altered diet [64]. The Sa and Sb components of CST are respectively analogous to Ia and Ib of the ι-toxin, as first determined by crossed-immunoelectrophoresis and neutralization studies with *C. perfringens* type E antiserum [1,11,35,70].

During the late 1970s it was thought that *C. perfringens* was causing colony outbreaks because type E antiserum neutralized the cytotoxic effects of cecal contents from diarrheic rabbits [62,71–74]. *C. perfringens* type E was however not isolated, and in 1983 a strong correlation was established between the presence of disease and *C. spiroforme* [1,62,64]. A clever selection process for clostridial spores was employed by Borriello and Carman involving heat (80 °C/10 min) or ethanol (50%/1 h at room temperature) treatment of cecal contents [1]. PCR-based detection methods for *C. spiroforme* (16S rRNA) as well as the toxin (Sa and Sb) genes have now been published [75]. A vaccine consisting of formalin-toxoided, *C. spiroforme* culture supernatant has been used experimentally but subsequent efforts are lacking in the literature [76]. The importance of vaccine development is further suggested, as antimicrobials used for treating rabbit colonies during a *C. spiroforme* outbreak have become less effective against this pathogen [77].

2.3. Clostridium difficile Toxin (CDT)

C. difficile was first described in 1935 and considered normal, healthy intestinal flora in infants [78]. Relative to *C. perfringens* and *C. botulinum* outlined in this review, the toxin-linked pathogenicity of *C. difficile* was determined relatively recent (~35 years ago, like that for *C. spiroforme*). In the United States, *C. difficile* infections cost the healthcare industry and patients billions of dollars every year [79]. The rise of *C. difficile* as a nosocomial and community-acquired pathogen can be attributed to multiple factors that include smoking, hospital design, highly dynamic bacterial genome, long-term care residency, advancing age (>65 years, involving decreased immunity and increased visits to healthcare facilities), as well as use of proton-pump inhibitors and antibiotics [79–81]. In fact, *C. difficile* is attributed to more nosocomial infections than the highly-heralded strains of methicillin-resistant *Staphylococcus aureus* [82]. Treatment options are few but include ironically antibiotics (*i.e.*, vancomycin, metronidazole, or more recently fidaxomicin) plus experimental methods such as fecal flora transplant as well as intravenous immunoglobulins and vaccines that target *C. difficile* toxins [83–85].

C. difficile produces various toxins (*i.e.*, toxins A and B) in addition to CDT, and represents a major cause of enterocolitis (*C. difficile* associated diarrhea, or CDAD) often nosocomially acquired following use of antibiotics [5,8,79,82,83,86]. It is possible for *C. difficile* to be pathogenic with only CDT, but not toxins A and B, further suggesting the pathogenic potential of CDT in humans [87]. However, in a common hamster model for *C. difficile* colitis, a toxins A/B negative-CDT positive strain colonizes but does not cause disease [88]. A pseudomembrane consisting of white blood cells, fibrin, mucin, dead cells (bacterial and host), as well as viable bacteria is the hallmark of severe disease caused by *C. difficile*. Recurring bouts of *C. difficile* colitis are particularly problematic for some unfortunate patients. There is a correlation between the presence of CDT (CDTa and CDTb) genes and increased disease severity (*i.e.*, mortality) elicited by different "epidemic" strains (*i.e.*, PCR ribotypes 023, 027, and 078) of *C. difficile* found throughout the world [86,89,90].

Besides humans, animals involved in food production (cattle, chickens, pigs, rabbits, sheep), wildlife (elephant), and even pets (cats, dogs) can be colonized by *C. difficile* and its transmission likely goes either human to animal or in reverse [91–93]. Furthermore, commercially available meats and vegetables can harbor *C. difficile* and thus represent other environmental sources for human colonization [94–96]. It is not only food, but water contaminated by *C. difficile* spores is also a concern before and after treatment [97,98]. Clearly, exposure to *C. difficile* can naturally occur throughout our everyday existence. One important issue not resolved is relative infectious dose of *C. difficile* spores needed to elicit human disease which hinges upon aforementioned factors, and likely others yet unknown.

To further understand relative amounts of CDT produced by *C. difficile* in the gastrointestinal tract and during culture, a monoclonal antibody (Mab)-based ELISA has been developed for detecting CDTb [99]. The findings suggest that *in vivo* production of CDT is ~20-fold higher in human feces *versus in vitro* culture, thus highlighting obvious (and unknown) differences existing between the intestinal tract and a broth tube. Furthermore, like ι-toxin and CST, CDT also shares much structural similarity (*i.e.*, 80% and 82% amino acid sequence identity of CDTa (48 kDa) and CDTb (75 kDa) with *C. perfringens* Ia and Ib, respectively) [8,13,22,23,100]. These three toxins represent the ι-family that does not include *C. perfringens* BEC, *C. botulinum* C2 toxin, or *B. cereus* VIP.

Additional structural commonalities of CDT with other ι-family toxins are highlighted by interchangeable protein components that result in biologically-active chimeras [11,22,100,101]. Interestingly, *C. difficile*, *C. perfringens* and *C. spiroforme* are all associated with gastrointestinal diseases in humans and/or animals [1,4,5,34,86,102], and the synthesis of common binary enterotoxins with interchangeable protein components likely reveals a shared evolutionary path for these ubiquitous pathogens in a common niche. Although not described in the literature, it is plausible that co-colonization by two binary enterotoxin-producing species could result in chimeric-toxin induced damage to the gastric epithelium *in vivo*.

Regarding CDT prevalence among hospital/patient isolates in the United Kingdom and United States, *C. difficile* strains analyzed in the early 2000s respectively revealed only 6% and 16% containing both CDTa and CDTb genes [5,86,103,104]. Barbut and colleagues found an 11% prevalence rate of CDT-producing strains in a French hospital from 2000–2004 [89]. In another study from the Netherlands (2005), a survey of 17 hospitals revealed a higher (30%) presence of both CDT genes among *C. difficile* isolates [105]. Furthermore, this same study revealed that 36% of patients with CDAD were community (not nosocomial) acquired cases. A very recent Romanian study of CDAD cases at one hospital in Bucharest (2011–2012) discovered that 69% of *C. difficile* isolates were CDT positive [106]. Finally, increasing prevalence of CDT-positive strains has been noted over time in Italy [107]. Overall, these and other studies suggest a disturbing trend revealing the importance of CDT in *C. difficile* pathogenesis and increasing prevalence around the world.

2.4. Clostridium botulinum C2 Toxin

C. botulinum (*Bacillus botulinus*) was first described in 1895 following a food poisoning incident in Belgium [108]. Like *C. perfringens*, the neurotoxin types (A-G) of *C. botulinum* are classically determined by mouse lethal assays with toxin-specific antisera [4,26,109]. In contrast to the classic botulinum neurotoxins, the C2 toxin produced by types C and D lacks neurotoxicity but induces vascular permeability, necrotic-hemorrhagic lesions, as well as a lethal fluid accumulation in lungs and intestinal tracts of various animals [7,110–118]. C2 toxin is synthesized by *C. botulinum* during sporulation [24] and incorporated into the spore coat [119], which is akin to CPE and BEC of *C. perfringens* [10,28,120]. Pioneering work on the cell binding and translocation component (C2II), as well as enzyme component (C2I), of C2 toxin was initiated in the late 1970s. This effort by Dr. Ohishi's laboratory was the first describing protein synergy employed by any *Clostridium* or *Bacillus* binary enterotoxin.

Trypsin activation of the 81-kDa C2II precursor into C2IIa (60 kDa) occurs between K181 and A182 [18], generating stable C2IIa homoheptamers in solution [121]. The C2IIa complex mediates biological effects on cells, in conjunction with an ADP-ribosyltransferase (C2I), that involve the formation of ion-permeable channels in lipid membranes [122]. Electron microscopy of C2IIa oligomers on lipid bilayers reveals annular heptameric structures with inner and outer diameters of 20–40 and 110–130 angstroms, respectively [121]. Although C2II precursor binds to cells, it will not dock with C2I or facilitate cytotoxicity [123,124].

There are intriguing physical (molecular weights and epitopes), as well as functional (cytotoxicity), variations between C2I and C2II components produced by different *C. botulinum* strains [125,126] that perhaps is not surprising from an evolutionary perspective. Comparable structural and functional data are lacking for the other *Clostridium* and *Bacillus* binary enterotoxins. Finally, following earlier reports that C2I possesses ADP-ribosyltransferase activity specific for arginine [127], the intracellular substrate of C2 toxin was identified in 1986 as actin via modification of R177 [128–130]. These results represent a ground-breaking discovery of a new family of bacterial ADP-ribosylating proteins that target the actin cytoskeleton.

2.5. Bacillus cereus Vegetative Insecticidal Proteins (VIPs)

In contrast to the aforementioned clostridial binary enterotoxins, the binary-based *B. cereus* VIPs target corn pests (*i.e.*, Northern and Western corn rootworms) but neither animals nor humans [3,12,131]. Additionally, lepidopterans are evidently not affected by the binary VIPs [12]. The *B. cereus* VIPs are composed of VIP1 (~86 kDa cell-binding component) and VIP2 (~54 kDa ADP-ribosyltransferase that targets actin) produced during the growth, not sporulation, phase [12,19]. VIP1 and VIP2 do not share structural similarity with VIP3 produced by *Bacillus thuringiensis*, a protein that does kill various lepidopteran species evidently through pore formation in the midgut epithelium [132]. In addition to its insect killing properties, *B. cereus* can cause human food poisoning via other protein toxins [3,133] and is considered a nonlethal intestinal symbiont of various soil-dwelling insects such as roaches, sow bugs, and termites [134].

Overall, the prevalence of binary enterotoxins in two different genera and diverse species suggests a shared evolutionary success. The striking similarities, and dissimilarities, in structure and function of these *Clostridium* and *Bacillus* proteins will now be addressed below.

3. Protein Structure and Function

Crystal structures are available for the A components of ι, CDT, C2, and VIP toxins (Figure 1). All of these enzymes contain two domains, in which the *N*-terminus promotes interactions with the B component and the *C*-terminus possesses enzymatic activity. Use of PDBeFold (Protein Data Bank in Europe, version 2.59) reveals that the secondary structure of Ia, *versus* CDTa, C2I and VIP2, is respectively 93%, 69% and 82% matched along with Z scores of 19.8, 12.4 and 11.9. In contrast, for the B components there is only a low quality structure available for C2II [135]. Additionally, various studies have further investigated the structure and function of the C2, ι, and more recently CDT toxins by employing various techniques as follows.

3.1. C. botulinum C2II and C2I

Like the other *Clostridium* and *Bacillus* binary enterotoxins in this review, the *C*-terminus of C2II facilitates receptor-mediated binding on the cell surface, as deletion of only seven *C*-terminal residues effectively prevents C2IIa interactions with cells [18]. Antisera specific for the *C*-terminus (domain 4; residues 592–721), but not domains 1 (residues 1–264) or 3 (residues 490–592), block C2IIa binding to cells as determined by Western blots and cytotoxicity [18]. Neutralizing epitopes are sterically hindered after C2IIa-cell interactions, as addition of domain-4 specific antiserum does not afford neutralization. Deletion studies focused upon the *N*-terminus of C2II reveal that residues 1–181, lost after proteolytic activation of the C2II precursor, may be important for proper folding of the molecule [18]. Mutagenesis of the C2II gene within a conserved region of domain II (amino acids 303–331) results a protein devoid of voltage-gating, but not chloroquine binding or translocation of C2I into the cytosol [139].

Figure 1. Ribbon plots of crystal structures for the A components from ɩ ([136]; Protein Data Bank (PDB) ID = 1GIQ), *C. difficile* toxin (CDT) ([137]; PDB ID = 2WN8), C2 ([135]; PDB ID = 2J3Z), and vegetative insecticidal protein (VIP) ([12]; PDB ID = 1QS1) toxins using Chimera (version 1.9) provided by the University of California, San Francisco, CA, USA [138]. Orange = alpha helix; Purple = beta sheet.

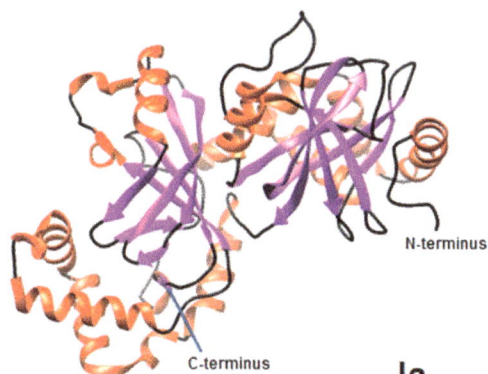

Ia

CDTa

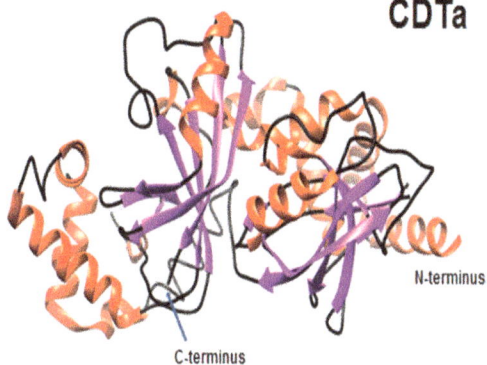

C2I

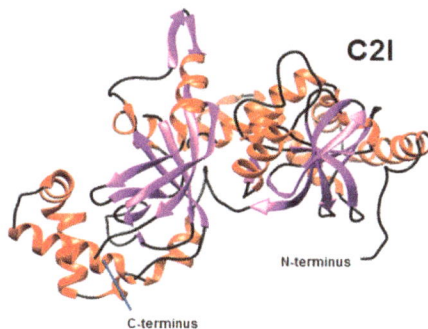

Figure 1. *Cont.*

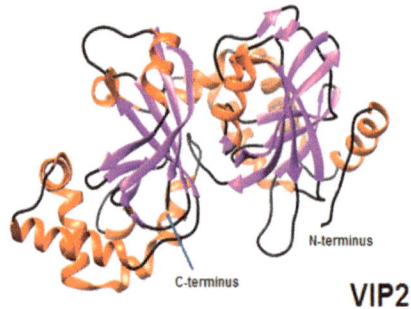

VIP2

For the C2I molecule, residues 1–87 primarily mediate binding to C2IIa heptamers and translocation across the endosomal membrane [140]. Alignment of C2I amino acids 1–225 with VIP2 residues 60–275 reveals common, surface-exposed α-helices. In particular, residues 12–29 of C2I are akin to the first α-helical structure encompassing residues 71–85 in VIP2 [12]. Further analysis of component A crystals from different *Clostridium* and *Bacillus* binary enterotoxins shows two structurally similar domains possessing the same folding patterns, perhaps a result of gene duplication (Figure 1).

X-ray crystallography of both C2 toxin components has been reported at varying resolutions, evaluating particularly the conformational effects of different pH [135]. For C2I, there are minimal effects upon structure at pH 3 *versus* 6.1, which mimics endosomal acidification necessary for translocation of C2I through a C2IIa-created pore into the cytosol. As C2I has a minimum diameter of 40 angstroms, and the inner pore diameter formed by oligomeric C2IIa is maximally 32 angstroms, there is likely unfolding of C2I facilitated by low pH and contact with C2IIa. Orientations of the *N*- and *C*-terminal domains of C2I are the same as that discovered for Ia and VIP2 [12,135,136]. For C2II there is also little change in shape at pH 4.3 and 6.0; however, the *C*-terminal domain used for binding cell-surface receptor was not readily resolved and suggests plasticity [135].

Crystallography of other bacterial ADP-ribosyltransferases like *Bordetella pertussis* pertussis toxin [141], *Corynebacterium diphtheriae* diphtheria toxin [142], *E. coli* heat-labile enterotoxin [143], *Pseudomonas aeruginosa* exotoxin A [144], as well as VIP2 [12] reveals within the *C*-terminus: (1) two antiparallel β-sheets flanked by a pair of α-helices; and (2) a highly conserved catalytic domain containing an EXE motif found in prokaryotic as well as eukaryotic ADP-ribosyltransferases [6,13,23,136,145]. In contrast, overall sequence similarity of these toxins within the *C*-terminus is low. Studies focused upon the EXE motif of *C. botulinum* C2I show that an E387Q mutation prevents ADP-ribosyltransferase, but not NAD-glycohydrolase, activity while the same alteration of E389 inhibits both [146].

3.2. C. perfringens Ib and Ia

Like C2II, four distinct domains on Ib have also been described via deletion mutagenesis and antibody studies [50,147]. For instance, cleaving just ten amino acids from the *C*-terminus (domain 4) prevents Ib binding to Vero cells, and Ib peptides containing ≥200 *C*-terminal residues are competitive inhibitors of the toxin [147]. On the other end of Ib, deletion of just 27 *N*-terminal Ib residues from domain 1 prevents Ia docking without affecting cell binding of this construct that also effectively competes with ι-toxin *in vitro* [147]. In this same study, three Mabs targeting an *N*-terminal epitope (residues 28–66) had no effect upon Ib binding or cytotoxicity. These immunoreagents might not occupy the Ib site necessary for Ia docking, or perhaps oligomerization of Ib and/or docking of Ia readily displace these antibodies. In contrast, two Mabs recognizing unique Ib epitopes within *C*-terminal residues 632–655 afford protection against ι cytotoxicity. One Mab prevents Ib binding to cells while the other does not, yet the latter efficiently prevents Ib oligomerization on the cell surface [50,147]. Results for the latter Mab further reveal the importance of Ib oligomerization on biological activity of ι-toxin, as do studies with Ibp, a molecule that remains as a cell-bound monomer [42,50]. None of these Mabs recognize Ib on the cell surface, suggesting that epitopes are not accessible after binding to receptor. Furthermore, oligomerization of Ib also does not occur on Vero cells at 4 °C, although there is binding to the cell surface, while ι-toxin resistant cells (MRC-5, human lung) bind Ib without subsequent oligomerization [42,50]. The cumulative data clearly support Ib oligomerization for biological activity of ι-toxin.

All aforementioned Mabs against Ib recognize Ibp or *C. spiroforme* Sb in an ELISA and Western blot [147]. C2II is also recognized in an ELISA by one Mab that prevents Ib binding to cells, but C2 cytotoxicity is not neutralized *in vitro*. C2II and Ib bind unique receptors via their *C*-terminus and share little sequence similarity within this region [13,15,18,42–44,147–149], thus these distinct biological characteristics make this finding of Mab cross-reactivity quite curious.

There is also a calcium binding motif (DXDXDXXXDXXE) found within the *N*-terminus of B components from these binary enterotoxins, as well as in the distantly-related protective antigen of *Bacillus anthracis* edema and lethal toxins [150,151]. The proposed role played by chelation of calcium involves maintenance of protein conformation that affects A docking, as evidenced by ι-toxin [151].

To date, there is no crystal structure for Ib but that for Ia is available [136,150,152] (Figure 1). Analysis of Ia reveals two domains that share conformational, but little sequence, similarity. Other A components in this binary-enterotoxin family have the same structure, typical of ADP-ribosyl transferases targeting actin [152]. Further highlighting structural commonalities among these proteins is the catalytic C domain of Ia (residues 211–413) [13] and VIP2 (266–461) [12] that are quite similar, possessing 40% sequence identity and equivalent surface charges. One obvious difference between Ia and VIP2 is the spatial orientation of the first glutamic acid found within the conserved catalytic motif, 378EXE380 of Ia. Like C2I [146], the first glutamic acid within the EXE motif of Ia facilitates ADP-ribosyltransferase, but not NAD-glycohydrolase, activity [153]. Further analysis of Ia reveals that R295 and E380, which are conserved residues among ADP-ribosyltransferases [154–157], are also important for Ia catalysis [145,153]. The 338STS340 motif is located near the active site of Ia and other ADP-ribosyltransferases. Alanine replacement of any residue in the STS motif of Ia, especially the first serine, decreases enzymatic activity [153]. Extensive mutagenesis studies of Ia that focus upon the NAD binding cavity reveal that Y246 and N255 are important for ADP-ribosyltransferase, but not NAD-glycohydrolase, activity unlike Y251 involvement in both [153]. The binding of actin and NAD by Ia is accomplished by five loop structures, via ionic and van der Waals interactions [152]. All ADP-ribosyltransferases within the *Clostridium* and *Bacillus* binary-enterotoxin family (BECa, C2I, CDTa, Ia, Sa, and VIP2) target G-actin, which is a commonly conserved protein found throughout nature that plays a pivotal role in the cytoskeleton and homeostasis [6,8,10,38,39,70,100,128,152,158].

3.3. C. difficile CDTa and CDTb

In contrast to the C2 and ι toxins, less biochemical work has been done with the CDT components. Crystal data at different pH (4.0, 8.5 and 9.0) reveal conformational shifts for CDTa, particularly within the active site at low pH [137]. Structure-function studies show that the same amino acids are also necessary for enzymatic activity, in which Ia shares respectively 40% and 84% overall sequence identity with C2I and CDTa [13,23,101,146]. Amino acids necessary for ADP-ribosylation are similar between Ia (R295, R296, R352, Q300, N335, E378, E380) and CDTa (R302, R303, R359, Q307, N342, E385, E387) [137,145]. NAD and NADPH, sources of ADP-ribose for enzymatic transfer to G-actin, also uniquely make direct contact with S345 in CDTa which suggests differences in substrate interactions *versus* Ia [137]. The EXE motif of ADP-ribosyltransferases, part of the ADP-ribosyl turn-turn (ARTT) loop important for stabilizing substrate-enzyme complexes, differ regarding substrate contact made by Ia and CDTa [137,152,153]. The E385 and E387 of CDTa do not make contact with NAD or NADPH, unlike the equivalent E378 and E380 of Ia [136,137]. The STS motif found in CDTa is important in ligand binding and possibly catalysis.

Unlike the *C*-terminal, enzyme-critical similarities of A components for the *Clostridium* and *Bacillus* binary enterotoxins, the *N*-terminal domains can differ. For instance, Ia (residues 1–210) and VIP2 (60–265) contain only 20% sequence identity, dissimilar surface charges, and different conformations as Ia contains an additional α helix (residues 61–66) [12,13,19,136]. Relative to enzymatic components of the other binary enterotoxins, the Ib docking region on Ia is more centrally located within the *N*-terminal domain (residues 129–257) [136,159], *versus* C2I residues 1–87 that interact with C2II (137) or CDTa residues 1–240 docking to CDTb [101]. Overall, these data reflect evolutionary variation within the *Clostridium* and *Bacillus* binary enterotoxins.

4. Intoxication Process

4.1. Toxin Binding to Cell

To access the cytosol and G-actin, the *Clostridium* and *Bacillus* binary-enterotoxin family must initiate intoxication via B components binding to a targeted cell via a receptor(s) to form a homooligomer complex. This acts as a platform for docking of A to the cell surface. To further understand the binding and oligomerization properties of B components on cell surfaces, studies have delved into the role(s) played by lipid rafts [59,60,160]. Lipid rafts are cholesterol-rich, detergent-insoluble (at 4 °C) "structures" or "microdomains" located on the cell membrane. These microdomains inadvertently serve as attachment, entry, and sometimes exit sites cleverly pirated by various bacteria, viruses, and toxins [161–164]. Results suggest that *C. perfringens* Ib, but not Ibp, localizes into these membrane microdomains on Vero cells that are susceptible to ι-toxin [59,60]. Comparable studies with C2IIa have been reported, and in fact the phosphatidylinositol 3-kinase pathway is necessary for C2 toxin internalization via lipid rafts [165]. Finally, the cell-surface receptor for CDT, CST and ι-toxin is lipolysis-stimulated lipoprotein receptor (LSR) which forms clusters in lipid rafts following binding of CDTb, precursor CDTb, or just the receptor binding domain (amino acids 677–876) of CDTb [43,160]. It appears that for ι, CDT and C2 toxins, lipid rafts are necessary for oligomer formation although monomeric B components can evidently bind to receptor located outside of these microdomains. To date, comparable studies have not been reported for the other *Clostridium* and *Bacillus* binary enterotoxins.

By using polarized CaCo-2 (human colon) cells, Blöcker *et al*. [40] discovered that the Ib receptor is primarily localized onto the basolateral membrane. Richard *et al*. [41] revealed that Ib traverses CaCo-2 cells from either the apical or basolateral surface and internalizes Ia found on the distal side, even when Ia is added 3 h after Ib. In this latter study, addition of Ib-neutralizing antiserum or Mabs with Ia on the opposite surface *versus* Ib does not affect ι cytotoxicity. Two different groups reveal that Ib rapidly binds to Vero cells at 37 °C and forms a large complex (>200 kDa) in less than 1 min that remains evident for at least 2 h [49,50]. Over time, Ib oligomers decrease on/in cells and tailing is evident in Western blots, suggesting lysosomal degradation of the oligomer into smaller protein species that appear on the plasma membrane [49,166].

These collective data [49,50] are relevant to earlier work by Sakurai and Kobayashi [36] showing that Ia injected intravenously into mice 2 h after Ib causes death, suggesting extended availability of Ib *in vitro* and *in vivo*. Furthermore, if toxin neutralizing antiserum targeting Ib is given just 5 min after an Ib + Ia injection, mice are not protected against lethality. In this same study it was also discovered that Ib given intradermally to guinea pigs, followed by intraperitoneal Ia, elicits a dermonecrotic lesion at the Ib injection site. Evidently, Ia remains in circulation for an extended period and is neutralized by antibody given 2 h after an Ia injection. In summary, Ia "locates" Ib in the body via general circulation, perhaps a common characteristic of not just ι but other *Clostridium* and *Bacillus* binary enterotoxins that could be exploited for therapeutic purposes? Comparable discoveries were reported by Simpson [118] for C2 toxin in mice and rats.

In regards to receptor binding, much work has been done in the past with C2II. The C2II precursor and proteolytically-activated C2IIa bind to intestinal cells and brush border membranes [167]; however, only C2IIa has hemagglutinating properties with human and animal erythrocytes, which is a process competitively inhibited by various sugars like *N*-acetylgalactosamine, *N*-acetylglucosamine, L-fucose, galactose, or mannose [168]. Trypsin or pronase pretreatment of human erythrocytes prevents C2IIa-induced hemagglutination, thus suggesting a glycoprotein receptor. Furthermore, chemical mutagenesis of CHO cells generated those resistant to C2 toxin because they lack *N*-acetylglucosaminyltransferase I activity necessary for synthesizing asparagine-linked carbohydrates [148,169]. These mutant cells are still susceptible to ι-toxin, providing evidence that C2IIa and Ib recognize unique receptors. Additionally, pretreatment of cells with various lectins or glycosidases does not affect Ib binding, suggesting that the receptor (or part of) is not a carbohydrate [42]. C2IIa and Ib form voltage-dependent channels in lipid membranes [48,122,139,170–172], likely employing hydrophobic and hydrophilic amino acids (*i.e.*, C2II residues 303–331) for insertion into the membrane [139].

In contrast to C2IIa, which binds and facilitates C2I-mediated cytotoxicity in all tested vertebrate cells [123,149,168,169,173,174], the cell-surface receptor for Ib is not as ubiquitous [42]. Recent studies by Papatheodorou *et al.* very nicely reveal LSR as the cell-surface receptor for the ι-family toxins (CDT, CST, and ι) [43,160,175]. LSR is a transmembrane lipoprotein found in various tissues and naturally facilitates lipoprotein clearance and tight junction formation, as well as plays a critical role in cell development [43]. Additional studies by Wigelsworth *et al.* provide evidence that CD44 also promotes intoxication of these same toxins [44]. Like LSR, the CD44 glycoprotein is also a single-pass transmembrane protein with multiple functions. CD44 acts as a receptor for multiple ligands, transduces signals, and is exploited by certain bacteria and viruses for cell entry. These independent discoveries of LSR and CD44 were made using different methodologies that respectively include single-gene knockouts of human, near-haploid leukemia cells (HAP1) and proteomic analysis of lipid rafts from Ib-treated *versus* untreated Vero cells [43,44,176]. How LSR and CD44 might interact to affect ι intoxication remains unresolved to date.

Moreover, it was recently shown in cancerous breast epithelial cells that reduced LSR levels significantly decrease ι-toxin sensitivity while overexpression of CD44 conveys toxin resistance [177]. Overexpression of CD44 in LSR-expressing cell lines correlates with decreased, toxin-stimulated formation of lysosomes and cytosolic levels of ι-toxin. These data suggest that CD44 drives ι-toxin resistance through inhibition of endocytosis in cancerous breast epithelial cells and highlights the importance of cell-type specificity during intoxication. Discovery of LSR and CD44 involvement in ι-family intoxications provides invaluable insight towards potential therapeutics targeting: (1) the cell-surface interactions of these toxins; and (2) breast cancer cells. To our knowledge, receptor-binding studies for *B. cereus* VIP1 have not been reported in the literature.

4.2. A Docking to B and Internalization

N-terminal domains within the A and B components of each *Clostridium* and *Bacillus* binary enterotoxin are intimately involved in forming an AB complex on the cell surface. After binding to a surface receptor, intracellular-acting bacterial toxins use two major pathways for gaining entry into the cytosol. There is retrograde routing through the Golgi apparatus and endoplasmic reticulum, employed by *S. dysenteriae* shiga [53,178] and *V. cholerae* cholera [56,179] toxins, which is inhibited by brefeldin A that subsequently causes protein accumulation within the endoplasmic reticulum [180]. The other route exploited by bacterial toxins involves translocation from acidified early endosomes into the cytosol, like that employed by diphtheria toxin [181], *B. anthracis* edema/lethal toxins [182], and the *Clostridium/Bacillus* binary enterotoxins described in this review. Subsequent transport of vesicles from early to late endosomes involves microtubules depolymerized by nocodazole, which inhibits trafficking into late endosomes [40,121,183,184]. Neither brefeldin A nor nocodazole influence the biological activity of C2 or ι toxins on cells. However, translocation of A components for C2, ι, or the edema and lethal toxins across the endosomal membrane is blocked by bafilomycin A, which inhibits vacuolar-type ATPases that acidify the endosome [40,121,185,186]. Decreased pH evidently induces conformational changes that promote membrane insertion of the heptameric B complex, followed by translocation of the A component(s) across the endosomal membrane. This is a process mimicked on the cell surface by simply lowering media pH [40,51,121,172,181,185,187]. It is not clear if B heptamers of these *Clostridium* and *Bacillus* binary enterotoxins enter the cytosol with the A components or remain attached to the endosomal membrane, possibly recycling to the cell surface in a degraded form [41,124,166].

Figure 2. Basic model showing cell-surface binding and internalization of *Clostridium* and *Bacillus* binary enterotoxins.

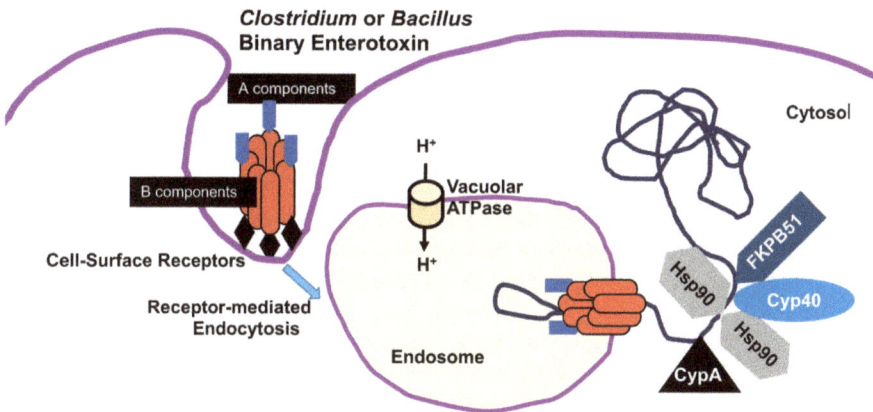

To leave the endosome and enter the cytosol, A components of C2, CDT, and ι toxins traverse the endosomal membrane via chaperones such as heat shock protein 90 (Hsp90) and protein-folding enzymes that include the peptidyl-prolyl *cis/trans* isomerases cyclophilin A [188–190], cyclophilin 40 [191], and FK506 binding protein (FKBP) 51 [192]. Hsp90 is a highly conserved ATPase produced by all eukaryotic cells that provides an essential housekeeping role by regulating various proteins associated with cell signaling [193]. Inhibitors of Hsp90 (geldanamycin, radicicol), cyclophilins (cyclosporine A), and FKBPs (FK506) effectively delay C2-, CDT-, or ι-induced cytotoxicity because they block the pH-dependent translocation of A components into the cytosol (Figure 2). The inhibitors have no effects upon ADP-ribosyltransferase activity, binding to cell-surface receptor(s), or endocytosis.

Although unproven, the A components of *Clostridium* and *Bacillus* binary enterotoxins likely unfold and thread through toxin-generated channels in the membrane into the cytosol as proposed for diphtheria toxin [194]. Ratts *et al.* [195] report that Hsp90 and thioredoxin reductase, found in a cytosolic complex, are both required to transport diphtheria toxin from the endosome. Intriguingly different is that geldanamycin and radicicol are necessary for inhibiting diphtheria cytotoxicity, whereas either drug alone inhibits CDT, C2, or ι cytotoxicity. Thioredoxin reductase might cleave the disulfide bond between the A and B chains of diphtheria toxin, like that for the *C. tetani* tetanus toxin and *C. botulinum* neurotoxin A [196]. However, disulfide bonds and reductive activation have never been described for any of the *Clostridium* and *Bacillus* binary enterotoxins.

4.3. ADP-Ribosylation: Destruction of the Actin Cytoskeleton, Intoxicated Cell, and Perhaps the Host

Once inside the cytosol, the A component can mono-ADP-ribosylate G-actin that subsequently disrupts F-actin formation and the cytoskeleton. The basic mechanism of ADP-ribosylation employed by BEC, C2, CDT, CST, ι and VIP toxins is remarkably conserved by diverse bacteria from many different genera. All known ADP-ribosylating toxins use NAD as a source of ADP-ribose. There are at least four bacterial groups of ADP-ribosylating toxins based upon intracellular targets, that include: (1) elongation factor 2 modified via histidine variant (diphthamide) by diphtheria toxin and exotoxin A through an *N*- and *C*-terminal active site, respectively; (2) heterotrimeric G-proteins modified via cysteine by pertussis toxin, or arginine by *E. coli* heat labile enterotoxin and cholera toxin through an *N*-terminal active site; (3) Rho and Ras GTPases respectively modified via asparagine by *C. botulinum* C3 exoenzyme and arginine by *P. aeruginosa* exoenzyme S through a *C*-terminal active site; and (4) G-actin modified via arginine through a *C*-terminal active site. This last group includes *B. cereus* VIP [12,19,131], *C. botulinum* C2 toxin [128–130,197], and the ι-toxin family of *C. difficile* CDT [100,101], *C. perfringens* ι-toxin [38,39,136,152,156,198] as well as *C. spiroforme* CST [70,199]. From a sequence similarity perspective, *C. perfringens* BEC appears to be distinct from all other *Clostridium/Bacillus* binary enterotoxins and the actin residue modified has not been identified to date [10].

Actin (~42 kDa, G monomer) is found in all eukaryotic cells and structurally conserved between diverse species [200,201]. Many important eukaryotic efforts depend upon actin and include maintenance of cell structure, homeostasis, as well as the immune system. However, some pathogenic viruses and bacteria exploit the actin cytoskeleton for cell entry, intra- and inter-cellular movement, and in the case of *Clostridium* and *Bacillus* binary enterotoxins, targeting of the actin cytoskeleton causing cell death [6,201].

The actin-ADP-ribosylating enterotoxins of *Clostridium* and *Bacillus* species can be subdivided into two groups. C2 toxin, the first bacterial toxin discovered to mono-ADP-ribosylate actin, exclusively modifies R177 on β/γ-nonmuscle, as well as γ-smooth muscle, isoforms [6,128–130,202–206]. The ι-family toxins are less discriminating and mono-ADP-ribosylate R177 found on all G-actin isoforms, that includes skeletal muscle [39,206]. Perhaps varying substrate specificities lie in CDTa, Ia, and Sa having an actin-binding sequence of LKDKE *versus* LKTKE for C2I [6,8,13,23,152,207]. Modification of actin by *C. perfringens* BEC likely occurs at R177, but this remains unconfirmed along with recognition of different actin isoforms [10]. F-actin does not represent a substrate target for any of these bacterial binary toxins, but ADP-ribosylation of G-actin inhibits assembly into F-actin strands and cytoskeletal development [6,129,136,197,202–204,208,209]. From a bacterium's perspective for survival, disruption of a eukaryote's cytoskeleton can prevent phagocytosis [210] and intracellular trafficking of vesicles necessary for homeostasis. Furthermore, the ι, CDT and C2 toxins induce microtubule-based protrusions from epithelial cells in just 90 min after toxin exposure [211,212]. Length of these cellular extensions is concentration dependent, as lower amounts of toxin cause longer protrusions. These actin-free protrusions respectively promote *C. difficile* adherence ~5- and 4-fold to intoxicated Caco-2 cells and intestinal tract of gnotobiotic mice, probably slowing bacterial elimination from the intestinal lumen as well as increasing toxin-based damage to the mucosa [211,212]. In particular, CDT reroutes vesicles containing fibronectin from the basolateral membrane to apical protrusions that ultimately promote bacterial adherence [212]. This was evident on cultured cells treated with CDT and in murine intestines (cecum and colon epithelium) following infection by CDT-producing *C. difficile* (ribotype 027). Correspondingly, intracellular levels of fibronectin are decreased in Caco-2 cells following CDT or C2 toxin exposure. The authors reveal that secretion of fibronectin from an intoxicated cell is dependent upon matrix metalloproteases and calcium signaling [212].

Ultimately, a susceptible cell exposed to a sufficient dose of *Clostridium* or *Bacillus* binary enterotoxin dies and releases valuable nutrients into the environment, that become readily available for bacterial consumption. From a research perspective, toxins that modify actin are invaluable tools for studying the cytoskeleton and numerous cell processes during both homeostasis and disease progression.

5. Conclusions

As presented in this review, the enteric-targeting binary toxins from various *Clostridium* and *Bacillus* species possess unique characteristics yet commonly target the cytoskeleton, specifically actin. Historically, the discovery of *C. perfringens* ι-toxin in 1943 was the first for any *Clostridium* or *Bacillus* binary enterotoxin [34]. Subsequently, the multi-component nature of *C. botulinum* C2 toxin, *C. perfringens* ι-toxin, *C. spiroforme* CST, *C. difficile* CDT, *B. cereus* VIP, and *C. perfringens* BEC were respectively elucidated in 1980 [7], 1986 [9,35], 1988 [70], 1997 [8], 1999 [12], and 2014 [10]. There have been many subsequent studies exploring the biochemistry and biological effects for some of these toxins. Clearly there is more work to be done, particularly involving the evolution and therapeutic potential of these enterotoxins.

Further exploration by researchers employing gene probes and specific toxin antibodies will likely unveil new binary toxins produced by other bacteria, and perhaps those from different genera. As one example of very recent discovery, *C. perfringens* BEC was isolated in Japan from human food-poisoning strains of type A [10]. This toxin, although employing a basic binary mechanism, is also structurally distinct from all of the other *Clostridium* and *Bacillus* binary enterotoxins described in this review. Further study will perhaps unveil intriguing insight involving the evolutionary paths taken by binary enterotoxin-producing *Clostridium* and *Bacillus* species. Finally, it is evident that a knowledge-based understanding of the past will foster additional creative efforts, by various international groups, targeting these fascinating proteins.

Acknowledgments

Kisha Pradhan and Bradley G. Stiles thank Wilson College for generous use of facilities that include computers, copiers, and library services.

Author Contributions

Bradley G. Stiles and Kisha Pradhan wrote the manuscript and provided figures. Jodie M. Fleming and Ramar Perumal Samy reviewed the manuscript and provided writing material via ideas. Holger Barth and Michel R. Popoff reviewed the manuscript, provided ideas, and facilitated figure formation.

Conflicts of Interest

The authors declare no conflict of interest.

References

1. Borriello, S.P.; Carman, R.J. Association of iota-like toxin and *Clostridium spiroforme* with both spontaneous and antibiotic-associated diarrhea and colitis in rabbits. *J. Clin. Microbiol.* **1983**, *17*, 414–418.
2. McDonel, J.L. Toxins of *Clostridium perfringens* types A, B, C, D and E. In *Pharmacology of Bacterial Toxins*; Dorner, F., Drews, J., Eds.; Pergamon Press: New York, NY, USA, 1986; pp. 477–517.
3. McKillip, J.L. Prevalence and expression of enterotoxins in *Bacillus cereus* and other *Bacillus* spp., a literature review. *Antonie Van Leeuwenhoek* **2000**, *77*, 393–399.
4. Songer, J.G. Clostridial enteric diseases of domestic animals. *Clin. Microbiol. Rev.* **1996**, *9*, 216–234.
5. Stoddart, B.; Wilcox, M.H. *Clostridium difficile*. *Curr. Opin. Infect. Dis.* **2002**, *15*, 513–518.
6. Aktories, K.; Schwan, C.; Papatheodorou, P.; Lang, A.E. Bidirectional attack on the actin cytoskeleton. Bacterial protein toxins causing polymerization or depolymerization of actin. *Toxicon* **2012**, *60*, 572–581.
7. Ohishi, I.; Iwasaki, M.; Sakaguchi, G. Purification and characterization of two components of botulinum C2 toxin. *Infect. Immun.* **1980**, *30*, 668–673.
8. Perelle, S.; Gibert, M.; Bourlioux, P.; Corthier, G.; Popoff, M.R. Production of a complete binary toxin (actin-specific ADP-ribosyltransferase) by *Clostridium difficile* CD196. *Infect. Immun.* **1997**, *65*, 1402–1407.
9. Stiles, B.G.; Wilkins, T.D. Purification and characterization of *Clostridium perfringens* iota toxin: Dependence on two nonlinked proteins for biological activity. *Infect. Immun.* **1986**, *54*, 683–688.
10. Yonogi, S.; Matsuda, S.; Kawai, T.; Yoda, T.; Harada, T.; Kumeda, Y.; Gotoh, K.; Hiyoshi, H.; Nakamura, S.; Kodama, T.; *et al.* BEC, a novel enterotoxin of *Clostridium perfringens* found in human clinical isolates from acute gastroenteritis outbreaks. *Infect. Immun.* **2014**, *82*, 2390–2399.
11. Popoff, M.R.; Milward, F.W.; Bancillon, B.; Boquet, P. Purification of the *Clostridium spiroforme* binary toxin and activity of the toxin on HEp-2 cells. *Infect. Immun.* **1989**, *57*, 2462–2469.

12. Han, S.; Craig, J.A.; Putnam, C.D.; Carozzi, N.B.; Tainer, J.A. Evolution and mechanism from structures of an ADP-ribosylating toxin and NAD complex. *Nature Struct. Biol.* **1999**, *6*, 932–936.

13. Perelle, S.; Gibert, M.; Boquet, P.; Popoff, M.R. Characterization of *Clostridium perfringens* iota toxin genes and expression in *Escherichia coli*. *Infect. Immun.* **1993**, *61*, 5147–5156.

14. Gibert, M.; Perelle, S.; Daube, G.; Popoff, M.R. *Clostridium spiroforme* toxin genes are related to *C. perfringens* iota toxin genes but have a different genomic localization. *Syst. Appl. Microbiol.* **1997**, *20*, 337–347.

15. Kimura, K.; Kubota, T.; Ohishi, I.; Isogai, E.; Isogai, H.; Fujii, N. The gene for component-II of botulinum C2 toxin. *Vet. Microbiol.* **1998**, *62*, 27–34.

16. Sakaguchi, Y.; Hayashi, T.; Yamamoto, Y.; Nakayama, K.; Zhang, K.; Ma, S.; Arimitsu, H.; Oguma, K. Molecular analysis of an extrachromosomal element containing the C2 toxin gene discovered in *Clostridium botulinum* type C. *J. Bacteriol.* **2009**, *191*, 3282–3291.

17. Fujii, N.; Kubota, T.; Shirakawa, S.; Kimura, K.; Ohishi, I.; Moriishi, K.; Isogai, E.; Isogai, H. Characterization of component-I gene of botulinum C2 toxin and PCR detection of its gene in clostridial species. *Biochem. Biophys. Res. Commun.* **1996**, *220*, 353–359.

18. Blöcker, D.; Barth, H.; Maier, E.; Benz, R.; Barbieri, J.T.; Aktories, K. The *C*-terminus of component C2II of *Clostridium botulinum* C2 toxin is essential for receptor binding. *Infect. Immun.* **2000**, *68*, 4566–4573.

19. Yu, X.; Liu, T.; Liang, X.; Tang, C.; Zhu, J.; Wang, S.; Li, S.; Deng, Q.; Wang, L.; Zheng, A.; Li, P. Rapid detection of *vip1*-type genes from *Bacillus cereus* and characterization of a novel *vip* binary toxin gene. *FEMS Microbiol. Lett.* **2011**, *325*, 30–36.

20. Warren, G.; Koziel, M.; Mullins, M.A.; Nye; Carr, B; Desai, N.; Kostichka, K.; Duck, N.; Estruch, J.J. Novel pesticidal proteins and strains. World Intellectual Property Organization. Patent WO96/10083, 1996.

21. Gibert, M.; Petit, L.; Raffestin, S.; Okabe, A.; Popoff, M.R. *Clostridium perfringens* iota toxin requires activation of both binding and enzymatic components for cytopathic activity. *Infect. Immun.* **2000**, *68*, 3848–3853.

22. Perelle, S.; Scalzo, S.; Kochi, S.; Mock, M.; Popoff, M.R. Immunological and functional comparison between *Clostridium perfringens* iota toxin, *C. spiroforme* toxin, and anthrax toxins. *FEMS Microbiol. Lett.* **1997**, *146*, 117–121.

23. Popoff, M.R. Molecular biology of actin-ADP-ribosylating toxins. In *Handbook of Experimental Pharmacology, Bacterial Protein Toxins*; Aktories, K., Just, I., Eds.; Springer-Verlag: Berlin, Germany, 2000; Volume 145, pp. 275–306.

24. Nakamura, S.; Serikawa, T.; Yamakawa, K.; Nishida, S.; Kozaki, S.; Sakaguchi, G. Sporulation and C2 toxin production by *Clostridium botulinum* type C strains producing no C1 toxin. *Microbiol. Immunol.* **1978**, *22*, 591–596.

25. Oakley, C.; Warrack, G. Routine typing of *Clostridium welchii*. *J. Hyg. Camb.* **1953**, *51*, 102–107.

26. *VPI Anaerobe Laboratory Manual*; Holdeman, L.V., Cato, E.P., Moore, W.E.C., Eds.; Southern Printing Co: Blacksburg, VA, USA, 1977; pp. 131–133.

27. Walker, P.; Batty, I.; Egerton, J. The typing of *C. perfringens* and the veterinary background. *Papua New Guinea Med. J.* **1979**, *22*, 50–56.

28. Sarker, M.R.; Singh, U.; McClane, B.A. An update on *Clostridium perfringens* enterotoxin. *J. Nat. Toxins.* **2000**, *9*, 251–266.

29. Niilo, L. Measurement of biological activities of purified and crude enterotoxin of *Clostridium perfringens*. *Infect. Immun.* **1975**, *12*, 440–442.

30. Daube, G.; Simon, P.; Limbourg, B.; Manteca, C.; Mainil, J.; Kaeckenbeeck, A. Hybridization of 2659 *Clostridium perfringens* isolates with gene probes for seven toxins (α, β, ε, ι, θ, μ and enterotoxin) and for sialidase. *Am. J. Vet. Res.* **1996**, *57*, 496–501.

31. Fach, P.; Popoff, M.R. Detection of enterotoxigenic *Clostridium perfringens* in food and fecal samples with a duplex PCR and the slide agglutination test. *Appl. Environ. Microbiol.* **1997**, *63*, 4232–4236.

32. Meer, R.R.; Songer, J.G. Multiplex polymerase chain reaction assay for genotyping *Clostridium perfringens*. *Am. J. Vet. Res.* **1997**, *58*, 702–705.

33. Uzal, F.A.; Plumb, J.J.; Blackall, L.L.; Kelly, W.R. PCR detection of *Clostridium perfringens* producing different toxins in faeces of goats. *Lett. Appl. Microbiol.* **1997**, *25*, 339–344.

34. Bosworth, T. On a new type of toxin produced by *Clostridium welchii*. *J. Comp. Path.* **1943**, *53*, 245–255.

35. Stiles, B.G.; Wilkins, T.D. *Clostridium perfringens* iota toxin: Synergism between two proteins. *Toxicon* **1986**, *24*, 767–773.

36. Sakurai, J.; Kobayashi, K. Lethal and dermonecrotic activities of *Clostridium perfringens* iota toxin: Biological activities induced by cooperation of two nonlinked components. *Microbiol. Immunol.* **1995**, *39*, 249–253.

37. Simpson, L.L.; Stiles, B.G.; Zepeda, H.H.; Wilkins, T.D. Molecular basis for the pathological actions of *Clostridium perfringens* iota toxin. *Infect. Immun.* **1987**, *55*, 118–122.

38. Schering, B.; Barmann, M.; Chhatwal, G.S.; Geipel, U.; Aktories, K. ADP-ribosylation of skeletal muscle and non-muscle actin by *Clostridium perfringens* iota toxin. *Eur. J. Biochem.* **1988**, *171*, 225–229.

39. Vandekerckhove, J.; Schering, B.; Bärmann, M.; Aktories, K. *Clostridium perfringens* iota toxin ADP-ribosylates skeletal muscle actin in Arg-177. *FEBS Lett.* **1987**, *225*, 48–52.

40. Blöcker, D.; Behelke, J.; Aktories, K.; Barth, H. Cellular uptake of the binary *Clostridium perfringens* iota toxin. *Infect. Immun.* **2001**, *69*, 2980–2987.

41. Richard, J.F.; Mainguy, G.; Gibert, M.; Marvaud, J.C.; Stiles, B.G.; Popoff, M.R. Transcytosis of iota toxin across polarized CaCo-2 cells. *Mol. Microbiol.* **2002**, *43*, 907–917.

42. Stiles, B.G.; Hale, M.L.; Marvaud, J.C.; Popoff, M.R. *Clostridium perfringens* iota toxin: Binding studies and characterization of cell surface receptor by fluorescence-activated cytometry. *Infect. Immun.* **2000**, *68*, 3475–3484.

43. Papatheodorou, P.; Carette, J.E.; Bell, G.W.; Schwan, C.; Guttenberg, G.; Brummelkamp, T.R.; Aktories, K. Lipolysis-stimulated lipoprotein receptor (LSR) is the host receptor for the binary toxin *Clostridium difficile* transferase (CDT). *Proc. Natl. Acad. Sci. USA* **2011**, *108*, 16422–16427.

44. Wigelsworth, D.J.; Ruthel, G.; Schnell, L.; Herrlich, P.; Blonder, J.; Veenstra, T.D.; Carman, R.J.; Wilkins, T.D.; van Tran Nhieu, G.; Pauillac, S.; *et al.* CD44 promotes intoxication by the clostridial iota-family toxins. *PLoS One* **2012**, *7*, e51356.

45. Billington, S.J.; Wieckowski, E.U.; Sarker, M.R.; Bueschel, D.; Songer, J.G.; McClane, B.A. *Clostridium perfringens* type E animal enteritis isolates with highly conserved, silent enterotoxin gene sequences. *Infect. Immun.* **1998**, *66*, 4531–4536.

46. Hart, B.; Hooper, P. Enterotoxaemia of calves due to *Clostridium welchii* type E. *Aust. Vet. J.* **1967**, *43*, 360–363.

47. Ross, H.E.; Warren, M.E.; Barnes, J.M. *Clostridium welchii* iota toxin: Its activation by trypsin. *J. Gen. Microbiol.* **1949**, *3*, 148–152.

48. Knapp, O.; Benz, R.; Gibert, M.; Marvaud, J.C.; Popoff, M.R. Interaction of *Clostridium perfringens* iota toxin with lipid bilayer membranes: Demonstration of channel formation by the activated binding component Ib and channel block by the enzyme component Ia. *J. Biol. Chem.* **2002**, *277*, 6143–6152.

49. Nagahama, M.; Nagayasu, K.; Kobayashi, K.; Sakurai, J. Binding component of *Clostridium perfringens* iota toxin induces endocytosis in Vero cells. *Infect. Immun.* **2002**, *70*, 1909–1914

50. Stiles, B.G.; Hale, M.L.; Marvaud, J.C.; Popoff, M.R. *Clostridium perfringens* iota toxin: Characterization of the cell-associated iota b complex. *Biochem. J.* **2002**, *367*, 801–808.

51. Simpson, L.L. The binary toxin produced by *Clostridium botulinum* enters cells by receptor-mediated endocytosis to exert its pharmacologic effects. *J. Pharmacol. Exp. Ther.* **1989**, *251*, 1223–1228.

52. Minami, J.; Katayama, S.; Matsushita, O.; Matsushita, C.; Okabe, A. Lambda-toxin of *Clostridium perfringens* activates the precursor of epsilon-toxin by releasing its *N*- and *C*-terminal peptides. *Microbiol. Immunol.* **1997**, *41*, 527–535.

53. Sandvig, K. The Shiga toxins: Properties and action on cells. In *The Comprehensive Sourcebook of Bacterial Protein Toxins*, 3rd ed.; Alouf, J.E., Popoff, M.R., Eds.; Academic Press: Amsterdam, The Netherlands, 2006; Chapter 17, pp. 310–322.

54. Garred, O.; Dubinina, E.; Polessakaya, A.; Olsnes, S.; Koslov, J.; Sandvig, K. Role of the disulfide bond in Shiga toxin A-chain for toxin entry into cells. *J. Biol. Chem.* **1997**, *272*, 11414–11419.

55. Garred, O.; van Deurs, B.; Sandvig, K. Furin-induced cleavage and activation of Shiga toxin. *J. Biol. Chem.* **1995**, *270*, 10817–10821.

56. Hirst, T.R.; D'Souza, J.M. *Vibrio cholerae* and *Escherichia coli* thermolabile enterotoxin. In *The Comprehensive Sourcebook of Bacterial Protein Toxins*, 3rd ed.; Alouf, J.E., Popoff, M.R., Eds.; Academic Press: Amsterdam, The Netherlands, 2006; Chapter 15, pp. 270–290.

57. Majoul, I.; Ferrari, D.; Soling, H.D. Reduction of protein disulfide bonds in an oxidizing environment—the disulfide bridge of cholera toxin A-subunit is reduced in the endoplasmic reticulum. *FEBS Lett.* **1997**, *401*, 104–108.

58. Nagahama, M.; Umezaki, M.; Oda, M.; Kobayashi, K.; Tone, S.; Suda, T.; Ishidoh, K.; Sakurai, J. *Clostridium perfringens* iota-toxin b induces rapid cell necrosis. *Infect. Immun.* **2011**, *79*, 4353–4360.

59. Hale, M.L.; Marvaud, J.C.; Popoff, M.R.; Stiles, B.G. Detergent-resistant membrane microdomains facilitate Ib oligomer formation and biological activity of *Clostridium perfringens* iota toxin. *Infect. Immun.* **2004**, *72*, 2186–2193.

60. Nagahama, M.; Yamaguchi, A.; Hagiyama, T.; Ohkubo, N.; Kobayashi, K.; Sakurai, J. Binding and internalization of *Clostridium perfringens* iota toxin in lipid rafts. *Infect. Immun.* **2004**, *72*, 3267–3275.

61. Borriello, S.P.; Davies, H.A.; Carman, R.J. Cellular morphology of *Clostridium spiroforme*. *Vet. Microbiol.* **1986**, *11*, 191–195.

62. Carman, R.J.; Borriello, S.P. Observations on an association between *Clostridium spiroforme* and *Clostridium perfringens* type E iota enterotoxaemia in rabbits. *Eur. J. Chemother. Antibiot.* **1982**, *2*, 143–144.

63. Carman, R.J.; Borriello, S.P. Infectious nature of *Clostridium spiroforme*-mediated rabbit enterotoxaemia. *Vet. Microbiol.* **1984**, *9*, 497–502.

64. Carman, R.J.; Evans, R.H. Experimental and spontaneous clostridial enteropathies of laboratory and free living lagomorphs. *Lab. Anim. Sci.* **1984**, *34*, 443–452.

65. Carman, R.J.; Perelle, S.; Popoff, M.R. Binary toxins from *Clostridium spiroforme* and *Clostridium perfringens*. In *The Clostridia: Molecular Biology and Pathogenesis*; Rood, J., McClane, B.A., Titball, R., Eds.; Academic Press: New York, NY, USA, 1997; Chapter 20, pp. 359–367.

66. Carman, R.J.; Wilkins, T.D. *In vitro* susceptibility of rabbit strains of *Clostridium spiroforme* to antimicrobial agents. *Vet. Microbiol.* **1991**, *28*, 391–397.

67. Peeters, J.E.; Geeroms, R.; Carman, R.J.; Wilkins, T.D. Significance of *Clostridium spiroforme* in the enteritis-complex of commercial rabbits. *Vet. Microbiol.* **1986**, *12*, 25–31.

68. Yonushonis, W.P.; Roy, M.J.; Carman, R.J.; Sims, R.E. Diagnosis of spontaneous *Clostridium spiroforme* iota enterotoxemia in a barrier rabbit breeding colony. *Lab. Anim. Sci.* **1987**, *37*, 69–71.

69. Kaneuchi, C.; Miyazato, T.; Shinjo, T.; Mitsuoka, T. Taxonomic study of helically coiled, sporeforming anaerobes isolated from the intestines of humans and other animals: *Clostridium cocleatum* sp. nov. and *Clostridium spiroforme* sp. nov. *Int. J. Syst. Bacteriol.* **1979**, *29*, 1–12.

70. Popoff, M.R.; Boquet, P. *Clostridium spiroforme* toxin is a binary toxin which ADP-ribosylates cellular actin. *Biochem. Biophys. Res. Commun.* **1988**, *152*, 1361–1368.

71. Butt, M.T.; Papendick, R.E.; Carbone, L.G.; Quimby, F.W. A cytotoxicity assay for *Clostridium spiroforme* enterotoxin in cecal fluid of rabbits. *Lab. Anim. Sci.* **1994**, *44*, 52–54.

72. Eaton, P.; Fernie, D.S. Enterotoxaemia involving *Clostridium perfringens* iota toxin in a hysterectomy-derived rabbit colony. *Lab. Anim.* **1980**, *14*, 347–351.

73. Katz, L.; Lamont, J.T.; Trier, J.S.; Sonnenblick, E.B.; Rothman, S.W.; Broitman, S.A.; Rieth, S. Experimental clindamycin associated colitis in rabbits. Evidence for toxin-mediated mucosal damage. *Gastroenterology* **1978**, *74*, 246–252.

74. Lamont, J.T.; Sonnenblick, E.B.; Rothman, S. Role of clostridial toxin in the pathogenesis of clindamycin colitis in rabbits. *Gastroenterology* **1979**, *76*, 356–361.

75. Drigo, I.; Bacchin, C.; Cocchi, M.; Bano, L.; Agnoletti, F. Development of PCR protocols for specific identification of *Clostridium spiroforme* and detection of *sas* and *sbs* genes. *Vet. Microbiol.* **2008**, *13*, 414–418.

76. Ellis, T.M.; Gregory, A.R.; Logue, G.D. Evaluation of a toxoid for protection of rabbits against enterotoxaemia experimentally induced by trypsin-activated supernatant of *Clostridium spiroforme*. *Vet. Microbiol.* **1991**, *28*, 93–102.

77. Agnoletti, F.; Ferro, T.; Guolo, A.; Marcon, B.; Cocchi, M.; Drigo, I.; Mazzolini, E.; Bano, L. A survey of *Clostridium spiroforme* antimicrobial susceptibility in rabbit breeding. *Vet. Microbiol.* **2009**, *136*, 188–191.

78. Hall, I.C.; O'Toole, E. Intestinal flora in new-born infants with a description of a new pathogenic anaerobe, *Bacillus difficilis*. *Am. J. Dis. Child.* **1935**, *49*, 390–402.

79. Jump, R.L. *Clostridium difficile* infection in older adults. *Aging Health* **2013**, *9*, 403–414.

80. He, M.; Sebaihia, M.; Lawley, T.D.; Stabler, R.A.; Dawson, L.F.; Martin, M.J.; Holt, K.E.; Seth-Smith, H.M.; Quail, M.A.; Rance, R.; *et al.* Evolutinary dynamics of *Clostridium difficile* over short and long time scales. *Proc. Natl. Acad. Sci. USA* **2010**, *107*, 7527–7532.

81. Loo, V.G.; Bourgault, A.-M.; Poirier, L.; Lamothe, F.; Michaud, S.; Turgeon, N.; Toye, B.; Beaudoin, A.; Frost, E.H.; Gilca, R.; *et al.* Host and pathogen factors for *Clostridium difficile* infection and colonization. *N. Eng. J. Med.* **2011**, *365*, 1693–1703.

82. Voelker, R. Increased *Clostridium difficile* virulence demands new treatment approach. *JAMA* **2010**, *303*, 2017–2019.

83. Tschudin-Sutter, S.; Widmer, A.F.; Perl, T.M. *Clostridium difficile*: Novel insights on an incessantly challenging disease. *Curr. Opin. Infect. Dis.* **2012**, *25*, 405–411.

84. Seekatz, A.M.; Aas, J.; Gessert, C.E.; Rubin, T.A.; Saman, D.M.; Bakken, J.S.; Young, V.B. Recovery of the gut microbiome following fecal microbiota transplantation. *mBio* **2014**, *5*, e00893–e00914.

85. Shah, N.; Shaaban, H.; Spira, R.; Slim, J.; Boghossian, J. Intravenous immunoglobulin in the treatment of severe *Clostridium difficile* colitis. *J. Glob. Infect. Dis.* **2014**, *6*, 82–85.

86. Gerding, D.N.; Johnson, S.; Rupnik, M.; Aktories, K. *Clostridium difficile* binary toxin CDT. Mechanism, epidemiology, and potential clinical importance. *Gut Microbes* **2014**, *5*, 1–13.

87. Elliott, B.; Reed, R.; Chang, B.J.; Riley, T.V. Bacteremia with a large clostridial toxin-negative, binary toxin-positive strain of *Clostridium difficile*. *Anaerobe* **2009**, *15*, 249–251.

88. Geric, B.; Carman, R.J.; Rupnik, M.; Genheimer, C.W.; Sambol, S.P.; Lyerly, D.M.; Gerding, D.N.; Johnson, S. Binary toxin-producing, large clostridial toxin-negative *Clostridium difficile* strains are enterotoxic but do not cause disease in hamsters. *J. Infect. Dis.* **2006**, *193*, 1143–1150.

89. Barbut, F.; Gariazzo, B.; Bonne, L.; Lalande, V.; Burghoffer, B.; Luiuz, R.; Petit, J.C. Clinical features of *Clostridium difficile*-associated infections and molecular characterization of strains: Results of a retrospective study, 2000–2004. *Infect. Cont. Hosp. Epidemiol.* **2007**, *28*, 131–139.

90. Bacci, S.; Molbak, K.; Kjeldsen, M.K.; Olsen, K.E. Binary toxin and death after *Clostridium difficile* infection. *Emerg. Infect. Dis.* **2011**, *17*, 976–982.

91. Keessen, E.C.; Gaastra, W.; Lipman, L.J. *Clostridium difficile* infection in humans and animals, differences and similarities. *Vet. Microbiol.* **2011**, *153*, 205–217.

92. Knight, D.R.; Riley, T.V. Prevalence of gastrointestinal *Clostridium difficile* carriage in Australian sheep and lambs. *Appl. Environ. Microbiol.* **2013**, *79*, 5689–5692.

93. Hensgens, M.P.; Keessen, E.C.; Squire, M.M.; Riley, T.V.; Koene, M.G.; de Boer, E.; Lipman, L.J.; Kuijper, E.J. European Society of Clinical Microbiology and Infectious Diseases Study Group for *Clostridium difficile*. *Clostridium difficile* infection in the community: A zoonotic disease? *Clin. Microbiol. Infect.* **2012**, *18*, 635–645.

94. Gould, L.H.; Limbago, B. *Clostridium difficile* in food and domestic animals: A new food-borne pathogen? *Clin. Infect. Dis.* **2010**, *51*, 577–582.

95. Metcalf, D.S.; Costa, M.C.; Dew, W.M.; Weese, J.S. *Clostridium difficile* in vegetables, Canada. *Lett. Appl. Microbiol.* **2010**, *51*, 600–602.

96. Harvey, R.B.; Norman, K.N.; Andrews, K.; Norby, B.; Hume, M.E.; Scanlan, C.M.; Hardin, M.D.; Scott, H.M. *Clostridium difficile* in retail meat and processing plants in Texas. *J. Vet. Diagn. Invest.* **2014**, *23*, 807–811.

97. Romano, V.; Pasquale, V.; Krovacek, K.; Mauri, F.; Demarta, A.; Dumontet, S. Toxigenic *Clostridium difficile* PCR ribotypes from wastewater treatment plants in southern Switzerland. *Appl. Environ. Microbiol.* **2012**, *78*, 6643–6646.

98. Xu, C.; Weese, J.S.; Flemming, C.; Odumeru, J.; Warriner, K. Fate of *Clostridium difficile* during wastewater treatment and incidence in Southern Ontario watersheds. *J. Appl. Microbiol.* **2014**, *117*, 891–904.

99. Carman, R.J.; Stevens, A.L.; Lyerly, M.W.; Hiltonsmith, M.F.; Stiles, B.G.; Wilkins, T.D. *Clostridium difficile* binary toxin (CDT) and diarrhea. *Anaerobe* **2011**, *17*, 161–165.

100. Popoff, M.R.; Rubin, E.J.; Gill, D.M.; Boquet, P. Actin-specific ADP-ribosyltransferase produced by a *Clostridium difficile* strain. *Infect. Immun.* **1988**, *56*, 2299–2306.

101. Gülke, I.; Pfeifer, G.; Liese, J.; Fritz, M.; Hofmann, F.; Aktories, K.; Barth, H. Characterization of the enzymatic component of the ADP-ribosyltransferase toxin CDTa from *Clostridium difficile*. *Infect. Immun.* **2001**, *69*, 6004–6011.

102. Braun, M.; Herholz, C.; Straub, R.; Choisat, B.; Frey, J.; Nicolet, J.; Kuhnert, P. Detection of the ADP-ribosyltransferase toxin gene (*cdt*A) and its activity in *Clostridium difficile* isolates from equidae. *FEMS Microbiol. Lett.* **2000**, *184*, 29–33.

103. Geric, B.; Johnson, S.; Gerding, D.N.; Grabnar, M.; Rupnik, M. Frequency of binary toxin genes among *Clostridium difficile* strains that do not produce large clostridial toxins. *J. Clin. Microbiol.* **2003**, *41*, 5227–5232.

104. Stubbs, S.; Rupnik, M.; Gibert, M.; Brazier, J.; Duerden, B.; Popoff, M. Production of actin-specific ADP-ribosyltransferase (binary toxin) by strains of *Clostridium difficile*. *FEMS Microbiol. Lett.* **2000**, *186*, 307–312.

105. Paltansing, S.; van den Berg, R.J.; Guseinova, R.A.; Visser, C.E.; van der Vorm, R.R.; Kuijper, E.J. Characteristics and incidence of *Clostridium difficile*-associated disease in The Netherlands, 2005. *Eur. Soc. Clin. Microbiol. Infect. Dis.* **2007**, *13*, 1058–1064.

106. Rafila, A.; Indra, A.; Popescu, G.A.; Wewalka, G.; Allerberger, F.; Benea, S.; Badicut, I.; Aschbacher, R.; Huhulescu, S. Occurrence of *Clostridium difficile* infections due to PCR ribotype 027 in Bucharest, Romania. *J. Infect. Dev. Ctries.* **2014**, *8*, 694–698.

107. Spigaglia, P.; Mastrantonio, P. Comparative analysis of *Clostridium difficile* clinical isolates belonging to different genetic lineages and time periods. *J. Med. Microbiol.* **2004**, *53*, 1129–1136.

108. Devriese, P.P. On the discovery of *Clostridium botulinum. J. Hist. Neurosci.* **1999**, *8*, 43–50.

109. Simpson, L.L. The origin, structure, and pharmacological activity of botulinum toxin. *Pharmacol. Rev.* **1981**, *33*, 155–187.

110. Ermert, L.; Bruckner, H.; Walmrath, D.; Grimminger, F.; Aktories, K.; Suttorp, N.; Duncker, H.R.; Seeger, W. Role of endothelial cytoskeleton in high-permeability edema due to botulinum C2 toxin in perfused rabbit lungs. *Am. J. Physiol.* **1995**, *268*, 753–761.

111. Iwasaki, M.; Ohishi, I.; Sakaguchi, G. Evidence that botulinum C2 toxin has two dissimilar components. *Infect. Immun.* **1980**, *29*, 390–394.

112. Jensen, W.I.; Duncan, R.M. The susceptibility of the mallard duck (*Anas platyrhynchos*) to *Clostridium botulinum* C2 toxin. *Jpn. J. Med. Sci. Biol.* **1980**, *33*, 81–86.

113. Kurazono, H.; Hosokawa, M.; Matsuda, H.; Sakaguchi, G. Fluid accumulation in the ligated intestinal loop and histopathological changes of the intestinal mucosa caused by *Clostridium botulinum* C2 toxin in the pheasant and chicken. *Res. Vet. Sci.* **1987**, *42*, 349–353.

114. Ohishi, I. Response of mouse intestinal loop to botulinum C2 toxin: Enterotoxic activity induced by cooperation of nonlinked protein components. *Infect. Immun.* **1983**, *40*, 691–695.

115. Ohishi, I. Lethal and vascular permeability activities of botulinum C2 toxin induced by separate injections of the two toxin components. *Infect. Immun.* **1983**, *40*, 336–339.

116. Ohishi, I.; Iwasaki, M.; Sakaguchi, G. Vascular permeability activity of botulinum C2 toxin elicited by cooperation of two dissimilar protein components. *Infect. Immun.* **1980**, *31*, 890–895.

117. Ohishi, I.; Sakaguchi, G. Oral toxicities of *Clostridium botulinum* type C and D toxins of different molecular sizes. *Infect. Immun.* **1980**, *28*, 303–309.

118. Simpson, L.L. A comparison of the pharmacological properties of *Clostridium botulinum* type C1 and C2 toxins. *J. Pharmacol. Exp. Ther.* **1982**, *223*, 695–701.

119. Yamakawa, K.; Nishida, S.; Nakamura, S. C2 toxicity in extract of *Clostridium botulinum* type C spores. *Infect. Immun.* **1983**, *41*, 858–860.

120. Frieben, W.R.; Duncan, C.L. Homology between enterotoxin protein and spore structural protein in *Clostridium perfringens* type A. *Eur. J. Biochem.* **1973**, *39*, 393–401.

121. Barth, H.; Blöcker, D.; Behlke, J.; Bergsma-Schutter, W.; Brisson, A.; Benz, R.; Aktories, K. Cellular uptake of *Clostridium botulinum* C2 toxin requires oligomerization and acidification. *J. Biol. Chem.* **2000**, *275*, 18704–18711.

122. Schmid, A.; Benz, R.; Just, I.; Aktories, K. Interaction of *Clostridium botulinum* C2 toxin with lipid bilayer membranes: Formation of cation-selective channels and inhibition of channel function by chloroquine and peptides. *J. Biol. Chem.* **1994**, *269*, 16706–16711.

123. Miyake, M.; Ohishi, I. Response of tissue-cultured cynomolgus monkey kidney cells to botulinum C2 toxin. *Microb. Pathog.* **1987**, *3*, 279–286.

124. Ohishi, I.; Yanagimoto, A. Visualizations of binding and internalization of two nonlinked protein components of botulinum C2 toxin in tissue culture cells. *Infect. Immun.* **1992**, *60*, 4648–4655.

125. Ohishi, I.; Hama, Y. Purification and characterization of heterologous component IIs of botulinum C2 toxin. *Microbiol. Immunol.* **1992**, *36*, 221–229.

126. Ohishi, I.; Okada, Y. Heterogeneities of two components of C2 toxin produced by *Clostridium botulinum* types C and D. *J. Gen. Microbiol.* **1986**, *132*, 125–131.

127. Simpson, L.L. Molecular basis for the pharmacological actions of *Clostridium botulinum* type C2 toxin. *J. Pharmacol. Exp. Ther.* **1984**, *230*, 665–669.

128. Aktories, K.; Bärmann, M.; Ohishi, I.; Tsuyama, S.; Jakobs, K.H.; Habermann, E. Botulinum C2 toxin ADP-ribosylates actin. *Nature* **1986**, *322*, 390–392.

129. Ohishi, I.; Tsuyama, S. ADP-ribosylation of nonmuscle actin with component I of C2 toxin. *Biochem. Biophys. Res. Commun.* **1986**, *136*, 802–806.

130. Aktories, K.; Ankenbauer, T.; Schering, B.; Jakobs, K.H. ADP-ribosylation of platelet actin by botulinum C2 toxin. *Eur. J. Biochem.* **1986**, *161*, 155–162.

131. Jucovic, M.; Walters, F.S.; Warren, G.W.; Palekar, N.V.; Chen, J.S. From enzyme to zymogen: Engineering Vip2, an ADP-ribosyltransferase from *Bacillus cereus*, for conditional toxicity. *Protein Eng. Des. Select.* **2008**, *21*, 631–638.

132. Fang, J.; Xu, X.; Wang, P.; Zhao, J.Z.; Shelton, A.M.; Cheng, J.; Feng, M.G.; Shen, Z. Characterization of chimeric *Bacillus thuringiensis* Vip3 toxins. *Appl. Env. Microbiol.* **2007**, *73*, 956–961.

133. Michelet, N.; Granum, P.E.; Mahillon, J. *Bacillus cereus* enterotoxins, bi- and tricomponent cytolysins, and other hemolysins. In *The Comprehensive Sourcebook of Bacterial Protein Toxins*, 3rd ed.; Alouf, J.E., Popoff, M.R., Eds.; Academic Press: Amsterdam, The Netherlands, 2006; Chapter 46, pp. 779–790.

134. Margulis, L.; Jorgensen, J.Z.; Dolan, S.; Kolchinsky, R.; Rainey, F.A.; Lo, S.C. The *Arthromitus* stage of *Bacillus cereus*: Intestinal symbionts of animals. *Proc. Natl. Acad. Sci. USA* **1998**, *95*, 1236–1241.

135. Schleberger, C.; Hochmann, H.; Barth, H.; Aktories, K.; Schulz, G.E. Structure and action of the binary C2 toxin from *Clostridium botulinum*. *J. Mol. Biol.* **2006**, *364*, 705–715.

136. Tsuge, H.; Nagahama, M.; Nishimura, H.; Hisatsune, J.; Sakaguchi, Y.; Itogawa, Y.; Katunuma, N; Sakurai, J. Crystal structure and site-directed mutagenesis of enzymatic components from *Clostridium perfringens* iota-toxin. *J. Mol. Biol.* **2003**, *325*, 471–483.

137. Sundriyal, A.; Roberts, A.K.; Shone, C.C.; Acharya, K.R. Structural basis for substrate recognition in the enzymatic component of ADP-ribosyltransferase toxin CDTa from *Clostridium difficile*. *J. Biol. Chem.* **2009**, *284*, 28713–28719.

138. Pettersen, E.F.; Goddard, T.D.; Huang, C.C.; Couch, G.S.; Greenblatt, D.M.; Meng, E.C.; Ferrin, T.E. UCSF Chimera—a visualization system for exploratory research and analysis. *J. Comput. Chem.* **2004**, *25*, 1605–1612.

139. Blöcker, D.; Bachmeyer, C.; Benz, R.; Aktories, K.; Barth, H. Channel formation by the binding component of *Clostridium botulinum* C2 toxin: Glutamate 307 of C2II affects channel properties *in vitro* and pH-dependent C2I translocation *in vivo*. *Biochemistry* **2003**, *42*, 5368–5377.

140. Barth, H.; Roebling, R.; Fritz, M.; Aktories, K. The binary *Clostridium botulinum* C2 toxin as a protein delivery system: Identification of the minimal protein region necessary for interaction of toxin components. *J. Biol. Chem.* **2002**, *277*, 5074–5081.

141. Stein, P.E.; Boodhoo, A.; Armstrong, G.D.; Cockle, S.A.; Klein, M.H.; Read, R.J. The crystal structure of pertussis toxin. *Structure* **1994**, *2*, 45–57.

142. Choe, S.; Bennett, M.J.; Fujii, G.; Curmi, P.M.; Kantardjieff, K.A.; Collier, R.J.; Eisenberg, D. The crystal structure of diphtheria toxin. *Nature* **1992**, *357*, 216–222.

143. Sixma, T.K.; Pronk, S.E.; Kalk, K.H.; Wartna, E.S.; van Zanten, B.A.; Witholt, B.; Hol, W.G. Crystal structure of a cholera toxin-related heat-labile enterotoxin from *E. coli. Nature* **1991**, *351*, 371–377.

144. Li, M.; Dyda, F.; Benhar, I.; Pastan, I.; Davies, D.R. Crystal structure of the catalytic domain of *Pseudomonas* exotoxin A complexed with a nicotinamide adenine dinucleotide analog: Implications for the activation process and for ADP ribosylation. *Proc. Natl. Acad. Sci. USA* **1995**, *93*, 6902–6906.

145. Van Damme, J.; Jung, M.; Hofmann, F.; Just, I.; Vandekerckhove, J.; Aktories, K. Analysis of the catalytic site of the actin ADP-ribosylating *Clostridium perfringens* iota toxin. *FEBS Lett.* **1996**, *380*, 291–295.

146. Barth, H.; Preiss, J.C.; Hofmann, F.; Aktories, K. Characterization of the catalytic site of the ADP-ribosyltransferase *Clostridium botulinum* C2 toxin by site-directed mutagenesis. *J. Biol. Chem.* **1998**, *273*, 29506–29511.

147. Marvaud, J.C.; Smith, T.; Hale, M.L.; Popoff, M.R.; Smith, L.A.; Stiles, B.G. *Clostridium perfringens* iota toxin: Mapping of receptor binding and Ia docking domains on Ib. *Infect. Immun.* **2001**, *69*, 2435–2441.

148. Fritz, G.; Schroeder, P.; Aktories, K. Isolation and characterization of a *Clostridium botulinum* C2 toxin-resistant cell line: Evidence for possible involvement of the cellular C2II receptor in growth regulation. *Infect. Immun.* **1995**, *63*, 2334–2340.

149. Stiles, B.G.; Blöcker, D.; Hale, M.L.; Guetthoff, M.A.; Barth, H. *Clostridium botulinum* C2 toxin: Binding studies with fluorescence-activated cytometry. *Toxicon* **2002**, *40*, 1135–1140.

150. Sakurai, J.; Nagahama, M.; Oda, M.; Tsuge, H.; Kobayashi, K. *Clostridium perfringens* iota toxin: Structure and function. *Toxins* **2009**, *1*, 208–228.

151. Kobayashi, K.; Nagahama, M.; Ohkubo, N.; Kojima, T.; Shirai, H.; Iwamoto, S.; Oda, M.; Sakurai, J. Role of Ca^{2+}-binding motif in cytotoxicity induced by *Clostridium perfringens* iota toxin. *Microb. Pathog.* **2008**, *44*, 265–270.

152. Tsuge, H.; Nagahama, M.; Oda, M.; Iwamoto, S.; Utsunomiya, H.; Marquez, V.E.; Katunuma, N.; Nishizawa, M.; Sakurai, J. Structural basis of actin recognition and arginine ADP-ribosylation by *Clostridium perfringens* iota toxin. *Proc. Natl. Acad. Sci. USA* **2008**, *105*, 7399–7404.

153. Nagahama, M.; Kihara, A.; Miyawaki, T.; Mukai, M.; Sakaguchi, Y.; Ochi, S.; Sakurai, J. Characterization of the enzymatic component of *Clostridium perfringens* iota toxin. *J. Bacteriol.* **2000**, *183*, 2096–2103.

154. Carroll, S.F.; Collier, R.J. NAD binding site of diphtheria toxin: Identification of a residue within the nicotinamide subsite by photochemical modification with NAD. *Proc. Natl. Acad. Sci. USA* **1984**, *81*, 3307–3311.

155. Jung, M.; Just, I.; van Damme, J.; Vandekerckhove, J.; Aktories, K. NAD-binding site of the C3-like ADP-ribosyltransferase from *Clostridium limosum*. *J. Biol. Chem.* **1993**, *268*, 23215–23218.

156. Sakurai, J.; Nagahama, M.; Hisatsune, J.; Katunuma, N.; Tsuge, H. *Clostridium perfringens* iota-toxin, ADP-ribosyltransferase: Structure and mechanism of action. *Adv. Enzyme Regul.* **2003**, *43*, 361–377.

157. Takada, T.; Iida, K.; Moss, J. Conservation of a common motif in enzymes catalyzing ADP-ribose transfer. *J. Biol. Chem.* **1995**, *270*, 541–544.

158. Chowdhury, H.H.; Popoff, M.R.; Zorec, R. Actin cytoskeleton depolymerization with *Clostridium spiroforme* toxin enhances the secretory activity of rat melanotrophs. *J. Physiol.* **1999**, *521*, 389–395.

159. Marvaud, J.C.; Stiles, B.G.; Chenal, A.; Gillet, D.; Gibert, M.; Smith, L.A.; Popoff, M.R. *Clostridium perfringens* iota toxin. Mapping of the Ia domain involved in docking with Ib and cellular internalization. *J. Biol. Chem.* **2002**, *277*, 43659–43666.

160. Papatheodorou, P.; Hornuss, D.; Nolke, T.; Hemmasi, S.; Castonguay, J.; Picchianti, M.; Aktories, K. *Clostridium difficile* binary toxin CDT induces clustering of the lipolysis-stimulated lipoprotein receptor into lipid rafts. *mBio* **2013**, *4*, e00244–13.

161. Fivaz, M.; Abrami, L.; Tsitrin, Y.; van der Goot, F.G. Not as simple as just punching a hole. *Toxicon* **2001**, *39*, 1637–1645.

162. Lafont, F.; van Tran Nhieu, G.; Hanada, K.; Sansonetti, P.; van der Goot, F.G. Initial steps of *Shigella* infection depend on the cholesterol/sphingolipid raft-mediated CD44-IpaB interaction. *EMBO J.* **2002**, *21*, 4449–4457.

163. Miyata, S.; Minami, J.; Tamai, E.; Matsushita, O.; Shimamota, S.; Okabe, A. *Clostridium perfringens* epsilon-toxin forms a heptameric pore within the detergent-insoluble microdomains of Madin-Darby canine kidney cells and rat synaptosomes. *J. Biol. Chem.* **2002**, *277*, 39463–39468.

164. Simons, K.; Ehehalt, R. Cholesterol, lipid rafts, and disease. *J. Clin. Investig.* **2002**, *110*, 597–603.

165. Nagahama, M.; Hagiyama, T.; Kojima, T.; Aoyanagi, K.; Takahashi, C.; Oda, M.; Sakaguchi, Y.; Oguma, K.; Sakurai, J. Binding and internalization of *Clostridium botulinum* C2 toxin. *Infect. Immun.* **2009**, *77*, 5139–5148.

166. Nagahama, M.; Umezaki, M.; Tashiro, R.; Oda, M.; Kobayashi, K.; Shibutani, M.; Takagishi, T.; Ishidoh, K.; Fukuda, M.; Sakurai, J. Intracellular trafficking of *Clostridium perfringens* iota-toxin b. *Infect. Immun.* **2012**, *80*, 3410–3416.

167. Ohishi, I.; Miyake, M. Binding of the two components of C2 toxin to epithelial cells and brush borders of mouse intestine. *Infect. Immun.* **1985**, *48*, 769–775.

168. Sugii, S.; Kozaki, S. Hemagglutinating and binding properties of botulinum C2 toxin. *Biochim. Biophys. Acta* **1990**, *1034*, 176–179.

169. Eckhardt, M.; Barth, H.; Blöcker, D.; Aktories, K. Binding of *Clostridium botulinum* C2 toxin to asparagine-linked complex and hybrid carbohydrates. *J. Biol. Chem.* **2000**, *275*, 2328–2334.

170. Bachmeyer, C.; Benz, R.; Barth, H.; Aktories, K.; Gibert, M.; Popoff, M. Interaction of *Clostridium botulinum* C2-toxin with lipid bilayer membranes and Vero cells: Inhibition of channel function by chloroquine and related compounds *in vitro* and intoxication *in vivo*. *FASEB J.* **2001**, *15*, 1658–1660.

171. Bachmeyer, C.; Orlik, F.; Barth, H.; Aktories, K.; Benz, R. Mechanism of C2-toxin inhibition by fluphenazine and related compounds: Investigation of their binding kinetics to the C2II-channel using the current noise analysis. *J. Mol. Biol.* **2003**, *333*, 527–540.

172. Blöcker, D.; Pohlamnn, K.; Haug, G.; Bachmeyer, C.; Benz, R.; Aktories, K.; Barth, H. *Clostridium botulinum* C2 toxin: Low pH-induced pore formation is required for translocation of the enzyme component C2I into the cytosol of host cells. *J. Biol. Chem.* **2003**, *278*, 37360–37367.

173. Barth, H.; Stiles, B.G. Binary actin-ADP-ribosylating toxins and their use as molecular Trojan horses for drug delivery into eukaryotic cells. *Curr. Med. Chem.* **2008**, *15*, 459–469.

174. Ohishi, I.; Miyake, M.; Ogura, H.; Nakamura, S. Cytopathic effect of botulinum C2 toxin on tissue-culture cells. *FEMS Microbiol. Lett.* **1984**, *23*, 281–284.

175. Papatheodorou, P.; Wilczek, C.; Nolke, T.; Guttenberg, G.; Hornuss, D.; Schwan, C.; Aktories, K. Identification of the cellular receptor of *Clostridium spiroforme* toxin. *Infect. Immun.* **2012**, *80*, 1418–1423.

176. Blonder, J.; Hale, M.L.; Chan, K.C.; Yu, L.R.; Lucas, D.A.; Conrads, T.P.; Zhou, M.; Popoff, M.R.; Issaq, H.J.; Stiles, B.G.; *et al.* Quantitative profiling of the detergent-resistant membrane proteome of iota-b toxin induced Vero cells. *J. Prot. Res.* **2005**, *4*, 523–531.

177. Fagan-Solis, K.; Reaves, D.K.; Rangel, M.C.; Popoff, M.R.; Stiles, B.G.; Fleming, J.M. Challenging the roles of CD44 and lipolysis stimulated lipoprotein receptor in conveying *Clostridium perfringens* iota toxin cytotoxicity in breast cancer. *Mol. Cancer* **2014**, *13*, 163–168.

178. Sandvig, K.; van Deurs, B. Endocytosis, intracellular transport, and cytotoxic action of shiga toxin and ricin. *Physiol. Rev.* **1996**, *76*, 949–966.

179. Orlandi, P.A.; Curran, P.K.; Fishman, P.H. Brefeldin A blocks the response of cultured cells to cholera toxin. Implications for intracellular trafficking in toxin action. *J. Biol. Chem.* **1993**, *8*, 12010–12016.

180. Chardin, P.; McCormick, F. Brefeldin A: The advantage of being uncompetitive. *Cell* **1999**, *97*, 153–155.

181. Madshus, I.H.; Stenmark, H.; Sandvig, K.; Olsnes, S. Entry of diphtheria toxin-protein A chimeras into cells. *J. Biol. Chem.* **1991**, *266*, 17446–17453.

182. Friedlander, A.M. Macrophages are sensitive to anthrax lethal toxin through an acid-dependent process. *J. Biol. Chem.* **1986**, *261*, 7123–7126.

183. Gruenberg, J.; Howell, K.E. Membrane traffic in endocytosis: Insights from cell-free assays. *Ann. Rev. Cell Biol.* **1989**, *5*, 453–481.

184. Sakai, T.; Yamashina, S.; Ohnishi, S. Microtubule-disrupting drugs blocked delivery of endocytosed transferrin to the cytocenter, but did not affect return of transferrin to plasma membrane. *J. Biochem.* **1991**, *109*, 528–533.

185. Menard, A.; Altendorf, K.D.; Berves, D.D.; Mock, M.; Montecucco, C. The vacuolar ATPase proton pump is required for the cytotoxicity of *Bacillus anthracis* lethal toxin. *FEBS Lett.* **1996**, *386*, 161–164.

186. Werner, G.; Hagenmaier, H.; Drautz, H.; Baumgartner, A.; Zahner, H. Metabolic products of microorganisms. 224. Bafilomycins, a new group of macrolide antibiotics. Production, isolation, chemical structure and biological activity. *J. Antibiotics* **1984**, *37*, 110–117.

187. Lord, J.M.; Smith, D.C.; Roberts, L.M. Toxin entry: How bacterial proteins get into mammalian cells. *Cell. Microbiol.* **1999**, *1*, 85–91.

188. Haug, G.; Leemhuis, J.; Tiemann, D.; Meyer, D.K.; Aktories, K.; Barth, H. The host cell chaperone Hsp90 is essential for translocation of the binary *Clostridium botulinum* C2 toxin into the cytosol. *J. Biol. Chem.* **2003**, *278*, 32266–32274.

189. Kaiser, E.; Pust, S.; Kroll, C.; Barth, H. Cyclophilin A facilitates translocation of the *Clostridium botulinum* C2 toxin across membranes of acidified endosomes into the cytosol of mammalian cells. *Cell. Microbiol.* **2009**, *11*, 780–795.

190. Kaiser, E.; Kroll, C.; Ernst, K.; Schwan, C.; Popoff, M.; Fischer, G.; Buchner, J.; Aktories, K.; Barth, H. Membrane translocation of binary actin-ADP-ribosylating toxins from *Clostridium difficile* and *Clostridium perfringens* is facilitated by cyclophilin A and Hsp90. *Infect. Immun.* **2011**, *79*, 3913–3921.

191. Lang, A.E.; Ernst, K.; Lee, H.; Papatheodorou, P.; Schwan, C.; Barth, H.; Aktories, K. The chaperone Hsp90 and PPIases of the cyclophilin and FKBP families facilitate membrane translocation of *Photorhabdus luminescens* ADP-ribosyltransferases. *Cell. Microbiol.* **2014**, *16*, 490–503.

192. Kaiser, E.; Bohm, N.; Ernst, K.; Langer, S.; Schwan, C.; Aktories, K.; Popoff, M.; Fischer, G.; Barth, H. FK506-binding protein 51 interacts with *Clostridium botulinum* C2 toxin and FK506 inhibits membrane translocation of the toxin in mammalian cells. *Cell. Microbiol.* **2012**, *14*, 1193–1205.

193. Pratt, W.B.; Toft, D.O. Regulation of signaling protein function and trafficking by the hsp90/hsp70-based chaperone machinery. *Exp. Biol. Med.* **2003**, *228*, 111–133.

194. Falnes, P.O.; Choe, S.; Madhus, I.H.; Wilson, B.A.; Olsnes, S. Inhibition of membrane translocation of diphtheria toxin A-fragment by internal disulfide bridges. *J. Biol. Chem.* **1994**, *296*, 8402–8407.

195. Ratts, R.; Zeng, H.; Berg, E.A.; Blue, C.; McComb, M.E.; Costello, C.E.; vanderSpek, J.C.; Murphy, J.R. The cytosolic entry of diphtheria toxin catalytic domain requires a host cell cytosolic translocation factor complex. *J. Cell Biol.* **2003**, *160*, 1139–1150.

196. Kistner, A.; Habermann, E. Reductive cleavage of tetanus toxin and botulinum neurotoxin A by the thioredoxin system from brain. Evidence for two redox isomers of tetanus toxin. *Naunyn Schmiedebergs Arch. Pharmacol.* **1992**, *345*, 227–234.

197. Reuner, K.H.; Presek, P.; Boschek, C.B.; Aktories, K. Botulinum C2 toxin ADP-ribosylates actin and disorganizes the microfilament network in intact cells. *Eur. J. Cell Biol.* **1987**, *43*, 134–140.

198. Perelle, S.; Domenighini, M.; Popoff, M.R. Evidence that Arg-295, Glu-378, and Glu-380 are active-site residues of the ADP-ribosyltransferase activity of iota toxin. *FEBS Lett.* **1996**, *395*, 191–194.

199. Simpson, L.L.; Stiles, B.G.; Zepeda, H.; Wilkins, T.D. Production by *Clostridium spiroforme* of an iota-like toxin that possesses mono(ADP-ribosyl)transferase activity: Identification of a novel class of ADP-ribosyltransferases. *Infect. Immun.* **1989**, *57*, 255–261.

200. Egelman, E.H. A tale of two polymers: New insights into helical filaments. *Nat. Rev. Mol. Cell Biol.* **2003**, *4*, 621–630.

201. Pollard, T.D.; Cooper, J.A. Actin, a central player in cell shape and movement. *Science* **2009**, *326*, 1208–1212.

202. Aktories, K.; Reuner, K.H.; Presek, P.; Barmann, M. Botulinum C2 toxin treatment increases the G-actin pool in intact chicken cells: A model for the cytopathic action of actin-ADP-ribosylating toxins. *Toxicon* **1989**, *27*, 989–993.

203. Just, I.; Wille, M.; Chaponnier, C.; Aktories, K. Gelsolin-actin complex is target for ADP-ribosylation by *Clostridium botulinum* C2 toxin in intact human neutrophils. *Eur. J. Pharmacol. Mol. Pharmacol.* **1993**, *246*, 293–297.

204. Just, I.; Hennessey, E.S.; Drummond, D.R.; Aktories, K.; Sparrow, J.C. ADP-ribosylation of *Drosophila* indirect-flight-muscle actin and arthrin by *Clostridium botulinum* C2 toxin and *Clostridium perfringens* iota toxin. *Biochem. J.* **1993**, *291*, 409–412.

205. Vandekerckhove, J.; Schering, B.; Bärmann, M.; Aktories, K. Botulinum C2 toxin ADP-ribosylates cytoplasmic β/γ-actin in arginine 177. *J. Biol. Chem.* **1988**, *263*, 696–700.

206. Mauss, S.; Chaponnier, C.; Just, I.; Aktories, K.; Gabbiani, G. ADP-ribosylation of actin isoforms by *Clostridium botulinum* C2 toxin and *Clostridium perfringens* iota toxin. *Eur. J. Biochem.* **1990**, *194*, 237–241.

207. Prekeris, R.; Mayhew, M.W.; Cooper, J.B.; Terrian, D.M. Identification and localization of an actin-binding motif that is unique to the epsilon isoform of protein kinase C and participates in the regulation of synaptic function. *J. Cell. Biol.* **1996**, *132*, 77–90.

208. Wegner, A.; Aktories, K. ADP-ribosylated actin caps the barbed ends of actin filaments. *J. Biol. Chem.* **1988**, *263*, 13739–13742.

209. Weigt, C.; Just, I.; Wegner, A.; Aktories, K. Nonmuscle actin ADP-ribosylated by botulinum C2 toxin caps actin filaments. *FEBS Lett.* **1989**, *246*, 181–184.

210. Al-Mohanna, F.A.; Ohishi, I.; Hallett, M.B. Botulinum C2 toxin potentiates activation of the neutrophil oxidase. Further evidence of a role for actin polymerization. *FEBS Lett.* **1987**, *219*, 40–44.

211. Schwan, C.; Stecher, B.; Tzivelekidis, T.; van Ham, M.; Rohde, M.; Hardt, W.D.; Wehland, J.; Aktories, K. *Clostridium difficile* toxin CDT induces formation of microtubule-based protrusions and increases adherence of bacteria. *PLoS Path.* **2009**, *5*, e1000626.

212. Schwan, C.; Kruppke, A.S.; Nolke, T.; Schumacher, L.; Koch-Nolte, F.; Kudryashev, M.; Stahlberg, H.; Aktories, K. *Clostridium difficile* toxin CDT hijacks microtubule organization and reroutes vesicle traffic to increase pathogen adherence. *Proc. Natl. Acad. Sci. USA* **2014**, *111*, 2313–2318.

The Combined Repetitive Oligopeptides of *Clostridium difficile* Toxin A Counteract Premature Cleavage of the Glucosyl-Transferase Domain by Stabilizing Protein Conformation

Alexandra Olling, Corinna Hüls, Sebastian Goy, Mirco Müller, Simon Krooss, Isa Rudolf, Helma Tatge and Ralf Gerhard

Abstract: Toxin A (TcdA) and B (TcdB) from *Clostridium difficile* enter host cells by receptor-mediated endocytosis. A prerequisite for proper toxin action is the intracellular release of the glucosyltransferase domain by an inherent cysteine protease, which is allosterically activated by inositol hexaphosphate (IP$_6$). We found that in *in vitro* assays, the *C*-terminally-truncated TcdA$^{1-1065}$ was more efficient at IP$_6$-induced cleavage compared with full-length TcdA. We hypothesized that the *C*-terminally-located combined repetitive oligopeptides (CROPs) interact with the *N*-terminal part of the toxin, thereby preventing autoproteolysis. Glutathione-*S*-transferase (GST) pull-down assays and microscale thermophoresis confirmed binding between the CROPs and the glucosyltransferase (TcdA$^{1-542}$) or intermediate (TcdA$^{1102-1847}$) domain of TcdA, respectively. This interaction between the *N*- and *C*-terminus was not found for TcdB. Functional assays revealed that TcdB was more susceptible to inactivation by extracellular IP$_6$-induced cleavage. *In vitro* autoprocessing and inactivation of TcdA, however, significantly increased, either by acidification of the surrounding milieu or following exchange of its CROP domain by the homologous CROP domain of TcdB. Thus, TcdA CROPs contribute to the stabilization and protection of toxin conformation in addition to function as the main receptor binding domain.

Reprinted from *Toxins*. Cite as: Olling, A.; Hüls, C.; Goy, S.; Müller, M.; Krooss, S.; Rudolf, I.; Tatge, H.; Gerhard, R. The Combined Repetitive Oligopeptides of *Clostridium difficile* Toxin A Counteract Premature Cleavage of the Glucosyl-Transferase Domain by Stabilizing Protein Conformation. *Toxins* **2014**, *6*, 2162-2176.

1. Introduction

Toxins A (TcdA) and B (TcdB) are the main virulence factors of *Clostridium difficile* and predominantly responsible for *C. difficile*-induced diseases, ranging from mild diarrhea to fulminant pseudomembranous colitis and toxic megacolon [1,2]. Following secretion into the gut, the toxins enter their target cells via receptor-mediated endocytosis after binding to the cell surface, at least by the *C*-terminally-located combined repetitive oligopeptides (CROPs) [3–5]. Endosomal acidification triggers conformational changes of TcdA and TcdB, resulting in vesicle membrane insertion and translocation of the *N*-terminus into the cytosolic compartment [6]. The outer *N*-terminal subunit harbors the glucosyltransferase (GT-) domain, which inactivates small Rho GTPases by mono-glucosylation. This leads to disruption of the actin cytoskeleton and, consequently, cell rounding [7]. As a prerequisite for substrate modification, the GT-domain is cleaved off and released into the cytosol by the action of the adjacent cysteine protease domain [8]. The function of the inherent protease is allosterically activated by reducing conditions and the binding of cytosolic inositol hexakisphosphate (IP$_6$) [9]. The importance of autocatalytic processing with regard to toxin action becomes evident by TcdA A^{541}G^{542}A^{543}, a cleavage-resistant mutant, which results in about a 75-fold reduction of cytotoxic potency compared to wild-type TcdA. Instead, extracellular cleavage prevents toxin-mediated cellular effects [10]. Although structural and functional elucidation of the individual toxin domains allowed insights into the multistep process of toxin uptake and toxicity, the interaction of the functional domains in the context of the full-length toxin is rarely investigated. Besides a low resolution analysis of TcdB-structure obtained by small-angle X-ray scattering (SAXS) [11], Pruitt and co-workers presented a structural model of full-length TcdA based on negative stain electron microscopy followed by 3D-reconstruction and mapping of the known functional domains [12]. These analyses revealed a closely-packed conformation of TcdA at neutral pH, assuming intramolecular contacts between the individual domains. Under acidic conditions, the molecule takes on a more elongated shape, reflecting toxin unfolding, necessary during endosomal translocation.

In order to understand the interaction of the functional toxin domains, which leads to a conformation that enables the protection of enzymatic domains and molecule flexibility, we analyzed a putative binding between the CROPs and other domains of the toxins and evaluated the functional consequences by *in vitro* cleavage assays. Here, we report that the CROPs of TcdA tightly interact with the residual molecule, which prevents premature autoproteolysis and, thus, inactivation of the toxin. In addition to the commonly accepted function in receptor binding, we therefore propose that the CROPs, at least of TcdA, play an important role in the conformation stability and protection of the toxin.

2. Results and Discussion

2.1. Efficiency of pH-Dependent Autoprocessing Differs between Full-Length TcdA and C-Terminally-Truncated Toxin Fragments

Intracellular autoproteolytic processing of TcdA and TcdB was emulated in a cell-free system by incubating the toxins in the presence of inositol hexakisphosphate (IP$_6$) and dithiothreitol (DTT). IP$_6$/DTT-incubation of fragment TcdA$^{1-1065}$, which lacks the intermediate and the CROP region of TcdA, resulted in about a 50% cleaved glucosyltransferase (GT-) domain at pH 7.4, whereas IP$_6$-induced cleavage of the full-length toxin was completely ineffective (Figure 1A). Interestingly, the reduction of pH to an acidic milieu stimulated autoprocessing of TcdA dramatically, as shown by western blot analysis targeting the 62-kDa GT-domain. We therefore systematically compared full-length TcdA and fragment TcdA$^{1-1065}$ with regard to the pH-dependency of cleavage (Figure 1B). Densitometric evaluation of western blots detecting the cleaved GT-domain illustrates that autoproteolytic processing of *C*-terminally-truncated TcdA$^{1-1065}$ is more efficient the more the surrounding milieu gets neutralized (right panel). Opposite results were obtained with full-length TcdA, whose cleavage efficacy continuously decreases with neutralization from pH 5 to pH 7 by a factor of five (left panel).

Figure 1. pH-dependent efficacy of autoproteolytic cleavage of toxin A (TcdA) and TcdA fragments. (**A**) Inositol-hexakisphosphate (IP$_6$)-induced cleavage of TcdA (308 kDa) and TcdA$^{1-1065}$ (120 kDa) at pH 5.0 and pH 7.4. The TcdA-specific immunoblot illustrates unprocessed toxins and the cleaved glucosyltransferase domain (62 kDa); (**B**) Western blot (**upper panel**) and densitometric analyses (**lower panel**) reflecting the cleaved GT-domain of TcdA and TcdA$^{1-1065}$ in dependence of pH. Data are presented as means $\pm$ SD ($n = 3$).

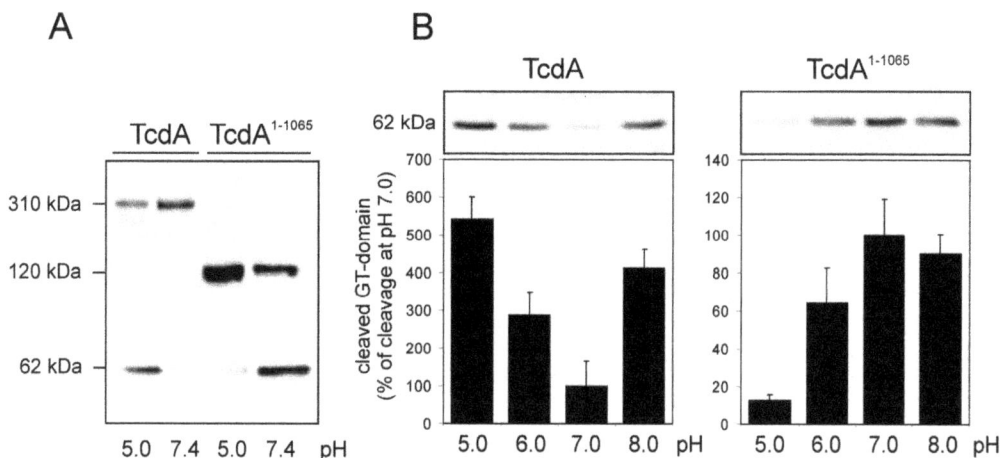

Assuming that the cleavage activity of partially-denatured toxin at pH 5 is solely determined by the cysteine protease activity, a cleavage efficiency of roughly 10% can be estimated for TcdA, as well as for TcdA$^{1-1065}$. From this, it can be extrapolated that at pH 7, about 2% of the holotoxin and about 70% of TcdA$^{1-1065}$ will be cleaved. These observations indicate that the intramolecular structures of the full-length toxin impede autoproteolytic cleavage at neutral pH, however, which are abrogated either under acidic conditions or in the case of toxin fragments lacking the C-terminus, respectively. The latter hypothesis indicates that the C-terminally-located CROPs are essentially involved in the maintenance of a closed, cleavage-protecting conformation and may even form an intramolecular bonding. Further flow cytometry experiments using HT-29 cells supported the assumption of the CROPs interacting with the rest of the toxin. Binding of fluorescently-labeled TcdA$^{1-1874}$, a mutant solely lacking the CROPs, was monitored and illustrated by a right shift in fluorescence (green curve) in Figure 2. To our surprise, previous saturation of the cell surface with the isolated TcdA CROP domain (amino acids 1875–2710) dramatically increased the intensity of fluorescence emitted from subsequent bound TcdA$^{1-1874}$ by a factor of 50 (red curve).

In an earlier study, we found that enhanced binding of truncated TcdA$^{1-1874}$ to HT29 cells pre-incubated with TcdA CROPs was associated with a faster internalization process, as determined by glucosylated Rac1 [4]. Interestingly, pre-incubation of HT29 cells with TcdA- or TcdB CROPs did not enhance the internalization of the homologous TcdB$^{1-1852}$ (Figure 3), which also indicates a lack of interaction of TcdB CROPS with the residual protein.

Figure 2. TcdA located combined repetitive oligopeptides (CROPs) facilitate the binding of TcdA$^{1-1874}$ to HT29 cells. Flow cytometry analysis shows the binding of fluorescently (Atto488)-labeled TcdA$^{1-1874}$ to HT29 cells (green curve) and to HT29 cells preloaded with TcdA CROPs (red curve). The blue peak with the black line represents the autofluorescence of HT29 cells in the absence of labeled TcdA$^{1-1874}$. The excitation and emission wavelengths of the fluorophore are 501 nm and 523 nm, respectively, setting the band-pass width to 10 nm.

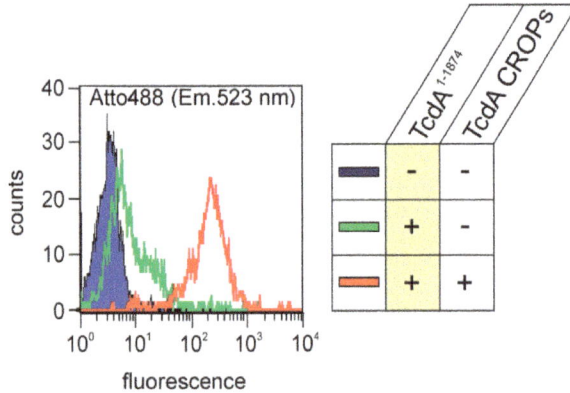

Figure 3. TcdB CROPs do not facilitate the uptake of TcdB$^{1-1852}$. Western blots showing the level of non-glucosylated Rac1 (**upper panel**) as a marker for the intracellular action of TcdB in HT-29 cells in triplicate. β-Actin (**lower panel**) served as the loading control. Rac1 glucosylation by CROP-truncated TcdB$^{1-1852}$ was not altered in dependence of the pre-incubation of cells with the CROPs of TcdA or TcdB, respectively.

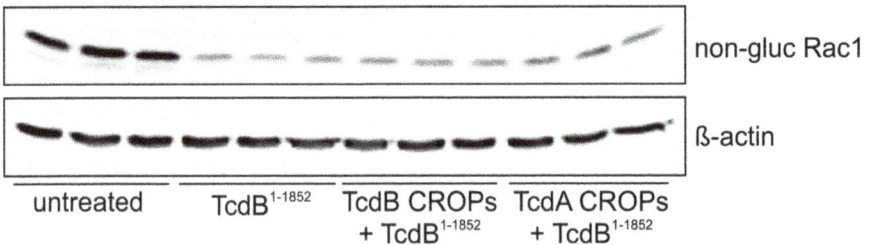

2.2. The TcdA CROPs, but Not the CROPs of TcdB, Interact with the Rest of the Respective Toxin

In order to analyze a potential interaction between the *C*-terminal CROP domain and the *N*-terminal part of the toxin, we performed glutathione-*S*-transferase (GST) pull-down assays using GST-fused CROPs as bait. Therefore, we coupled either GST or GST-CROPs to GSH beads, followed by incubation with the TcdA CROP deletion mutant, TcdA$^{1-1874}$. Toxin fragments used for pull-down experiments are shown in Figure 4a.

Figure 4. TcdA CROPs interact with TcdA$^{1-1874}$. (**A**) TcdA/TcdB are multi-domain toxins with a glucosyltransferase domain (GTD), cysteine protease domain (CPD), a hydrophobic region (HR), an intermediate domain (IMD) and the CROPs; (**B**) Glutathione-S-transferase (GST) (mock) or GST-TcdA CROPs were coupled to GSH beads and used as bait to precipitate CROP-truncated TcdA$^{1-1874}$. Coomassie staining proves equal applied amounts of GST and GST-TcdA CROPs (**left panel**). Immunoblot with α-TcdA$^{1-1065}$ shows the input and precipitation of truncated TcdA (**right panels**); (**C**) Binding of intermediate TcdA$^{1102-1847}$, the glucosyltransferase domain, TcdA$^{1-542}$, and the CROP-deletion mutant, TcdA$^{1-1874}$, to GST-fused TcdA CROPs was analyzed. Immunoblot targeting GST reflects equal amounts of bait (**upper panel**); immunoblot against the Penta-His tag displays the input and bound fraction of the N-terminal toxin fragments (**lower panel**).

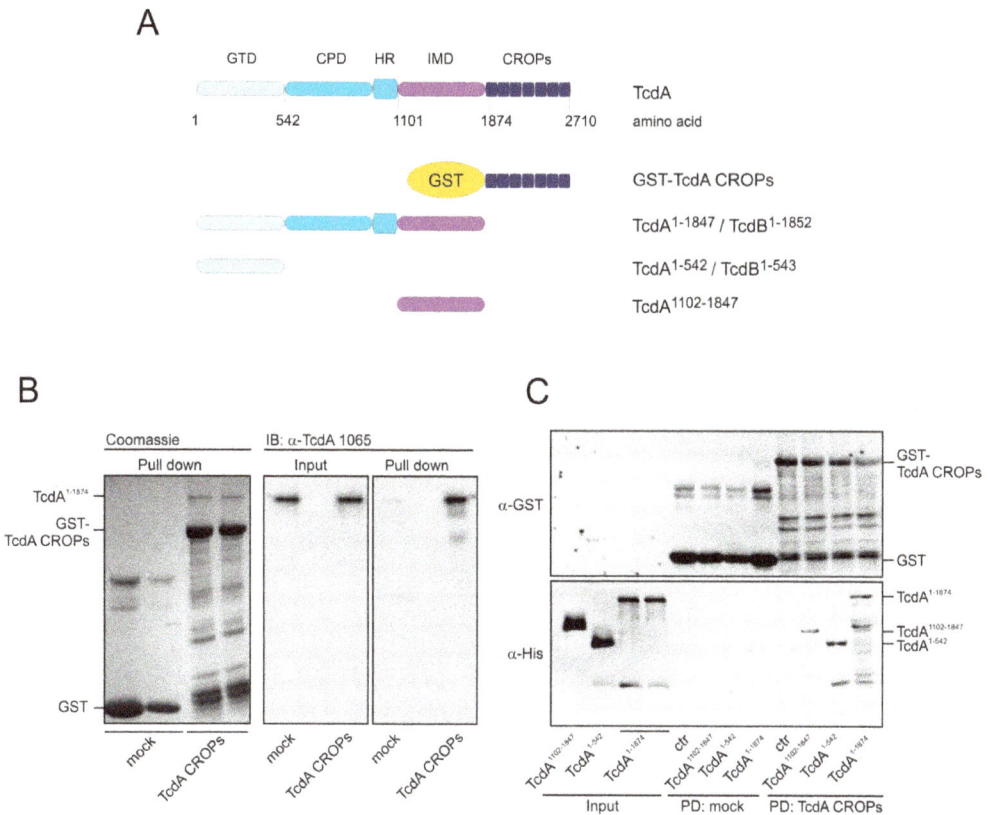

Immunoblot analyses with an antibody targeting the N-terminal part of TcdA (α-TcdA$^{1-1065}$) revealed that TcdA$^{1-1874}$ was specifically precipitated by the CROPs of TcdA, whereas almost no signal was detected when applying GST-coupled beads (mock) as negative bait (Figure 4B). Coomassie staining proved equal amounts of input. However, the question was which domain of TcdA interacts with the CROPs. Referring to the structural model of Pruitt and co-workers, which is based on negative stain electron microscopy followed by 3D reconstruction [12,13], we applied the

N-terminal GT-domain (TcdA$^{1-542}$) or the poorly-characterized intermediate region of TcdA (TcdA$^{1102-1847}$) to beads and checked for binding to the CROPs (Figure 4C). An immunoblot against GST ensured that equal amounts of beads were used (upper panel). Input and bound toxin fragments were detected through a *C*-terminal histidine epitope (lower panel). The GT-domain, as well as the intermediate region were precipitated by the CROPs, though with less affinity than mutant TcdA$^{1-1874}$, which comprises all toxin domains, except the CROPs. Moreover, binding affinity neither increased by simultaneous incubation of all isolated domains nor by using a fragment consisting of amino acids 543–1847 (Figure S1). Thus, a distinct domain responsible for the observed CROP-interaction could not be identified. This phenomenon was specific to TcdA fragments, since TcdB CROPs immobilized at beads were not capable of precipitating the homologous *N*-terminal TcdB fragments (Figure 5).

Figure 5. TcdB CROPs do not associate with TcdB$^{1-1852}$. Immunoblots showing pull-downs of GST-fused TcdB CROPs and TcdB fragments (TcdB$^{1-543}$ and TcdB$^{1-185}$). Precipitation of TcdA$^{1-1874}$ by GST-TcdA CROPs served as the positive control.

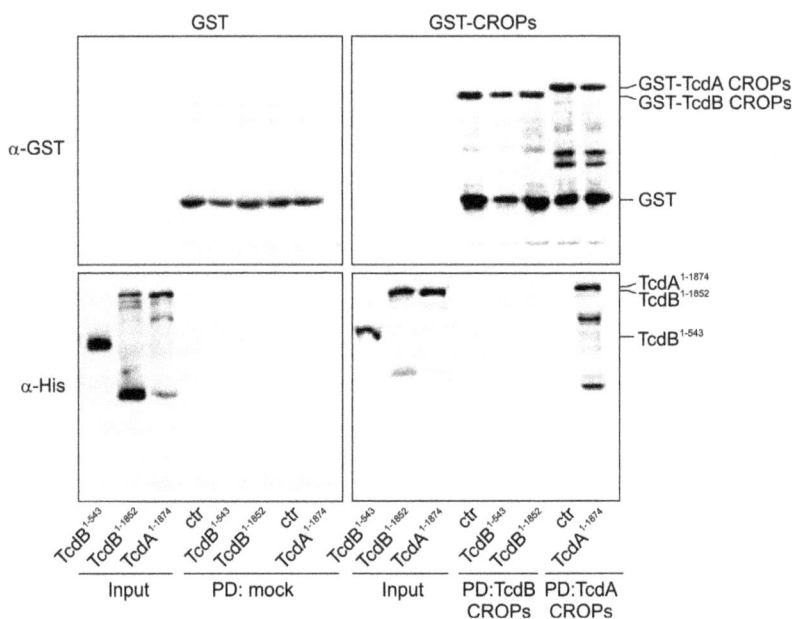

While TcdA$^{1-1874}$ was confirmed to be precipitated by TcdA CROPs, neither the isolated glucosyltransferase domain of TcdB (TcdB$^{1-543}$) nor the respective CROP-deletion mutant, TcdB$^{1-1852}$, was detected in the pull-down approach with the CROPs of TcdB (lower right panel).

In order to verify the results and to quantify the binding affinities, we took advantage of microscale thermophoresis (MST), a novel and sensitive method that enables the monitoring of the complex formation between fluorescent and non-fluorescent proteins under close-to-native conditions. We therefore titrated EGFP-fused toxin fragments (300 nM) with increasing concentrations of unlabeled TcdA or TcdB CROPs and plotted the resulting thermophoresis signals against the respective CROP concentration (Figure 6).

Figure 6. The determination of binding affinities using microscale thermophoresis. Microscale thermophoresis measurements were done to quantify the binding of EGFP-fused TcdA$^{1-1874}$ (♦), TcdA$^{1-542}$ (■) and TcdA$^{1102-1847}$ (○) to unlabeled CROPs. EGFP alone was used as the negative control (mock). The resulting thermophoresis signal was normalized and plotted against the respective CROP concentration. Data were fitted by the Hill slope, and the equilibrium binding constant K_D was obtained. Data are presented as means ± SEM ($n \geq 4$). The R-squared (R^2) reflects the goodness of the respective fit. (**A**) The substrate Rac 1 was used as the positive control for binding to TcdA$^{1-1874}$ ($K_D = 3.32 \pm 0.36$ μM); (**B**) Binding of TcdA$^{1-1874}$ to TcdA CROPs resulted in a binding constant of 1.44 ± 0.07 μM ($R^2 = 0.98$). The affinity of the shorter toxin fragments, TcdA$^{1-542}$ and TcdA$^{1102-1847}$, was less, with K_D values of 2.96 ± 0.18 μM ($R^2 = 0.96$) and 3.06 ± 0.18 μM ($R^2 = 0.89$); (**C**) In contrast to the CROP domain of TcdA, the TcdB CROPs did not interact with the toxin fragments.

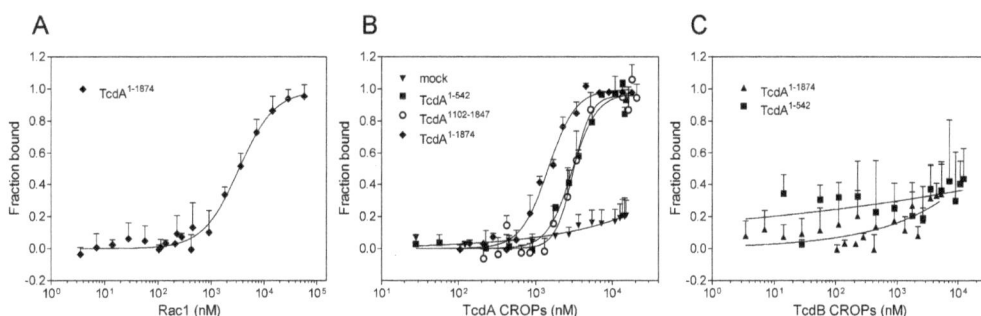

Since Rac1 is a target substrate of the clostridial toxins, it was used as the positive control and titrated to the fluorescent CROP-deletion mutant, TcdA$^{1-1874}$ ($K_D = 3.32 \pm 0.36$ μM). In addition to a standard cytotoxicity assay (Figure S2), the observed dose-dependent binding to Rac1 ensures the correct folding and functionality of the respective EGFP-fused toxin fragments. Similar to the pull-down assays, MST revealed almost identical binding affinities of the glucosyltransferase (TcdA$^{1-542}$) and the intermediate domain (TcdA$^{1102-1847}$) of TcdA with the CROPs, resulting in K_D values of 2.96 ± 0.18 μM (■) and 3.06 ± 0.18 μM (○), respectively (Figure 6B). In fact, TcdA$^{1-1874}$, comprising all toxin domains, except the CROPs, bound the TcdA CROP domain with enhanced affinity ($K_D = 1.44 \pm 0.07$ μM) compared to the shorter toxin fragments. In contrast, none of the analyzed TcdA mutants showed concentration-dependent binding to the CROPs of homologous TcdB (Figure 6C).

2.3. Intramolecular Interactions of TcdA Protect from Its Premature Autoproteolytic Cleavage

As indirect proof of intramolecular bonding, IP$_6$-induced cleavage of different toxins and chimeras was performed. Based on the previous results, we assume that the close conformation observed at neutral pH for TcdA originates from its CROP domain and sterically hinders the binding of IP$_6$ and subsequent activation of the inherent cysteine protease. This might explain the inefficient processing of TcdA at neutral pH, which is observed following the application of the co-factors, IP₆ and DTT (Figure 7A, first panel). The amounts of the cleaved GT-domain were semi-quantified by western blotting to $4.52\% \pm 1.17\%$ (Figure 7B). Under the given conditions, the cleavage of a chimera comprising the amino acids 1–1874 of TcdA and the CROPs of TcdB differs significantly, as this mutant is characterized by a ten-fold increase in autocatalysis compared to wild-type TcdA ($41.06\% \pm 1.77\%$). This behavior reflects an impaired conformation that allows IP₆-binding and cleavage induction, which shows that toxin structure and inhibition of premature cysteine protease activation is predominantly determined by the CROP domain. Interestingly, this seems to be true only for TcdA, since TcdB was efficiently processed at neutral pH ($31.71\% \pm 6.86\%$), though being exposed to 100-fold less IP₆. However, an inhibitory function of the TcdA CROPs only becomes important towards the N-terminus of TcdA, rather than the N-terminus of TcdB. This conclusion was drawn from a chimera of TcdB$^{1-1852}$ and the TcdA CROP domain, which was as efficiently processed as wild-type TcdB ($42.04\% \pm 4.07\%$). The linkage between toxin conformation and function was further investigated by a functional assay. As previously shown, extracellular cleavage of the toxins prevents the cytopathic effect of TcdA and TcdB [10]. Rounding of cells treated with TcdA or TcdB, which were previously incubated with IP₆/DTT, correlates well with the cleavage efficacies described above (Figure 8). Incubation of toxin with IP₆/DTT at pH 7 hardly affected the potency of TcdA towards 3T3 fibroblasts, due to an ineffective toxin cleavage. In addition to the exchange or deletion of the CROP domain, the protective conformation of TcdA is also impaired during acidification, allowing IP₆-induced cleavage, and results in the abrogation of the cytopathic effect. In contrast, due to quantitative processing, even at pH 7, TcdB is functionally inactivated in the presence of IP₆/DTT, and thus, does not significantly affect cell morphology. Cleavage efficiency at pH 5, however, was even less than at pH 7, since the cysteine protease activity itself is reduced under acidic conditions, as could be seen for TcdA$^{1-1065}$ (compare Figure 1A). IP₆ as an ingredient of dietary fiber is physiologically present in the human large intestine at concentrations reaching 4 mM [14]. Therefore, the necessity for the protection of the cysteine protease by, e.g., intramolecular structures, is obvious. This begs the question of how the TcdB structure is stabilized *in vivo* and protected from premature autoproteolytic cleavage. It is conceivable that external factors, hence intermolecular interactions, rather than an intramolecular structure, stabilize TcdB. This might ensure faster conformational alterations of the toxin and, consequently, a quicker translocation process. Whether both toxins adapt to different niches with TcdA, as the less susceptible molecule, ensuring basic cytotoxicity and TcdB predominantly focusing on efficacy, needs to be elucidated.

Figure 7. The CROPs of TcdA protect from premature autoproteolytic toxin inactivation. (**A**) Autocatalytic processing of TcdA, TcdB and the chimeras, TcdA$^{1-1874}$-TcdB CROPs and TcdB$^{1-1852}$-TcdA CROPs, respectively. Cleavage was induced by the addition of IP$_6$ and DTT at pH 7.0. Specific antibody directed against the glucosyltransferase domain (GTD) of TcdA (indicated by the arrow) was applied in the case of TcdA and TcdA$^{1-1874}$-TcdB CROPs (α-TcdA 542) and against the homologous domain of TcdB in the case of TcdB and TcdB$^{1-1852}$-TcdA CROPs (α-TcdB 543); (**B**) The bar chart shows the densitometrical evaluation of the cleaved glucosyltransferase domain. The cleavage efficacy of TcdA differs significantly from that of chimera TcdA$^{1-1874}$-TcdB CROPs (*** $p < 0.0001$) and TcdB (* $p = 0.017$), respectively; ns = not significant.

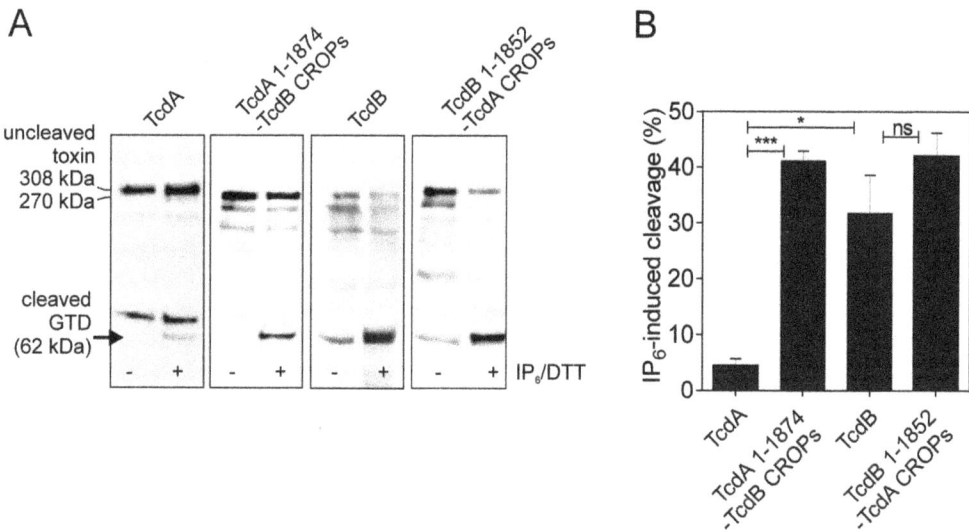

3. Experimental Section

3.1. Antibodies and Reagents

Monoclonal anti-Rac1 antibody recognizing non-glucosylated Rac1 (clone 102, BD PharMingen, Heidelberg, Germany); β-actin antibody (clone AC15, Sigma-Aldrich, Hamburg, Germany); GAPDH antibody, Penta-His antibody (Qiagen, Hamburg, Germany), HRP-conjugated secondary mouse antibody (Rockland, Gilbertsville, PA, USA); *Bacillus megaterium* expression system (MoBiTec, Göttingen, Germany); Inositol hexakisphosphate (Calbiochem/Merck, Darmstadt, Germany).

3.2. Expression and Purification of Recombinant Toxins

The *C. difficile* toxins (strain VPI 10463, GenBank Accession No. X51797) were recombinantly expressed in the *B. megaterium* expression system as *C*-terminally His-tagged fusion proteins, unless otherwise noted. Expression and purification was performed following the standard protocol, as described previously [15]. Cloning of recombinant TcdA and TcdB, the isolated TcdA CROPs (TcdA$^{1875-2710}$), the CROP deletion mutants TcdA$^{1-1874}$ and TcdB$^{1-1852}$, as well as fragment

TcdA$^{1-1065}$ is described elsewhere [3,16]. The *C*-terminally EGFP-fused constructs, TcdA$^{1-1874}$-EGFP and TcdA$^{1-542}$ $^{D285/287N}$-EGFP, were generated by mobilization of *tcdA* 1–5622 bp and *tcdA* 1–1626 bp from the host plasmids by the *SpeI* or *SpeI* and *BamHI* restriction sites, respectively, and ligation into the modified *B. megaterium* expression vector pHIS1522 harboring the *egfp* gene (pHIS1522-EGFP). The construct encoding *N*-terminally EGFP-labeled TcdA$^{1102-1847}$ was cloned by amplification of *egfp* from vector pEGFP-C1 (BD Biosciences Clontech, Heidelberg, Germany) and insertion through *BamHI* restriction sites into pQE30 plasmid harboring base pairs *tcdA* 3304-5541. The GST-tagged CROPs of TcdA and TcdB were generated by amplification of base pairs *tcdA* 5623-8130 (sense: 5'-AGCTAGATCTTATAAAATTATTAATGGTAAAC; antisense: 5'-AGTCGGATCCGCCATATATCCCAGGGGCTTTTAC) and *tcdB* 5542-7098 (sense: 5'-AGCTGGATCCCCAGTAAATAATTTGATAA; antisense: 5'-AGCTGAATTCCTTCACTAA TCACTAATTG) using the vector, pWH-TcdA, or genomic DNA of *C. difficile* strain VPI 10463 as the template, respectively. The resulting amplicons were digested with *BamHI* and *EcoRI* in the case of TcdB and restriction enzymes *BglII* and *BamHI* in the case of TcdA and were ligated into the prepared pGEX-2T vectors (GE Healthcare, Hamburg, Germany). Expression of the respective GST-tagged gene products occurred in TG1 *E. coli* following IPTG induction.

Figure 8. Premature autoproteolysis affects cytotoxicity. (**A**) Representative phase contrast microscopy of 3T3 fibroblasts treated with TcdA or TcdB at pH 7 or pH 5 in the absence or presence of IP$_6$/DTT. Cell rounding confirms the successful internalization of the glucosyltransferase domain, which is prohibited by extracellular IP$_6$/DTT-induced cleavage. Scale bars represent 50 μm; (**B**) Quantification of relative non-glucosylated Rac1 (in relation to GAPDH) by specific antibody (mean ± SD, $n = 3$). Immunoblots were performed from the samples shown under (**A**); Representative immunoblots are shown in the inserts above the bars.

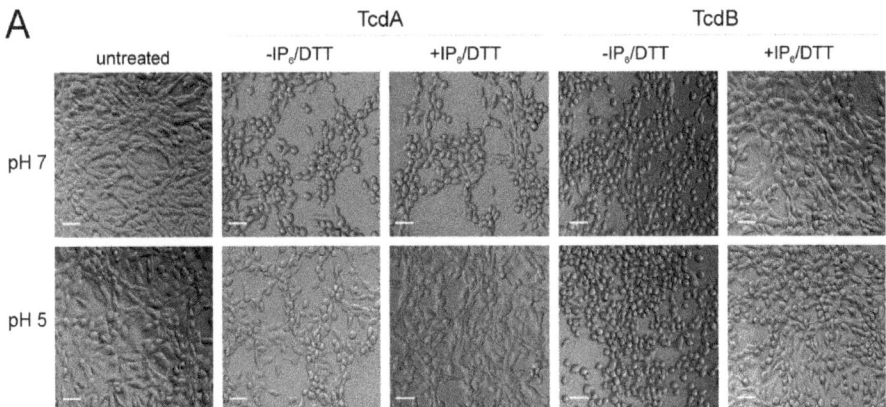

Figure 8. *Cont.*

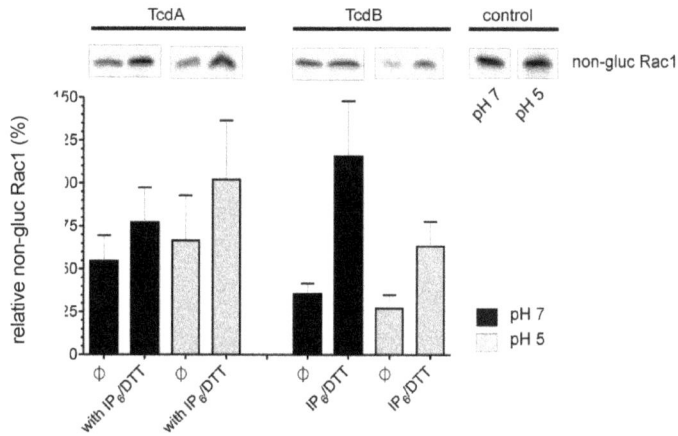

Purification of the His-tagged toxins and fragments was achieved by Ni^{2+} affinity chromatography. For pull-down experiments, the cleared supernatants of bacterial lysates harboring GST-fused TcdA or TcdB CROPs were incubated with GSH beads overnight at 4 °C in order to obtain the bead-coupled CROP domain. Afterwards, the beads were collected by centrifugation at 1000× g, washed three times with 50 mM Tris (pH 8.0), 80 mM NaCl and 0.5% Triton X-100 and used for precipitation experiments.

Chimera TcdB$^{1-1852}$-TcdA$^{1875-2710}$ consists of the TcdA CROP domain fused to the CROP-deletion mutant of TcdB. It was generated by amplification of *tcdA* 5623–8130 using primers 5'-AGCTGGATCCTTTATAAAATTATTAATGGTAAACAC (sense) and 5'-AGTCGCATGCCC GCCATATATCCCAGGGGCTTTTAC (antisense) and ligation into plasmid pHIS1522-TcdB 1–1852 through *BamHI* and *SphI* restriction sites. The reciprocal chimera harboring CROP-truncated TcdA$^{1-1874}$ and the TcdB CROPs was cloned by mobilization of *tcdA* 1–5622 from plasmid pWH-TcdA by *SpeI* restriction and ligation into the respective *SpeI* digested vector, pHIS-TcdB 1848–2366. All constructs are listed in Table S1, showing the location of tags and providing data about linker and additional amino acids as carry-over from the cloning strategy.

3.3. Generation of Specific Antibody

For the generation of specific antibodies, either TcdA fragment 1–1065 or the glucosyltransferase-inactive mutant (D286/288N) of TcdB fragment 1–543 were expressed in *B. megaterium* and purified by affinity chromatography and gel extraction. First, immunization of a New Zealand rabbit was performed after the standard protocol with 100 µg of protein followed by a single boost after four weeks. Blood was collected four weeks after boost immunization. The specificity of anti-serum was checked by western blot using the antigens, as well as full-length TcdA and TcdB as positive and negative controls.

3.4. Flow Cytometry

Mutant $TcdA^{1-1874}$ was fluorescently labeled with Lightning-Link™ Atto488 as previously described [3], and binding to HT29 cells in dependence of TcdA CROP pre-incubation was analyzed by flow cytometry. Therefore, adherent cells were suspended by Accutase treatment, and 500,000 cells were incubated at 4 °C for 30 min, either directly with 4 nM of fluorescent labeled $TcdA^{1-1874}$ or after pre-incubation for 30 min with non-labeled TcdA CROPs ($TcdA^{1875-2710}$) at 4 °C. Cells were washed twice with ice-cold PBS by centrifugation at $200\times$ g for 5 min at 4 °C to eliminate non-bound toxins and finally subjected to flow cytometry (FACScan flow cytometer; Becton Dickinson, Heidelberg, Germany). Ten thousand events were monitored per condition.

3.5. Pull-Down

Pull-down experiments were performed in binding buffer (50 mM Tris, 80 mM NaCl, 5 mM KCl, 1 mM $CaCl_2$, 1 mM $MgCl_2$, 1% (v/v) NP-40, pH 7.4) at 4 °C for 2 h. As bait, GST-fused TcdA or TcdB CROPs coupled to GSH beads were used, previously blocked for 2 h at 4 °C in 50 M Tris (pH 8.0), 80 mM NaCl, 0.1% NP-40 (v/v) and 1% BSA (w/v). Precipitation was done in a total volume of 200 μL with 180 pmol of GST-tagged CROPs and 50 pmol of the respective toxin fragment or H_2O as the control. Previous to incubation, a sample of each approach was taken as the input control. Following binding, beads were pelleted by centrifugation at $1000\times$ g for 5 min at 4 °C and washed three times in 20 mM Tris (pH 8.0), 80 mM NaCl, 0.5% Triton X-100. Finally, toxin fragment-bound beads were resuspended in Laemmli buffer; proteins were denaturized for 10 min at 95 °C, and the soluble fraction was subjected to SDS-PAGE and western blotting.

3.6. Microscale Thermophoresis (MST)

Thermophoresis was used to determine the binding affinities between the TcdA or TcdB CROP domain and N-terminal TcdA fragments. Therefore, a fixed concentration of 300 nM EGFP-fused $TcdA^{1-542}$, $TcdA^{1102-1847}$ or $TcdA^{1-1874}$, respectively, was titrated with 20 μM to 0.01 μM of TcdA or TcdB CROPs in 20 mM Tris, 50 mM NaCl (pH 7.4). As a positive control, a concentration series of recombinant Rac1 was applied. In order to allow binding, samples were incubated at least 30 min at room temperature followed by centrifugation for 5 min at $15,000\times$ g to eliminate potential precipitates. Experiments were performed in standard or hydrophilic capillaries using a NanoTemper Monolith™ NT.115 instrument for green dye fluorescence according to Duhr, Braun and co-workers [17]. Thermophoresis signals for each of the 16 capillaries were monitored, which harbor different ratios of binding partners. The normalized fluorescence at a given time point was plotted against the concentration of unlabeled CROPs. The resulting sigmoidal curves were normalized, and the means $\pm$ SEM ($n \geq 4$) to each data point were determined. Data points were finally fitted by the Hill slope, and K_D values were obtained.

3.7. IP6-Induced Cleavage

Two hundred sixty nanomoles of toxins were incubated for 2 h at 37 °C in 50 mM Hepes (pH range 5.0–8.0, as indicated) and 2 mM dithiothreitol supplemented with 10 µM inositol hexakisphosphate (IP6) in the case of TcdB and TcdB$^{1-1852}$-TcdA$^{1875-2710}$ or 1 mM IP6 in the case of TcdA, TcdA$^{1-1065}$ and TcdA$^{1-1874}$-TcdB$^{1848-2366}$, respectively. For cytotoxicity assays, samples were taken, neutralized to pH 7.4 and applied to 3T3 fibroblasts at final concentrations of 30 to 900 pM. In order to evaluate cleavage efficacies, cleavage reactions were stopped by boiling at 95 °C for 5 min in Laemmli sample buffer, and after neutralization, samples were subjected to SDS-PAGE and western blot analyses. Antibodies were directed against the N-terminus of TcdA (α-TcdA 542) or TcdB (α-TcdB 543), respectively. Densitometrical evaluation is illustrated as a bar diagram showing only IP6/DTT-induced cleavage minus the background signal in the absence of the inducers. Data are presented as the means ± SEM ($n = 3$).

3.8. Cell Culture and Cytotoxicity Assay

3T3 mouse fibroblasts were cultivated under standard conditions in Dulbecco's modified Eagle's medium (DMEM) supplemented with 10% fetal bovine serum (FBS), 100 µM penicillin, 100 µg/mL streptomycin. For cytotoxicity assays, cells were seeded in 24-well chambers and grown for 48 h to sub-confluence. Cleaved and non-cleaved samples were neutralized to pH 7.4, diluted in medium and applied to the cells with final toxin concentrations of 30 pM of TcdB or TcdB$^{1-1852}$-TcdA$^{1875-2710}$ and 900 pM of TcdA or TcdA$^{1-1874}$-TcdB$^{1848-2366}$, respectively. Toxin-induced cell rounding was monitored by light microscopy after 3 h of incubation.

3.9. Western Blotting

Protein samples were separated by SDS-PAGE and transferred onto a nitrocellulose membrane. After blocking with 3% (w/v) BSA and 2% (w/v) nonfat dry milk in TBST (50 mM Tris HCl pH 7.2, 150 mM NaCl, 0.05% (v/v) Tween-20), the membrane was incubated overnight with the primary antibody at 4 °C. Following washing with TBST, it was incubated for 45 min at room temperature with horseradish peroxidase-conjugated secondary antibody in TBST. Detection was performed by means of chemiluminescence. Rac1-glucosylation as a direct marker for intracellular toxin action was determined as described earlier [18].

3.10. Statistical Analysis

Two-tailed t-tests were performed using GraphPad Prism 5.02 (GraphPad Software, San Diego, CA, USA, 2008) to evaluate statistical significance. Significance was set at a p-value of <0.05. Data are presented as the mean ± standard error of the mean (SEM).

4. Conclusions

The present study describes the CROPs of TcdA to shield toxin conformation. Beyond the role as the main receptor binding domain, the C-terminal CROPs interact with the N-terminal part of the toxin to prevent premature cleavage and, thus, inactivation of the toxin. In line with this, a lack of interaction between CROPs and the *N*-terminus in TcdB correlated with more efficient autoprocessing. We conclude that the *C. difficile* toxins complement one another, together providing full pathogenic potential under any given condition: TcdA is less potent, but robust against different milieu circumstances, and TcdB, though being more susceptible, ensures efficient cytotoxicity, due to its high potency.

Acknowledgments

Parts of this work were performed using resources of the Hannover Medical School (MHH) Research Core Facility for Structural Analysis and of the "Laboratory Animal Science". This work was funded by the Federal State of Lower Saxony, Niedersächsisches Vorab (VWZN2889), by the Deutsche Forschungsgemeinschaft, SFB621 (Project B5), and by the Hannover Medical School, Hochschulinterne Leistungsförderung (HiLF) project.

Author Contributions

Alexandra Olling, concept of the study, performed experiments, wrote the manuscript; Corinna Hüls, performed experiments; Mirco Müller, data analysis; Sebastian Goy, generated constructs, performed experiments; Simon Krooss, generated constructs, performed experiments; Isa Rudolf, performed experiments, established the method; Helma Tatge, established the method, generated and provided constructs and proteins; Ralf Gerhard, performed experiments, concept of the study, wrote the manuscript.

Conflicts of Interest

The authors declare no conflict of interest.

References

1. Just, I.; Gerhard, R. Large clostridial cytotoxins. *Rev. Physiol. Biochem. Pharmacol.* **2004**, *152*, 23–47.
2. Hookman, P.; Barkin, J.S. Clostridium difficile-associated disorder/diarrhea and Clostridium difficile colitis: The emergence of a more virulent era. *Dig. Dis. Sci.* **2007**, *52*, 1071–1075.
3. Olling, A.; Goy, S.; Hoffmann, F.; Tatge, H.; Just, I.; Gerhard, R. The repetitive oligopeptide sequences modulate cytopathic potency but are not crucial for cellular uptake of Clostridium difficile toxin A. *PLoS One* **2011**, *6*, e17623.
4. Gerhard, R.; Frenzel, E.; Goy, S.; Olling, A. Cellular uptake of *Clostridium difficile* TcdA and truncated TcdA lacking the receptor binding domain. *J. Med. Microbiol.* **2013**, *62*, 1414–1422.
5. Papatheodorou, P.; Zamboglou, C.; Genisyuerek, S.; Guttenberg, G.; Aktories, K. Clostridial glucosylating toxins enter cells via clathrin-mediated endocytosis. *PLoS One* **2010**, *5*, e10673.
6. Qa'Dan, M.; Spyres, L.M.; Ballard, J.D. pH-induced conformational changes in *Clostridium difficile* toxin B. *Infect. Immun.* **2000**, *68*, 2470–2474.

7. Just, I.; Selzer, J.; Wilm, M.; von Eichel-Streiber, C.; Mann, M.; Aktories, K. Glucosylation of Rho proteins by *Clostridium difficile* toxin B. *Nature* **1995**, *375*, 500–503.

8. Egerer, M.; Giesemann, T.; Jank, T.; Satchell, K.J.; Aktories, K. Auto-catalytic cleavage of Clostridium difficile toxins A and B depends on cysteine protease activity. *J. Biol. Chem.* **2007**, *282*, 25314–25321.

9. Reineke, J.; Tenzer, S.; Rupnik, M.; Koschinski, A.; Hasselmayer, O.; Schrattenholz, A.; Schild, H.; von Eichel-Streiber, C. Autocatalytic cleavage of Clostridium difficile toxin B. *Nature* **2007**, *446*, 415–419.

10. Kreimeyer, I.; Euler, F.; Marckscheffel, A.; Tatge, H.; Pich, A.; Olling, A.; Schwarz, J.; Just, I.; Gerhard, R. Autoproteolytic cleavage mediates cytotoxicity of Clostridium difficile toxin A. *Naunyn Schmiedebergs Arch. Pharmacol.* **2011**, *383*, 253–262.

11. Albesa-Jové, D.; Bertrand, T.; Carpenter, E.P.; Swain, G.V.; Lim, J.; Zhang, J.; Haire, L.F.; Vasisht, N.; Braun, V.; Lange, A.; *et al.* Four distinct structural domains in Clostridium difficile toxin B visualized using SAXS. *J. Mol. Biol.* **2010**, *396*, 1260–1270.

12. Pruitt, R.N.; Chambers, M.G.; Ng, K.K.; Ohi, M.D.; Lacy, D.B. Structural organization of the functional domains of Clostridium difficile toxins A and B. *Proc. Natl. Acad. Sci. USA* **2010**, *107*, 13467–13472.

13. Pruitt, R.N.; Lacy, D.B. Toward a structural understanding of Clostridium difficile toxins A and B. *Front. Cell. Infect. Microbiol.* **2012**, *2*, 1–14.

14. Owen, R.W.; Weisgerber, U.M.; Spiegelhalder, B.; Bartsch, H. Faecal phytic acid and its relation to other putative markers of risk for colorectal cancer. *Gut* **1996**, *38*, 591–597.

15. Burger, S.; Tatge, H.; Hofmann, F.; Just, I.; Gerhard, R. Expression of recombinant Clostridium difficile toxin A using the Bacillus megaterium system. *Biochem. Biophys. Res. Commun.* **2003**, *307*, 584–588.

16. Teichert, M.; Tatge, H.; Schoentaube, J.; Just, I.; Gerhard, R. Application of mutated Clostridium difficile Toxin A for determination of glucosyltransferase-dependent effects. *Infect. Immun.* **2006**, *74*, 6006–6010.

17. Wienken, C.; Baske, P.; Rothbauer, U.; Braun, D.; Duhr, S. Protein-binding assays in biological liquids using microscale thermophoresis. *Nat. Commun.* **2010**, *1*, doi:10.1038/ncomms1093.

18. Genth, H.; Huelsenbeck, J.; Hartmann, B.; Hofmann, F.; Just, I.; Gerhard, R. Cellular stability of Rho-GTPases glucosylated by Clostridium difficile toxin B. *FEBS Lett.* **2006**, *580*, 3565–3569.

Clostridium Perfringens Enterotoxin (CPE) and CPE-Binding Domain (*c*-CPE) for the Detection and Treatment of Gynecologic Cancers

Jonathan D. Black, Salvatore Lopez, Emiliano Cocco, Carlton L. Schwab, Diana P. English and Alessandro D. Santin

Abstract: *Clostridium perfringens* enterotoxin (CPE) is a three-domain polypeptide, which binds to Claudin-3 and Claudin-4 with high affinity. Because these receptors are highly differentially expressed in many human tumors, claudin-3 and claudin-4 may provide an efficient molecular tool to specifically identify and target biologically aggressive human cancer cells for CPE-specific binding and cytolysis. In this review we will discuss these surface proteins as targets for the detection and treatment of chemotherapy-resistant gynecologic malignancies overexpressing claudin-3 and -4 using CPE-based theranostic agents. We will also discuss the use of fluorescent *c*-CPE peptide in the operative setting for real time detection of micro-metastatic tumors during surgery and review the potential role of CPE in other medical applications.

Reprinted from Toxins. Cite as: Black, J.D.; Lopez, S.; Cocco, E.; Schwab, C.L.; English, D.; Santin, A.D. Clostridium Perfringens Enterotoxin (CPE) and CPE-Binding Domain (c-CPE) for the Detection and Treatment of Gynecologic Cancers. *Toxins* **2015**, *7*, 1116-1125.

1. Introduction

Clostridium perfringens enterotoxin (CPE) is a single polypeptide of 35 kDa. It consists of three domains; (1) *C*-terminal domain I, which is responsible for receptor binding; (2) domain II, which is responsible for oligomerization and membrane insertion; and (3) domain III, which may participate in physical changes when the CPE protein inserts into membranes. CPE is comprised of 319 amino acids. It is the virulence factor responsible for *Clostridium perfringens* type A food poisoning, which is the second most commonly reported food-borne illness in the United States [1]. The structure and function of CPE has been investigated through the characterization of the functional properties of enterotoxin fragments and point mutations [2–6]. The $CPE_{290-319}$ COOH-terminus fragment is adequate for high-affinity binding to target cell receptors. However, this fragment is not capable of initiating large complex formation or causing cytolysis [7]. Of note, the $CPE_{290-319}$ COOH-terminus fragment inhibits cytolysis of target cells by full-length CPE [8]. Residues CPE_{45-116} are essential for large complex formation and cytotoxicity. When the NH_2 terminus is deleted, a CPE_{45-319} fragment is generated, which facilitates enhanced large membrane complex formation and cytotoxic activity [9]. Claudin-3, claudin-4, claudin-6, claudin-8 and claudin-14 [10,11] have been shown to represent natural receptors for CPE; they are the members of the transmembrane tissue-specific claudin proteins capable of facilitating CPE binding and cytolysis [12,13].

2. Role of Novel Treatments in Chemotherapy Resistant Gynecologic Cancer

Although many patients with gynecologic malignancies may initially respond to cytoreductive surgery plus platinum/taxane chemotherapy, and/or radiation, many experience recurrences and a poor-prognosis [14–16]. The development of cisplatin resistance reduces therapeutic effectiveness. Resistance mechanisms are often a result of intrinsic pathway activation used during development or as a defense mechanism against environmental toxins [17].

Patients with platinum-resistant/refractory ovarian cancer are known to have a poor prognosis and are classified as having chemotherapy-resistant/refractory ovarian cancer [18]. In these cases, single-agent therapies are used to treat this subset of patients. These include paclitaxel, doxorubicin, and topotecan. The response rate is low, at approximately 10%–15%, and overall survival is approximately 12 months [19]. Trials of combination chemotherapy in platinum-resistant ovarian cancer have failed to improve overall survival. Notably, it increases toxicity [20]. An encouraging discovery in ovarian cancer research has been that chemotherapy-resistant ovarian tumors express claudin-3 and -4 genes at considerably higher levels when compared with chemotherapy-naive ovarian tumors [21,22]. Preclinical studies have investigated the utility of this finding and have found that chemotherapy-resistant ovarian cancer cell lines continue to display substantial sensitivity to CPE *in vitro* and *in vivo* despite their resistance to multiple different chemotherapeutic agents [23].

For patients with endometrial cancer who failed first line therapy, a combination regimen is the most effective. However, there is no established and universally recommended second-line agent in this disease. In patients with measurable disease, second-line agents produce a response in only 50% of patients and a complete response is infrequent. While progression free survival (PFS) and overall survival (OS) times are improving, the 5-year survival rate with advanced/recurrent measurable disease is still <10% [16]. Low initial complete response rates and the high rate of recurrence and/or progression suggest de novo and/or rapidly developing drug resistance. The underlying causes of drug resistance are multifactorial. In endometrial cancer, the resistance is mostly related to the overexpression of β-tubulin subtypes and/or the multidrug-resistance gene (*MDR-1*). Inhibition of apoptosis via alterations in the extrinsic and intrinsic apoptosis pathways and the PI3KCA pathway are also known to occur [16].

Patients with recurrent cervical cancer, which is not amenable to surgery or radiation, have a poor prognosis. The most common treatment for recurrent or metastatic cervical cancer is a combination of paclitaxel and cisplatin or paclitaxel, cisplatin and bevacizumab [24–26]. Although not curative, these treatments result in median survival times of 1–1.5 years. When a patient progresses after this initial therapy for a recurrence or metastases, there are no FDA approved or NCCN level 1 or 2A therapies available and therefore, options are limited [24–27].

2.1. Claudins

Before Mikio Furuse and Shoichiro Tsukita discovered claudins-1 and -2 in 1998, it was thought that tight junctions were mediated by occludins alone [28]. The discovery of this new class of proteins within tight junctions led to an investigation into the proteomics of tight junctions. Ultimately, this led to the discovery of additional claudins, most notably, claudins-3 and -4 [29]. Today there are

27 members of the claudin family of proteins. This family of proteins is primarily known for the role they play in mediating tight junctions between cells; they create a physiologic tight barrier, which prevents unregulated diffusion of water, lipids, proteins and anions. Many claudins (claudins-1, -3, -5, -11, -14, -19, and tricellulin) have sealing functions as well. Other claudins form channels across tight junctions which feature selectivity for cations (claudins-2, -10b, and -15), selectivity for anions (claudin-10a and -17), or are permeable to water (claudin-2). For several claudins (claudins-4, -7, -8, -16, and occludin), we still do not understand their exact function, as their effects on epithelial barriers are inconsistent [30]. What we know, however, is that the importance of claudins in maintaining a normal physiologic state is clear and the dysregulation of their expression and/or interaction plays an important role in carcinogenesis.

Claudins-3 and -4 are upregulated in a number of cancers, including ovarian, uterine, breast, prostate and pancreatic [9,31–34]. Additionally, claudin-3 overexpression is associated with increased malignant potential in colorectal cancer and breast cancer [35,36]. Claudin-3 and -4 have also been studied in uterine serous carcinoma, carcinosarcoma, and high-grade serous ovarian cancer, all of which are known for their aggressive behavior. Claudin-4 in particular is expressed in approximately 70%–90% of ovarian cancers and is differentially expressed across subtypes with the lowest expression seen in the clear cell subtype. In these tumors, differential expression of claudin-3 and -4 between cancerous and normal tissue suggest that these proteins may be exploited as either markers for early detection of carcinogenesis or as targets for tumor directed therapy [32,37–40].

2.2. Claudins as a Target of CPE

Since claudin-3 and -4 genes are highly and differentially expressed in biologically aggressive malignancies (including ovarian carcinoma), they represent a viable detection and treatment target [41]. Claudin-3 and -4 are natural receptors for *Clostridium perfringens* enterotoxin (CPE). The use of CPE for clinical benefit in claudin-3 and -4 expressing tumors has been evaluated. For example, in breast cancer, CPE-mediated toxicity was achieved by using claudin-3 and -4 as targets [42]. Furthermore, in prostate cancer, CPE-mediated cytotoxicity was identified when using claudin-4 as a target [43]. Lastly, claudin-4 was used as the target for CPE-mediated cytotoxicity in pancreatic cancer [44].

The expression of claudin-6 has been reported in multiple human cancers, including rhabdoid tumors, breast cancers and gastric cancers [45]. Claudin-6 is also expressed in ovarian cancer and may represent a novel functional receptor for CPE [10]. Interestingly, knock out claudin-6, as was done with UCI-101, an ovarian cancer cell line that is highly sensitive to CPE, decreased sensitivity to the toxin. Furthermore, when claudin-6 is overexpressed, different ovarian cell lines that are resistant to the effects of CPE become sensitive [10]. Lastly, binding assays reveal that CPE can bind claudin-6 in cells and this binding is associated with CPE cytotoxicity [10]. As a result of this research, claudin-6 has been established as a receptor for CPE and is now considered a therapeutic target in ovarian cancer and other cancers.

2.3. Claudin-3 and -4 Are Potential Targets for CPE-Based Theranostics

As natural receptors for CPE, claudin-3 and -4 are the members of the trans-membrane tissue-specific claudin family most capable of mediating CPE binding and cytolysis [12]. When the CPE toxin binds to cells, small pores form in the plasma membrane and osmotic equilibrium is lost. As a result, rapid cell death occurs [13]. The C-terminal fragment of CPE (c-CPE peptide) is composed of the CPE amino acids 184 to 319 with the receptor-binding region being in amino acids 290 to 319. c-CPE is a non-toxic ligand of claudin-3 and -4 and has fewer antigenic determinants. c-CPE can cause disruption of the tight junction barrier and thereby enhancing drug absorption through mucosal surfaces in a reversible and concentration-dependent manner [46]. A relatively low dose of c-CPE sensitizes epithelial ovarian cancer cells to the cytotoxic effects of carboplatin and paclitaxel in a claudin-4 dependent manner. Compared with single agent paclitaxel or carboplatin, the addition of c-CPE to paclitaxel is able to significantly suppress large tumor burdens by inhibiting tumor cell proliferation and accelerating apoptosis [47]. Furthermore, c-CPE can target TNFα to ovarian cancer cells and has been used as a carrier for other bacterial toxins aimed at claudin-4 positive tumor cells [48]. Yuan et al., linked the c-CPE to TNFα. With c-CPE-TNF, they demonstrated 6.7-fold greater toxicity against the claudin-3 and -4 positive human ovarian carcinoma cells than free TNF, indicating that specific targeting to tumor cells expressing high levels of claudin-3 and -4 was achieved [48]. Unfortunately, thus far, this clinical application has been limited due to its failure to concentrate at the site of tumors and the associated side effects.

Several research groups have reported using claudins-3 or -4 as targets for fluorescent molecule binding as a means to assist with the localization of ovarian and breast cancer cells [40,49]. c-CPE, when used as an intraperitoneal (i.p.) chemosensitizer, decreases tumor density and improves drug penetration by surface diffusion in ovarian cancer cells [50]. Since ovarian carcinoma is largely a disease of the peritoneal cavity and claudin-3 and -4 are not expressed in mesothelial cells, the utilization of i.p. treatment with full length CPE may have great potential for claudin-4 expressing ovarian tumors. Pharmacologic studies in ovarian cancer patients have demonstrated a therapeutic advantage to i.p. CPE and a significant reduction in systemic toxicity when compared with an identical dose of the drug given intravenously. Strategies to limit CPE toxicity to normal tissues already exist in vivo and are based on the local delivery of the blocking CPE peptide fragment to gut and lung via enteral and inhalation routes, respectively. These observations, combined with the finding that ovarian cancer remains confined to the peritoneal cavity for much of its natural history, suggest that i.p. administration of CPE in human patients harboring chemotherapy-resistant/residual disease may result in reduced toxicity and better therapeutic responses compared with identical doses of CPE given intravenously. Consistent with this view, multiple i.p. injections of sublethal doses of CPE every three days significantly inhibited tumor growth in 100% of mice harboring claudin-3 and -4 positive, chemotherapy resistant, ovarian tumor xenografts [23].

Other work in the area of chemoresistant ovarian cancer has demonstrated that CD44+ ovarian cancer stem cells represent a proportion of cancer cells capable of sustaining tumor growth and chemo-resistance. These cancer stem cells highly express genes encoding claudin-4. Casagrande et al. showed that multiple doses of i.p CPE were effective for the eradication of claudin-4 expressing chemoresistant ovarian cancer stem cells in mice harboring these xenografts. They showed a 100% reduction in tumor burden in half of the mice treated ($p < 0.0001$) [21]. In this study, recombinant CPE was not found to induce toxin-associated side effects. It should be noted, however, that repeated i.p. administrations were required in order to attain a therapeutic effect [51].

Another novel approach of safely targeting claudin-3 and -4 expressing ovarian tumor cells is through gene therapy. Intra-tumoral gene transfer of CPE-expressing vectors can be employed for selective suicide gene therapy of claudin-3 and -4 positive tumors. In one study, 72 hours after gene transfer, up to 100% cytotoxicity was noted in claudin-3 and -4 expressing tumor lines. The *in vivo* data from this study also revealed inhibition of ovarian cancer xenograft growth in SCID mice [52].

2.4. c-CPE-Based Diagnostic Agents in Gynecologic Cancers

Imaging of claudin-3 and -4 represents a promising diagnostic approach for identifying metastatic chemotherapy-resistant ovarian cancer [40]. When injected intravenously in mice harboring i.p. chemotherapy-resistant ovarian cancer, the non-toxic carboxy-terminal fragment of CPE (*c*-CPE) conjugated to the fluorescent FITC molecule revealed a preferential accumulation in the ovarian tumor when compared to the normal surrounding tissue. More importantly, the fluorescent *c*-CPE identified ovarian tumor spheroids isolated from the mouse ascites [40]. This research demonstrated that using a FITC-conjugated CPE peptide *in vitro* and *in vivo* revealed binding to multiple primary chemoresistant ovarian carcinoma cell lines and xenografts. This suggests that CPE peptide is a good candidate for tumor therapy and/or the development of new diagnostic tracers (*i.e.*, radioisotopes), including the use of near-infrared fluorescent imaging to identify microscopic metastatic ovarian cancer preoperatively or at the time of tumor recurrence [40]. It also suggests that *c*-CPE can be complexed to biocompatible nanocarriers for the delivery of contrast agents (*i.e.*, iron oxide) or therapeutics (*i.e.*, chemotherapy, gene therapy) [8].

Another interesting approach is *c*-CPE conjugated to fluorescent dyes (*i.e.*, FITC or IRDye CW800). It has been employed as an intra-operative imaging system aimed at guiding surgeons to identify residual disease during real-time debulking surgery [40]. Real-time intra-operative approach imaging has recently been reported in humans undergoing surgery for ovarian cancer using folic acid conjugated to the FITC molecule [53]. The success of such a strategy resides on the fact that ovarian cancer metastases are commonly found attached to the organs of the peritoneal cavity and therefore easily exposed to the illumination source [54]. Uterine Serous Carcinoma (USC), a biologically aggressive variant of endometrial cancer, spreads in a pattern similar to high-grade ovarian cancer [55]. Furthermore, USC expresses high levels of claudin-3 and -4 and *in vitro* and *in vivo* is extremely sensitive to CPE [32]. This suggests that fluorescent *c*-CPE may be useful intra-operatively for the management of USC. Claudin-3 and -4 are also overexpressed in other solid tumors such as pancreatic, breast and prostate cancer [9,34,51,56]. The work of Kominsky *et al.* in particular showed that claudin-3 and -4 were consistently expressed in breast cancer brain metastasis, while negligibly expressed in healthy regions of the Central Nervous System (CNS) [51]. In line with this, intracranial administration of the full length CPE in mice harboring metastatic breast cancer in the brain significantly prolonged their survival compared to control animals. This suggests that *c*-CPE-based theranostic approaches may be useful for breast cancer brain metastasis identification.

2.5. Future Directions

Currently, there are no early detection systems used to identify microscopic, recurrent and/or chemotherapy resistant gynecologic tumors. There are also no established or clinically effective treatment regimens available. Experimental results suggest that by using claudin-3 and -4 as targets and also as receptors for *c*-CPE conjugated to fluorescent dyes, toxins, and radioisotopes or combined with biodegradable/biocompatible nanoparticles for use in imaging, we will be able to diagnose and treat disease earlier and on orders of magnitude smaller than what is currently detected with imaging modalities today [40]. This novel approach is desperately needed to identify and ultimately achieve a cure for uterine and ovarian cancer that is recurrent and/or chemotherapy resistant.

Acknowledgments

This work was supported in part by grants from NIH R01 CA154460-01A1 and U01 CA176067-01A1, the Honorable Tina Brozman Foundation, the Deborah Bunn Alley Ovarian Cancer Research Foundation, the Guido Berlucchi Research Foundation and the Discovery to Cure Foundation to ADS. This investigation was also supported by NIH Research Grant CA-16359 from the NCI.

Author Contributions

J.B. and A.S. designed the study. All authors contributed equally to the literature search and writing of the review article.

Conflicts of Interest

The authors declare no conflict of interest.

References

1. McClane, B.A. An overview of *Clostridium perfringens* enterotoxin. *Toxicon* **1996**, *34*, 1335–1343.
2. Kokai-Kun, J.F.; McClane, B.A. Evidence that a region(s) of the *Clostridium perfringens* enterotoxin molecule remains exposed on the external surface of the mammalian plasma membrane when the toxin is sequestered in small or large complexes. *Infect. Immun.* **1996**, *64*, 1020–1025.
3. Kokai-Kun, J.F.; McClane, B.A. Determination of functional regions of *Clostridium perfringens* enterotoxin through deletion analysis. *Clin. Infect. Dis.* **1997**, *25*, S165–S167.
4. Kokai-Kun, J.F.; McClane, B.A. Deletion analysis of the *Clostridium perfringens* enterotoxin. *Infect. Immun.* **1997**, *65*, 1014–1022.
5. Miyamoto, K.; Fisher, D.J.; Li, J.; Sayeed, S.; Akimoto, S.; McClane, B.A. Complete sequencing and diversity analysis of the enterotoxin-encoding plasmids in *Clostridium perfringens* type A non-food-borne human gastrointestinal disease isolates. *J. Bacteriol.* **2006**, *188*, 1585–1598.

6. McClane, B.A.; Chakrabarti, G. New insights into the cytotoxic mechanisms of *Clostridium perfringens* enterotoxin. *Anaerobe* **2004**, *10*, 107–114.

7. Hanna, P.C.; Mietzner, T.A.; Schoolnik, G.K.; McClane, B.A. Localization of the receptor-binding region of *Clostridium perfringens* enterotoxin utilizing cloned toxin fragments and synthetic peptides. The 30 *C*-terminal amino acids define a functional binding region. *J. Biol. Chem.* **1991**, *266*, 11037–11043.

8. English, D.P.; Santin, A.D. Claudins overexpression in ovarian cancer: Potential targets for *Clostridium perfringens* enterotoxin (CPE) based diagnosis and therapy. *Int. J. Mol. Sci.* **2013**, *14*, 10412–10437.

9. Long, H.; Crean, C.D.; Lee, W.H.; Cummings, O.W.; Gabig, T.G. Expression of *Clostridium perfringens* enterotoxin receptors claudin-3 and claudin-4 in prostate cancer epithelium. *Cancer Res.* **2001**, *61*, 7878–7881.

10. Lal-Nag, M.; Battis, M.; Santin, A.D.; Morin, P.J. Claudin-6: A novel receptor for CPE-mediated cytotoxicity in ovarian cancer. *Oncogenesis* **2012**, *1*, e33.

11. Shrestha, A.; McClane, B.A. Human claudin-8 and -14 are receptors capable of conveying the cytotoxic effects of *Clostridium perfringens* enterotoxin. *MBio* **2013**, doi:10.1128/mBio.00594-12/

12. Katahira, J.; Sugiyama, H.; Inoue, N.; Horiguchi, Y.; Matsuda, M.; Sugimoto, N. *Clostridium perfringens* enterotoxin utilizes two structurally related membrane proteins as functional receptors *in vivo*. *J. Biol. Chem.* **1997**, *272*, 26652–26658.

13. Katahira, J.; Inoue, N.; Horiguchi, Y.; Matsuda, M.; Sugimoto, N. Molecular cloning and functional characterization of the receptor for *Clostridium perfringens* enterotoxin. *J. Cell. Biol.* **1997**, *136*, 1239–1247.

14. Du Bois, A.; Lück, H.J.; Meier, W.; Adams, H.P.; Möbus, V.; Costa, S.; Bauknecht, T.; Richter, B.; Warm, M.; Schröder, W.; *et al.* A randomized clinical trial of cisplatin/paclitaxel *versus* carboplatin/paclitaxel as first-line treatment of ovarian cancer. *J. Natl. Cancer Inst.* **2003**, *95*, 1320–1329.

15. International Collaborative Ovarian Neoplasm Group. Paclitaxel plus carboplatin *versus* standard chemotherapy with either single-agent carboplatin or cyclophosphamide, doxorubicin, and cisplatin in women with ovarian cancer: The ICON3 randomised trial. *Lancet* **2002**, *360*, 505–515.

16. Moxley, K.M.; McMeekin, D.S. Endometrial carcinoma: A review of chemotherapy, drug resistance, and the search for new agents. *Oncologist* **2010**, *15*, 1026–1033.

17. Shen, D.W.; Pouliot, L.M.; Hall, M.D.; Gottesman, M.M. Cisplatin resistance: A cellular self-defense mechanism resulting from multiple epigenetic and genetic changes. *Pharmacol. Rev.* **2012**, *64*, 706–721.

18. Markman, M.; Bookman, M.A. Second-line treatment of ovarian cancer. *Oncologist* **2000**, *5*, 26–35.

19. Naumann, R.W.; Coleman, R.L. Management strategies for recurrent platinum-resistant ovarian cancer. *Drugs* **2011**, *71*, 1397–1412.

20. Lortholary, A.; Largillier, R.; Weber, B.; Gladieff, L.; Alexandre, J.; Durando, X.; Slama, B.; Dauba, J.; Paraiso, D.; Pujade-Lauraine, E. Weekly paclitaxel as a single agent or in combination with carboplatin or weekly topotecan in patients with resistant ovarian cancer: The CARTAXHY randomized phase II trial from Groupe d'Investigateurs Nationaux pour l'Etude des Cancers Ovariens (GINECO). *Ann. Oncol.* **2012**, *23*, 346–352.

21. Casagrande, F.; Cocco, E.; Bellone, S.; Richter, C.E.; Bellone, M.; Todeschini, P.; Siegel, E.; Varughese, J.; Arin-Silasi, D.; Azodi, M.; *et al.* Eradication of chemotherapy-resistant CD44+ human ovarian cancer stem cells in mice by intraperitoneal administration of *Clostridium perfringens* enterotoxin. *Cancer* **2011**, *117*, 5519–5528.

22. Stewart, J.J.; White, J.T.; Yan, X.; Collins, S.; Drescher, C.W.; Urban, N.D.; Hood, L.; Lin, B. Proteins associated with Cisplatin resistance in ovarian cancer cells identified by quantitative proteomic technology and integrated with mRNA expression levels. *Mol. Cell. Proteomics* **2006**, *5*, 433–443.

23. Santin, A.D.; Cané, S.; Bellone, S.; Palmieri, M.; Siegel, E.R.; Thomas, M.; Roman, J.J.; Burnett, A.; Cannon, M.J.; Pecorelli, S. Treatment of chemotherapy-resistant human ovarian cancer xenografts in C.B-17/SCID mice by intraperitoneal administration of *Clostridium perfringens* enterotoxin. *Cancer Res.* **2005**, *65*, 4334–4342.

24. Coleman, R.E.; Harper, P.G.; Gallagher, C.; Osborne, R.; Rankin, E.M.; Silverstone, A.C.; Slevin, M.L.; Souhami, R.L.; Tobias, J.S.; Trask, C.W. A phase II study of ifosfamide in advanced and relapsed carcinoma of the cervix. *Cancer Chemother. Pharmacol.* **1986**, *18*, 280–283.

25. Sutton, G.P.; Blessing, J.A.; McGuire, W.P.; Patton, T.; Look, K.Y. Phase II trial of ifosfamide and mesna in patients with advanced or recurrent squamous carcinoma of the cervix who had never received chemotherapy: A Gynecologic Oncology Group study. *Am. J. Obstet. Gynecol.* **1993**, *168*, 805–807.

26. Monk, B.J.; Sill, M.W.; Burger, R.A.; Gray, H.J.; Buekers, T.E.; Roman, L.D. Phase II trial of bevacizumab in the treatment of persistent or recurrent squamous cell carcinoma of the cervix: A gynecologic oncology group study. *J. Clin. Oncol.* **2009**, *27*, 1069–1074.

27. Poolkerd, S.; Leelahakorn, S.; Manusirivithaya, S.; Tangjitgamol, S.; Thavaramara, T.; Sukwattana, P.; Pataradule, K. Survival rate of recurrent cervical cancer patients. *J. Med. Assoc. Thail.* **2006**, *89*, 275–282.

28. Furuse, M.; Fujita, K.; Hiiragi, T.; Fujimoto, K.; Tsukita, S. Claudin-1 and -2: novel integral membrane proteins localizing at tight junctions with no sequence similarity to occludin. *J. Cell. Biol.* **1998**, *141*, 1539–1550.

29. Morita, K.; Furuse, M.; Fujimoto, K.; Tsukita, S. Claudin multigene family encoding four-transmembrane domain protein components of tight junction strands. *Proc. Natl. Acad. Sci. USA* **1999**, *96*, 511–516.

30. Günzel, D.; Fromm, M. Claudins and other tight junction proteins. *Compr. Physiol.* **2012**, *2*, 1819–1852.

31. Hough, C.D.; Sherman-Baust, C.A.; Pizer, E.S.; Montz, F.J.; Im, D.D.; Rosenshein, N.B.; Cho, K.R.; Riggins, G.J.; Morin, P.J. Large-scale serial analysis of gene expression reveals genes differentially expressed in ovarian cancer. *Cancer Res.* **2000**, *60*, 6281–6287.

32. Santin, A.D.; Zhan, F.; Cane, S.; Bellone, S.; Palmieri, M.; Thomas, M.; Burnett, A.; Roman, J.J.; Cannon, M.J.; Shaughnessy, J., Jr.; *et al.* Gene expression fingerprint of uterine serous papillary carcinoma: Identification of novel molecular markers for uterine serous cancer diagnosis and therapy. *Br. J. Cancer* **2005**, *92*, 1561–1573.

33. Soini, Y. Claudins 2, 3, 4, and 5 in Paget's disease and breast carcinoma. *Hum. Pathol.* **2004**, *35*, 1531–1536.

34. Michl, P.; Barth, C.; Buchholz, M.; Lerch, M.M.; Rolke, M.; Holzmann, K.H.; Menke, A.; Fensterer, H.; Giehl, K.; Löhr, M.; *et al.* Claudin-4 expression decreases invasiveness and metastatic potential of pancreatic cancer. *Cancer Res.* **2003**, *63*, 6265–6271.

35. De Souza, W.F.; Fortunato-Miranda, N.; Robbs, B.K.; de Araujo, W.M.; de-Freitas-Junior, J.C.; Bastos, L.G.; Viola, J.P.B.; Morgado-Díaz, J.A. Claudin-3 overexpression increases the malignant potential of colorectal cancer cells: Roles of ERK1/2 and PI3K-Akt as modulators of EGFR signaling. *PLoS One* **2013**, *8*, e74994.

36. Kolokytha, P.; Yiannou, P.; Keramopoulos, D.; Kolokythas, A.; Nonni, A.; Patsouris, E.; Pavlakis, K. Claudin-3 and claudin-4: Distinct prognostic significance in triple-negative and luminal breast cancer. *Appl. Immunohistochem. Mol. Morphol.* **2014**, *22*, 125–131.

37. Santin, A.D.; Bellone, S.; Siegel, E.R.; McKenney, J.K.; Thomas, M.; Roman, J.J.; Burnett, A.; Tognon, G.; Bandiera, E.; Pecorelli, S. Overexpression of *Clostridium perfringens* enterotoxin receptors claudin-3 and claudin-4 in uterine carcinosarcomas. *Clin. Cancer Res.* **2007**, *13*, 3339–3346.

38. Santin, A.D.; Zhan, F.; Bellone, S.; Palmieri, M.; Cane, S.; Bignotti, E.; Anfossi, S.; Gokden, M.; Dunn, D.; Roman, J.J.; *et al.* Gene expression profiles in primary ovarian serous papillary tumors and normal ovarian epithelium: Identification of candidate molecular markers for ovarian cancer diagnosis and therapy. *Int. J. Cancer* **2004**, *112*, 14–25.

39. Fujita, K.; Katahira, J.; Horiguchi, Y.; Sonoda, N.; Furuse, M.; Tsukita, S. *Clostridium perfringens* enterotoxin binds to the second extracellular loop of claudin-3, a tight junction integral membrane protein. *FEBS Lett.* **2000**, *476*, 258–261.

40. Cocco, E.; Casagrande, F.; Bellone, S.; Richter, C.E.; Bellone, M.; Todeschini, P.; Holmberg, J.C.; Fu, H.H.; Montagna, M.K.; Mor, G.; *et al.* *Clostridium perfringens* enterotoxin carboxy-terminal fragment is a novel tumor-homing peptide for human ovarian cancer. *BMC Cancer* **2010**, *10*, 349.

41. Morin, P.J. Claudin proteins in human cancer: Promising new targets for diagnosis and therapy. *Cancer Res.* **2005**, *65*, 9603–9606.

42. Kominsky, S.L.; Vali, M.; Korz, D.; Gabig, T.G.; Weitzman, S.A.; Argani, P.; Sukumar, S. *Clostridium perfringens* enterotoxin elicits rapid and specific cytolysis of breast carcinoma cells mediated through tight junction proteins claudin 3 and 4. *Am. J. Pathol.* **2004**, *164*, 1627–1633.

43. Maeda, T.; Murata, M.; Chiba, H.; Takasawa, A.; Tanaka, S.; Kojima, T.; Masumori, N.; Tsukamoto, T.; Sawada, N. Claudin-4-targeted therapy using *Clostridium perfringens* enterotoxin for prostate cancer. *Prostate* **2012**, *72*, 351–360.

44. Michl, P.; Buchholz, M.; Rolke, M.; Kunsch, S.; Löhr, M.; McClane, B.; Tsukita, S.; Leder, G.; Adler, G.; Gress, T.M. Claudin-4: A new target for pancreatic cancer treatment using *Clostridium perfringens* enterotoxin. *Gastroenterology* **2001**, *121*, 678–684.

45. Birks, D.K.; Kleinschmidt-DeMasters, B.K.; Donson, A.M.; Barton, V.N.; McNatt, S.A.; Foreman, N.K.; Handler, M.H. Claudin 6 is a positive marker for atypical teratoid/rhabdoid tumors. *Brain Pathol.* **2010**, *20*, 140–150.

46. Kondoh, M.; Takahashi, A.; Fujii, M.; Yagi, K.; Watanabe, Y. A novel strategy for a drug delivery system using a claudin modulator. *Biol. Pharm. Bull.* **2006**, *29*, 1783–1789.

47. Gao, Z.; Xu, X.; McClane, B.; Zeng, Q.; Litkouhi, B.; Welch, W.R.; Berkowitz, R.S.; Mok, S.C.; Garner, E.I.O. C terminus of *Clostridium perfringens* enterotoxin downregulates CLDN4 and sensitizes ovarian cancer cells to Taxol and Carboplatin. *Clin. Cancer Res.* **2011**, *17*, 1065–1074.

48. Yuan, X.; Lin, X.; Manorek, G.; Kanatani, I.; Cheung, L.H.; Rosenblum, M.G.; Howell, S.B. Recombinant CPE fused to tumor necrosis factor targets human ovarian cancer cells expressing the claudin-3 and claudin-4 receptors. *Mol. Cancer Ther.* **2009**, *8*, 1906–1915.

49. Yoshida, H.; Sumi, T.; Zhi, X.; Yasui, T.; Honda, K.I.; Ishiko, O. Claudin-4: A potential therapeutic target in chemotherapy-resistant ovarian cancer. *Anticancer Res.* **2011**, *31*, 1271–1277.

50. Litkouhi, B.; Kwong, J.; Lo, C.M.; Smedley, J.G., 3rd.; McClane, B.A.; Aponte, M.; Gao, Z.; Sarno, J.L.; Hinners, J.; *et al.* Claudin-4 overexpression in epithelial ovarian cancer is associated with hypomethylation and is a potential target for modulation of tight junction barrier function using a *C*-terminal fragment of *Clostridium perfringens* enterotoxin. *Neoplasia* **2007**, *9*, 304–314.

51. Kominsky, S.L.; Tyler, B.; Sosnowski, J.; Brady, K.; Doucet, M.; Nell, D.; Smedley, J.G.; McClane, B.; Brem, H.; Sukumar, S. *Clostridium perfringens* enterotoxin as a novel-targeted therapeutic for brain metastasis. *Cancer Res.* **2007**, *67*, doi:10.1158/0008-5472.can-07-1314.

52. Walther, W.; Petkov, S.; Kuvardina, O.N.; Aumann, J.; Kobelt, D.; Fichtner, I.; Lemm, M.; Piontek, J.; Blasig, I.E.; Stein, U. Novel *Clostridium perfringens* enterotoxin suicide gene therapy for selective treatment of claudin-3- and -4-overexpressing tumors. *Gene Ther.* **2012**, *19*, 494–503.

53. Van Dam, G.M.; Themelis, G.; Crane, L.M.A.; Harlaar, N.J.; Pleijhuis, R.G.; Kelder, W.; Sarantopoulos, A.; de Jong, J.S.; Arts, H.J.G.; van der Zee, A.G.J.; *et al.* Intraoperative tumor-specific fluorescence imaging in ovarian cancer by folate receptor-α targeting: First in-human results. *Nat. Med.* **2011**, *17*, 1315–1319.

54. Boente, M.P.; Godwin, A.K.; Hogan, W.M. Screening, imaging, and early diagnosis of ovarian cancer. *Clin. Obstet. Gynecol.* **1994**, *37*, 377–391.

55. Schwartz, P.E. The management of serous papillary uterine cancer. *Curr. Opin. Oncol.* **2006**, *18*, 494–499.

56. Facchetti, F.; Lonardi, S.; Gentili, F.; Bercich, L.; Falchetti, M.; Tardanico, R.; Baronchelli, C.; Lucini, L.; Santin, A.; Murer, B. Claudin 4 identifies a wide spectrum of epithelial neoplasms and represents a very useful marker for carcinoma *versus* mesothelioma diagnosis in pleural and peritoneal biopsies and effusions. *Virchows. Arch.* **2007**, *451*, 669–680.

Cholera Toxin B: One Subunit with Many Pharmaceutical Applications

Keegan J. Baldauf, Joshua M. Royal, Krystal Teasley Hamorsky and Nobuyuki Matoba

Abstract: Cholera, a waterborne acute diarrheal disease caused by *Vibrio cholerae*, remains prevalent in underdeveloped countries and is a serious health threat to those living in unsanitary conditions. The major virulence factor is cholera toxin (CT), which consists of two subunits: the A subunit (CTA) and the B subunit (CTB). CTB is a 55 kD homopentameric, non-toxic protein binding to the GM1 ganglioside on mammalian cells with high affinity. Currently, recombinantly produced CTB is used as a component of an internationally licensed oral cholera vaccine, as the protein induces potent humoral immunity that can neutralize CT in the gut. Additionally, recent studies have revealed that CTB administration leads to the induction of anti-inflammatory mechanisms *in vivo*. This review will cover the potential of CTB as an immunomodulatory and anti-inflammatory agent. We will also summarize various recombinant expression systems available for recombinant CTB bioproduction.

Reprinted from *Toxins*. Cite as: Baldauf, K.J.; Royal, J.M.; Hamorsky, K.T.; Matoba, N. Cholera Toxin B: One Subunit with Many Pharmaceutical Applications. *Toxins* **2015**, *7*, 974-996.

1. Introduction

1.1. Cholera

Cholera is a highly contagious acute dehydrating diarrheal disease caused by *Vibrio cholerae*. There are over 200 serogroups of *V. cholerae* known to date; however, only two (O1 and 139 serotypes) are responsible for the vast majority of outbreaks [1,2]. The pathology of cholera results from *V. cholerae* colonization in the small intestine and subsequent production of the cholera toxin (CT).

V. cholerae are found in coastal waters and deltas due to their preference for salinity in water; however under proper conditions (warm and sufficient nutrients), *V. cholerae* can grow in low salinity environments [3]. Natural disasters (e.g., floods, monsoons, and earthquakes) and poor sanitation are major players in the spread of cholera epidemics. Symptomatic individuals can shed the organism from 2 days to 2 weeks after infection and recently shed organisms (5–24 h after shedding) have hyperinfectivity; in this state the infectious dose is 10 to 100 times lower than non-shed organisms ($\sim 10^6$ bacteria) [4,5]. This can lead to the rapid spread of cholera in densely populated areas without proper management of patients and their waste.

The most common symptom of cholera is a life-threatening amount of watery diarrhea, causing an extreme loss of water, up to 1 L per hour, which can lead to death within hours of the first onset of symptoms if left untreated [3]. The diarrhea is usually painless and not accompanied by the urge to evacuate the bowels. Early in the illness, vomiting can be a common symptom as well.

Cholera is considered endemic in over 50 countries, but it can manifest as an epidemic, as has recently been the case in Haiti (2010–present), a country previously not exposed to cholera [6–8]. Reported world incidences of cholera increased from 2007 until a peak of approximately 600,000 cases in 2011 [9]. In 2012, the number of reported cases decreased to approximately 245,000 with 49% of the cases resulting from the ongoing outbreak in Haiti and the Dominican Republic. However, the World Health Organization (WHO) estimates the actual global burden of the disease to be between 3 and 5 million cases per year and 100,000 to 130,000 deaths per year [10]. Additionally, a more virulent strain of *V. cholerae* O1 is making inroads in Africa and Asia [11]. The WHO suggests there should also be concern for the spread of antibiotic-resistant strains of *V. cholerae*. This has already been shown with *V. cholerae* O139 and some isolates from *V. cholerae* O1 El Tor, which have acquired resistance traits for co-trimoxazole and streptomycin [3]. It is clear that cholera, despite its long history, is still an emerging disease that is necessary to combat.

1.2. CT

CT produced by *V. cholerae*, is the main virulence factor in the development of cholera. The molecular characteristics of CT and its toxic effects in humans have been well characterized [12–14]. CT is an 84 kD protein made up of two major subunits, CTA and CTB [15,16] (Figure 1). The CTA subunit is responsible for the disease phenotype while CTB provides a vehicle to deliver CTA to target cells. CTA is a 28 kD subunit consisting of two primary domains, CTA1 and CTA2, with the toxin activity residing in the former and the latter acting as an anchor into the CTB subunit [17]. The CTB subunit consists of a homopentameric structure that is approximately 55 kD (11.6 kD monomers) and binds to the GM1-ganglioside; found in lipid rafts, on the surface of intestinal epithelial cells [13]. The exact mechanism of delivering CTA1 into the intracellular space is still not fully resolved; however, the current understanding is that CT is endocytosed and travels through a retrograde transport pathway from the Golgi apparatus to the endoplasmic reticulum (ER) [12–14,17,18]. Recently, it has been shown that CT can also move from the apical to basolateral surface of epithelial cells via transcytosis, enabling transport of whole CT through the intestinal barrier [19]. CTA is dissociated from CTB after the toxin reaches the ER and translocated to the cytosol via the ER-associated degradation pathway [15]. Intoxication occurs when CTA1 enters the cell cytosol and catalyzes the ADP ribosylation of adenylate cyclase, which leads to increased intracellular cAMP. This increase in intracellular cAMP results in impaired sodium uptake and increased chloride outflow, causing water secretion and diarrhea [12,17].

1.3. Current Vaccines

The emergence of a more virulent strain of *V. cholerae*, coupled with the increasing number of endemic and newly exposed countries suggests a growing need for a consistent vaccination strategy. Currently, there are two WHO pre-qualified vaccines for cholera: Dukoral® (SBL Vaccin AB, Stockholm, Sweden) and Shanchol® (Shantha Biotechnics Limited, Basheerbagh, India). Dukoral® contains killed *V. cholerae* (Inaba and Ogawa serotypes of *V. cholerae* O1) and recombinant (r) CTB, while Shanchol® contains the killed *V. cholerae* (serogroups O1 and O139) [20]. Due to the cross-reactivity of anti-CTB antibodies to heat labile enterotoxin (LTB), Dukoral® is also effective against enterotoxigenic *Escherichia coli* (ETEC), an advantage not offered by Shanchol®. On the other hand, Shanchol® is a less expensive cholera vaccine than Dukoral® because the latter includes costs related to rCTB, *i.e.*, recombinant production, a buffer to neutralize stomach acid to prevent rCTB degradation and additional storage space and logistics. In a vaccination cost analysis study performed in 2012, it was found to cost approximately US$10 to purchase two doses of Dukoral® and approximately US$3 to deliver those doses [21]. However, these costs could be reduced by developing cost-effective rCTB production methods (see below) and formulating the vaccine in a solid oral dosage form able to pass through the stomach and dissolve in the small intestine [22].

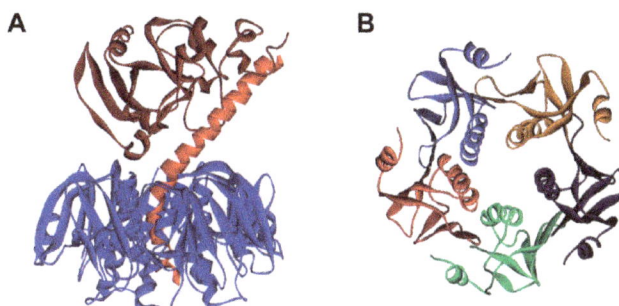

Figure 1. Cholera toxin (CT) crystal structure. (**A**) CT (side view; Protein Data Bank [PDB] ID: 1XTC). The CTA subunit is shown in red (CTA1 in dark red and CTA2 in light red) and the CTB subunit is shown in blue; (**B**) CTB (top view; PDB ID: 1XTC with CTA subunit removed). Each monomer of the B subunit is show in a different color. Images were created in Accelrys Discovery Studio Visualizer 2.5.

Interestingly, a field trial performed in 1985 suggests that a whole cell-killed vaccine with CTB (WCB) may be more efficacious than a whole cell-killed vaccine without CTB (WC) [23]. Children 2 to 10 years old were almost completely and significantly protected (92%) from cholera after 3 vaccinations with WCB compared to a non-significant 53% protection for WC for the first six months after vaccination. Hence, children were far better protected with the CTB-containing vaccine. In older populations (>10 years old) both vaccines showed similar protective efficacy over 6 months; the WCB vaccine protected 77% of the adults compared to 62% with the WC vaccine. Additionally, perhaps most importantly, the WCB vaccine significantly protected against severe cholera episodes (89% protective) *versus* no significant protection by the WC vaccine (44% protective). Lastly, within approximately the first 6 months following vaccination, the WCB vaccine significantly protected the recipients while WC vaccine recipients lost protective efficacy approximately 3 months after vaccination. This short-term enhanced protection could provide a significant implication for a reactive vaccination strategy to contain outbreaks.

The same population was also tracked for three years following vaccination and differences between WCB and WC vaccination were further elucidated [24]. Again, it was found that 2–5 year old children, who received all three vaccine doses, were significantly protected when receiving the

WCB vaccine for up to 2 years following vaccination when compared to the placebo group. At no point was WC vaccine significantly protective of the 2–5 year old cohort in this study. For up to 3 years following vaccination both WCB and WC protected study participants over the age of 5. Additionally, the number of doses needed to see strong protection against cholera was another point of differentiation. WCB vaccination required two doses to provide significant protection while the same level of protection was not achieved with the WC vaccine until a third dose was administered. It should be noted that WCB contains non-recombinant CTB (purified from CT) and thus should not be confused with the currently available Dukoral®, which contains rCTB.

In this regard, a more recent work has been performed to evaluate the protective efficacy of Dukoral® in adults and children [25]. The study by Alam *et al.*, divided children into 2 groups: young (median age 5) and older (median age 10) and had an adult group with a median age of 32. Significant antibody responses in all groups were seen 3 days following the first dose in all study groups and continued to day 42 in all groups. However at day 90, the next time point in the study, both groups of children lost the antibody response while the adult antibody response persisted until at least 270 days following the second vaccination. Additionally, a 2005 study in Mozambique showed that an rCTB whole cell-killed vaccine was able to protect at similar levels of the WCB vaccine used in Bangladesh [26]. The results from this study also confirmed that the vaccine containing rCTB may have improved protection in severe cases of cholera. Confounding these results, a field trial performed in Peru in 1994 is often reported as having negative results (increased cholera infection) in rCTB vaccine recipients [27]. However, the study did report positive protection after a booster third dose was given just prior to the start of the next cholera outbreak season in Peru. Additionally, this study evaluated only two time points, 1 year and 2 year protection, which could have overlooked the early protection (<6 months after vaccination) observed previously with WCB [28]. Lastly, the fact that a single booster provided protection during the second year of the study suggests that an rCTB containing vaccine does in fact protect against cholera outbreaks.

Shanchol® has been studied in both Bangladesh and Haiti; participants in both studies showed strong immune responses to the two dose vaccine regimen [20,29]. In 2012, Shanchol® was used in an outbreak in Guinea and found to be effective in protecting adults from cholera infection [30]. These findings were thought to be in line with results seen with Dukoral®, but there was no rCTB vaccine group in this study to compare to. An advantage to Shanchol® is that it has been tested in children as young as 1 year old and protection has been noted in this young population [29]. The lack of a large scale study comparing Shanchol® and Dukoral® makes any comparison difficult.

A recent paper may help elucidate the potential benefit of including rCTB in any cholera vaccine. Although mice do not develop cholera, a model of pulmonary *V. cholerae* infection has recently been established [31]. In this model, severe pneumonia was induced in mice and was found to be fatal within several days of inoculation with *V. cholerae*. Interestingly, mice vaccinated intranasally, twice with Dukoral® prior to *V. cholerae* challenge, were significantly protected compared to controls. Unvaccinated animals died within 24 h of the challenge while none of the mice vaccinated died for up to 7 days following challenge. Notably, Dukoral® without rCTB showed no protection in this model, while protection was restored upon inclusion of rCTB. These results provide unequivocal evidence that rCTB is essential in protecting mice from the lethal pneumonia induced by *V. cholerae* infection.

Coupled with the earlier findings with WCB vaccines in the field trial, it is suggested that, in the case of cholera outbreaks, vaccines containing rCTB may provide immediate benefit to vaccine recipients that would not be seen in rCTB-free vaccines.

2. CTB as a Vaccine Adjuvant

In addition to its toxic properties, CT is also known to have strong mucosal immunogenic properties that have been investigated for beneficial use as well as inducing an allergic response in animal models [32–37]. CT has also been shown previously to have adjuvant potential when incorporated into mucosal vaccines [38–40]. However, the toxicity of CT made its use in humans undesirable and work now focuses on removing the toxicity from the molecule while maintaining the adjuvant effect. The CTB subunit was previously shown to induce an immune response without the toxicity associated with the CTA subunit [41]. CTB has proven to be a strong adjuvant to uncoupled antigens when administered via the nasal route but less so when administered orally [15,42,43]. However, the nasal route of administration is not preferred due to the potential risk for developing Bell's palsy [44–46]. Fortunately, it was found that by coupling the antigen to CTB, a much stronger response is achieved via the oral administration route [47]. We should also point out that the adjuvant potential of CTB has also been shown in large animal models, indicating that the adjuvant potential is scalable to higher species [48–50]. The utility of CTB becomes apparent when looking at the various disease states in which it has been used as an adjuvant: bacterial and viral infections, allergy, and diabetes have been targeted [51–53]. Also, an interesting approach to resolving cocaine addiction has been attempted by binding rCTB to succinylnorcocaine, which has been tested in a Phase IIb randomized double-blind placebo-controlled trial [54,55]. The hypothesis behind the vaccine was that the anti-cocaine antibodies may block the uptake of cocaine in the brain from the blood. While the results were inconclusive, with only ~40% of participants achieving inhibitory antibody concentrations in the blood, this study shows potential utility of CTB-based vaccines in addiction therapy.

For a general overview of the work on CTB as a vaccine adjuvant, readers are referred to thorough reviews published previously [41,56,57]. For this review we will focus on some findings not addressed in these previous reviews.

2.1. CTB-Based Immunogens against Bacterial Pathogens

Development of vaccines against several bacterial pathogens has been attempted recently by conjugating antigens to CTB to induce immune responses against the bacteria. *Helicobacter pylori* is a bacterium that infects greater than 50% of the world population and can cause a variety of gastrointestinal diseases [58]. Specifically, *H. pylori* urease, a two subunit enzyme, has been targeted by linking both subunits (UreA and UreB) of the enzyme to CTB. Guo *et al.* described a fusion protein of rCTB with the B cell epitope of UreA (denoted rCTB-UA) that was expressed in *E. coli* [58]. In a mouse immunization experiment they found that rCTB-UA could induce antibodies to UreA and UreB proteins, which inhibited the activity of *H. pylori* urease. In a follow up paper, the group showed prophylactic and therapeutic dosing with rCTB-UA could protect mice from *H. pylori* infection [47]. This work has resulted in a second generation epitope vaccine (rCTB-UE) which not only consists of the original B cell epitope but a T helper cell epitope from both UreA and UreB [51,59]. In a Mongolian gerbil model of *H. pylori* infection, rCTB-UE protected against infection and decreased inflammation in the gastric tissue (inflammatory cytokines and histology) [59]. Additionally, the paper showed that the immune-protective mechanism of rCTB-UE was related to the upregulation of microRNA-155, which led to the activation of T helper (Th)1 and B cell immune responses against *H. pylori* infection. Meanwhile, Kono *et al.* showed protection from a fatal systemic infection of *Streptococcus pneumonia* in 10 day old mouse pups immunized via breast milk from mothers [60]. The mothers were intranasally immunized with Pneumococcal surface protein A (PSPA) and CTB and the anti-PSPA antibodies were present in serum and breast milk of the mothers. Through breast feeding, the offspring were protected from *S. pneumonia* infection. This study provided an important finding that mucosal immunization of a female population with vaccines containing CTB may be able to protect their offspring during early stages of life, when they are most vulnerable to respiratory diseases.

2.2. CTB-Based Immunogens against HIV

Viral pathogens have also been targeted by CTB-based vaccine development research. Given that CTB has the ability to induce potent mucosal humoral immune responses, perhaps the best opportunity to exploit CTB may be found in vaccines against mucosally transmitting viruses, such as human immunodeficiency virus (HIV-1). Indeed, a number of studies have used CTB as a mucosal adjuvant component of experimental HIV-1 vaccines [61–67].

Over the past decade, we reported a series of studies demonstrating that $rCTB\text{-}MPR_{649\text{-}684}$, a rCTB fusion protein displaying a peptide spanning the HIV-1 gp41 membrane proximal region, is capable of inducing gp41-binding antibodies in mice and rabbits [61,68–71]. These antibodies efficiently blocked transcytosis of primary HIV-1 isolates in a human tight epithelial model, suggesting that $rCTB\text{-}MPR_{649\text{-}684}$ protein may provide an effective prophylactic vaccine preventing HIV-1 mucosal transmission [61,69,70]. In a separate study, CTB was co-administered with a plasmid generated from an envelope protein ($gp145_{5m}$) of HIV-1 intramuscularly to mice [64]. The immune response by intramuscular dosing with $gp145_{5m}$ and CTB was significantly enhanced when compared to $gp145_{5m}$ alone. This study confirms that CTB, while an effective adjuvant via the nasal or oral administration routes, can also be considered for intramuscular dosing vaccine regimens to enhance the immune response. Meanwhile, Maeto *et al.* evaluated if supplementing a DNA plasmid expressing an HIV-1 Env and Interleukin-12 (IL-12) with CTB could enhance the immune response after intranasal immunization in mice [63]. IL-12 had previously been reported to enhance an antigen-specific immune response by the intranasal vaccination route [72]. In this study, not only did the combination enhance the immune response to the HIV-1 Env antigen but also significantly decreased the concentration needed to trigger Interferon (IFN)-γ, a Th1 cytokine, production by 3 times. HIV-specific CD8 responses in spleen and genital tract and genito-rectal draining lymph nodes were effectively improved, showing cytotoxic T cell responses with higher avidity, polyfunctionality and cytolytic activity. Hence, the results indicate that a greater adjuvant effect can be achieved when CTB is co-administered with another adjuvant.

2.3. Novel CTB-Based Vaccine Delivery and Antigen Conjugation Methods

In the majority of previous studies, CTB has been administered directly to mucosal surfaces via the intranasal or oral routes. In contrast, Hu *et al.* recently reported a novel approach of delivering CTB to the mucosa. In this study, they orally administered genetically engineered *Bacillus subtilis* to mice and guinea pigs, which expressed multiple epitopes of the foot-and-mouth disease virus and rCTB [73]. This method induced a significantly stronger immune response compared to the commercially available vaccine in the gut and lung, although upon viral challenge, the commercial vaccine provided slightly better protection in immunized animals.

In addition to mucosal routes of administration, CTB has been used as a component of a skin patch to vaccinate against hepatitis B virus in mice. The study was aimed at showing that transcutaneous immunization, involving microneedles which penetrate the stratum corneum without contacting nerves followed by applying a medicated patch to the area, could effectively produce antibodies against the hepatitis B surface antigen (HBsAg). CTB showed the ability to not only enhance the immune response against HBsAg but also extend the duration of protection through the transcutaneous immunization route [74]. Combined with results of other studies using a similar strategy [75–78], there is now a compelling reason to explore the development of transcutaneous vaccines including CTB as an adjuvant.

While antigen-CTB coupling has been most commonly achieved by chemical crosslinking to specific functional groups of amino acid residues or genetic fusion to the *N*- or *C*-terminus of CTB, an alternative approach has been seen in the literature that uses the CTA2 domain to link antigens to CTB [52,79,80]. For example, this approach was used for a vaccine against West Nile virus, in which the domain III (DIII) region of the virus was used as the antigen genetically fused to the CTA2 domain (see Figure 1). The DIII-CTA2 protein was co-expressed with rCTB to form a chimeric CT-like molecule, DIII-CTA2/B [52]. Intranasal delivery of DIII-CTA2/B in mice produced DIII-specific antibodies that could trigger complement-mediated killing. Although not as heavily studied as conventional CTB *C/N*-terminal fusion methods, the CTA2/B strategy may provide a useful means to develop a vaccine comprising a relatively large antigen.

Lastly, CTB has been incorporated into other alternative drug delivery systems such as liposomes, microspheres and nanoparticles. Harokopakis and colleagues found that coating liposomes with rCTB enhanced the immune response against the saliva-binding region of *S. mutans* AgI/II adhesin [81]. O'Hagan *et al.*, encapsulated rCTB in poly(lactide-co-glycolide) microparticles, which showed comparable humoral immunogenicity with CTB admixed with CT upon oral administration in mice [82]. In a more recent example, a DNA vaccine for cholera (pVAX-ctxB) encapsulated in microspheres, allowing the vaccine to pass through the acidic environment of the stomach, has shown the ability to generate an immune response in mice [83].

3. CTB in Inflammation

Besides the mucosal vaccine adjuvant activity summarized above, recent studies have revealed that CTB can also induce anti-inflammatory and regulatory T cell responses. Indeed, the protein was shown to suppress immunopathological reactions in allergy and autoimmune diseases (reviewed in: [57]). In a mouse model, the airway administration of CTB ameliorated experimental asthma [84]. Furthermore, the anti-inflammatory and immunoregulatory effects of CTB are effectively conferred on bystander protein antigens that are chemically or genetically linked to CTB; oral administration of rCTB chemically cross-linked to a peptide from the human 60 kD heat shock protein was shown to mitigate uveitis of Behcet's disease in a Phase I/II clinical trial [85]. Meanwhile, rCTB was also shown to mitigate the intestinal inflammation of Crohn's disease in mice and humans [57]. Below, we will highlight some of these and a few other recent findings regarding CTB as an anti-inflammatory agent.

3.1. CTB's Anti-Inflammatory Activity in Various Inflammatory Diseases

Type 1 Diabetes Mellitus induces cellular oxidative stress which leads to chronic inflammation and secondary effects such as: atherosclerosis, blindness, and stroke [86]. CTB has been used to target multiple anti-inflammatory agents that alone were either short lived or could not effectively induce an immune response. An example of this comes from Odumosu et al., who fused glutamic acid decarboxylase (GAD) to rCTB (GAD-rCTB) and showed suppression of dendritic cell activation in human umbilical cord blood isolated dendritic cells [87]. Dendritic cells are often implicated in islet β-cell loss in Type 1 Diabetes so this presents an attractive therapeutic option. Additionally, the group showed that pro-inflammatory cytokines, IL-12 and IL-6, were down-regulated while IL-10 was significantly increased in vitro using dendritic cells. Another study was performed incorporating GAD with rCTB and a recombinant vaccinia virus (rVV) by Denes et al., which co-administered the rVV-rCTB-GAD generated in their lab with Complete Freund's adjuvant (CFA) to see if multiple adjuvants could further enhance the immune response to the vaccine [88]. Vaccination with both rVV-rCTB-GAD alone and CFA alone showed some measureable protection in the NOD mouse model of diabetes compared to control animals given PBS at approximately 39 weeks of age. However, when rVV-rCTB-GAD and CFA were combined, hyperglycemia was delayed further to 43 weeks of age. Overall, the study showed by combining the vaccines, NOD mice could be protected from hyperglycemia and pancreatic islet inflammation better than either vaccine alone.

CTB had previously been shown to protect against uveitis resulting from Behcet's disease in a clinical trial performed in 2004 [85]. This work linked a T cell proliferative peptide (p336–351) to rCTB, which conferred protection on 5 of 8 patients following withdrawal of all immunosuppressive drugs. Other CTB conjugates have also been evaluated in a mouse model of uveitis and shown promise more recently [89]. Shil and colleagues delivered two components of the Renin-angiotensin system (RAS) to the retina, ACE2 and Ang-(1–7) by fusing them to rCTB and administering them orally to mice. Protection was noted by decreased inflammatory cytokines (e.g., IL-6, IL-1β, and TNF-α) and inflammation scoring. Additionally, these components were significantly elevated in the retina of the mice. This study showed that CTB can also be used as a delivery system to inflamed tissue and not just to enhance an immune response.

Atherosclerosis, an inflammatory condition, has recently become a target for rCTB fusion proteins [90–92]. In 2010, a mouse model of atherosclerosis showed protection by nasal administration of an rCTB fusion protein (p210-CTB) [91]. The p210 portion is derived from the apolipoprotein B-100 (ApoB100) peptide sequence as an alternative to a low density lipoprotein.

Indeed this vaccination strategy reduced atherosclerotic lesion formation and provided some clues to mechanism. IL-10 was significantly upregulated by p210-CTB, while transforming growth factor-β (TGF-β) was not, which led the authors to hypothesize that T regulatory 1 (T$_R$1) cells may be responsible for the protection. However, FoxP3 was upregulated thus the authors could not rule out some level of protection from the FoxP3$^+$ T regulatory cell population as well. Interestingly, T$_R$1 cells are believed to play a more important role when immunity is conferred through nasal administration [93]. A second rCTB-linked protein targeting both ApoB100 and cholesteryl ester transfer protein (implicated in atherosclerosis pathogenesis) was explored more recently, in a proof of concept study, in which antibodies were detected in mouse serum to the target proteins [92]. In this study, the route of administration was by foot pad injection, so it will be interesting to see if altering the route of administration will have impacts on the efficacy and/or mechanism of protection from atherosclerosis.

Liver inflammation and fibrosis were also significantly blunted by an intranasal administration of a rCTB-Sm-p40 egg antigen immunodominant peptide fusion in mice following infection with *Schistosoma mansoni*, which results in schistosomiasis [94]. This protection was associated with a significant increase in TGF-β in the mesenteric lymph node (MLN) CD4 T cells and granuloma cells. The studies on atherosclerosis and this study suggest that CTB may have a compartmentalized effect on TGF-β production in tissues, since both conjugates were administered intranasally, yet only the MLN CD4 T cells and liver granuloma cells showed elevated TGF-β.

Organ transplantation can lead to rejection through inflammation. In a rat model of kidney transplantation, an anti-inflammatory D-amino acid decapeptide, RDP58, chemically conjugated to CTB was shown to enhance the survival time compared to the therapeutic compound alone [95]. Allergic inflammation in mouse airways has also been shown to be reduced by CTB administration, not only in a preventative sense but also in mice that have already been sensitized to airway inflammation [84].

Lastly, CTB has shown in animal models as well as clinical trials to be effective in decreasing inflammation in Inflammatory Bowel Disease (IBD). IBD is subcategorized into Crohn's disease and ulcerative colitis. In 2001, Boirivant *et al.* showed that oral administration of rCTB protected against Trinitrobenzene Sulfonic Acid (TNBS) induced intestinal inflammation, which is a mouse model resembling Crohn's disease [96]. This finding was further explored to reveal that IL-12 and IFN-γ were significantly downregulated by rCTB administration in TNBS induced colitis [97]. In addition, rCTB inhibited both STAT-4 and STAT-1 activation and downregulated T-bet expression. These results showed a possible mechanism for protecting against inflammation by inhibiting Th1 cell signaling. The protection seen in the TNBS colitis model was confirmed in a human clinical trial, in which rCTB significantly decreased inflammation in mild to moderately active Crohn's disease [98]. However, IFN-γ did not correlate with the reductions in Crohn's disease activity index in the patients. This might suggest that CTB reduced inflammation in humans through more than inhibition of Th1 cell signaling. On the other hand CTB's effect in ulcerative colitis, which is another form of IBD involving inflammatory signaling and pathogenesis that is different from that of Crohn's disease, is currently not known. As noted earlier in the atherosclerosis and liver fibrosis studies, CTB's anti-inflammatory potential seems to be mediated by different pathways despite having the same route of administration. In this regard, it is of particular interest to investigate whether oral administration of CTB may have therapeutic potential in both Crohn's disease and ulcerative colitis.

3.2. Recombinant or Non-Recombinant CTB: Conflicting Results of CTB's Anti-Inflammatory Activity in in Vitro Experiments

While a number of studies have reported the anti-inflammatory activity of CTB *in vitro* and *in vivo*, the quality of the CTB used in those studies has not been consistent, which may have had a significant impact on the results of some of those studies. Hence, before concluding this section, we would like to point out the potential influence that the quality of the CTB may have on the outcome of anti-inflammatory studies, particularly those using cell culture experiments.

Many of the early studies have used non-recombinant CTB obtained from a commercial source, which is prepared from the CT holotoxin by chemical dissociation of CTA and CTB subunits. As a result, there is a trace amount of CT and CTA subunit remaining in the CTB product [99]. In a conventional *in vitro* assay using the murine macrophage cell line RAW264.7, we found that a commercial CTB product (Sigma-Aldrich, St. Louis, MO, USA; C9903), which contains ≤0.5% of CT according to the datasheet provided, significantly inhibited the production of TNFα induced by lipopolysaccharides (LPS), while rCTB produced in *E. coli* (purified to >95% homogeneous pentamer, with <0.003 endotoxin unit/μg) failed to show such an effect (Figure 2A) [100]. Notably, in this assay picomolar concentrations (<10 ng/mL) of CT exerted strong anti-inflammatory activity (Figure 2B). These results indicate that the trace amount of CT contamination in non-recombinant CTB products could have a major impact on results generated in similar assay systems. Hence, care should be taken when choosing the source of CTB for anti-inflammatory studies. It should be noted that some of the groundbreaking studies showing CTB's anti-inflammatory activity outlined above, including human clinical studies, have used rCTB. Consequently, there is compelling evidence for the immunotherapeutic potential of rCTB in various inflammatory disorders.

4. rCTB Production Methods

Given that CTB exerts strong mucosal immunomodulatory effects and rCTB is currently used in the WHO-prequalified oral cholera vaccine Dukoral® (see above), the protein has provided an attractive target for various recombinant production platforms. These include prokaryotic cells such as genetically modified *V. cholerae, E. coli, Bacillus* and *Lactobacillus*, as well as eukaryotes ranging from yeast cells to multicellular organisms such as silkworms and plants (Table 1) [100–126]. In cell culture systems rCTB is produced in fermenters and bioreactors [102–108]. Alternatively, in plant expression systems, rCTB is expressed in whole plants grown in controlled growth rooms or greenhouses [100,101,112–126].

Plant-based production of rCTB has been approached from two different angles. One approach is to vaccinate individuals with raw or minimally processed edible tissues of transgenic plants expressing rCTB (edible vaccines). For example, carrots, rice, tomatoes, potatoes and maize have been engineered to produce rCTB using transgenic technologies [101,112–119,121–125]. Among these, rice has provided the most advanced platform thus far towards an edible cholera vaccine. Yuki and colleagues have developed a transgenic rice expressing rCTB in the seed endosperm and showed that oral administration of the rice seeds induced CT holotoxin-neutralizing antibodies in mice and non-human primates [126]. No major side effects, including an IgE response to rice endogenous proteins, were observed. Interestingly, however, rCTB was shown to be N-glycosylated upon expression in plant cells. To avoid this unique post-translational modification, the same group created a mutant of CTB by replacing the corresponding Asn residue to Gln, and showed that the mutant expressed in transgenic rice endosperm was similarly effective to the original rice-based vaccine in mice and macaques [115]. These studies suggest that the rice-based experimental vaccine may provide a cost-effective oral cholera vaccine. It remains to be seen whether the approach of using edible plant tissue to deliver vaccines could be feasible from regulatory and public acceptance standpoints.

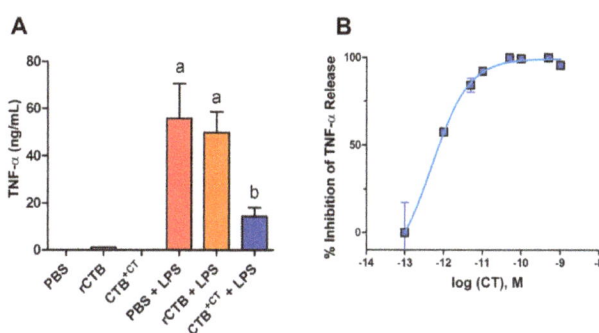

Figure 2. CT, not rCTB, inhibits the release of TNF-α by Raw 264.7 cells stimulated with LPS. (**A**) Commercial non-recombinant CTB containing a trace amount of CT (CTB^{+CT}) significantly reduces the production of TNF-α due to LPS stimulation. Raw 264.7 cells were pretreated with 10 µg/mL rCTB (produced in *E. coli* [100]), CTB^{+CT} (Sigma-Aldrich, St. Louis, MO, USA; catalog no. C9903), or PBS, and a final concentration of 1 µg/mL LPS was added and incubated for 24 h. TNF-α levels in cell supernatants were determined using a commercial ELISA kit (eBioscience, San Diego, CA, USA). Data represent the mean ± SEM ($n = 4$). a: $p < 0.001$, compared to PBS; b: $p < 0.05$, compared to PBS + LPS and rCTB + LPS (one-way ANOVA with Bonferroni multiple comparison tests); (**B**) Picomolar levels of CT inhibit the production of TNF-α. Raw 264.7 cells were pretreated for 2 h with varying concentration of CT, and a final concentration of 0.1 µg/mL LPS was added and incubated for 6 h. The 50% inhibitory concentration (IC$_{50}$) of CT was determined by non-linear regression analysis (GraphPad Prism 5.0, GraphPad Software, Inc., La Jolla, CA, USA) to be 0.49 pM. Data represent the mean ± SEM ($n = 2$). The TNF-α level of PBS + LPS was 4516.8 ± 791.1 pg/mL (mean ± SEM; $n = 2$).

Table 1. rCTB Production Systems.

System	Expression Host	Mode of Expression	Functional Evaluation	CTB Yield	Purification	Reference
	V. cholerae	Expression plasmid: (pML-LCTBtac2) transformation	Affinity for GM1-ganglioside confirmed (GM1-ELISA) and immunogenic in mice	1g/L culture	Affinity chromatography (lyso-GM1 ganglioside Spherosil column)	[102]
		Expression plasmid: pQE30 transformation	Affinity for GM1-ganglioside confirmed (GM1-ELISA)	9 mg/L culture	IMAC* Purification and membrane-filtration	[103]
	E. coli	Expression plasmid: pAE_ctxB transformation	Detected by anti-CT antibody (Western Blot)	1.2g/L culture	Centrifugation	[104]
Bacterial fermentation		Expression plasmid: pTG8148 transformation	Affinity for GM1-ganglioside confirmed (GM1-ELISA)	1 g/L culture	Cation exchange Chromatography (S-Sepharose FF column)	[105]
		Expression plasmid: pGEM-T-ctxB transformation	Detected by anti-CT antibody (Western Blot)	80 mg/L culture	Centrifugation	[106]
	Lactobacilli	Expression plasmid: (pLDH-CTB-His-Term) transformation	Affinity for GM1-ganglioside confirmed (GM1-ELISA) and immunogenic in mice	1 mg/L culture	IMAC Purification	[107]
	Bacillus brevis	Expression plasmid: (pNU212-CTB) transformation	Affinity for GM1-ganglioside confirmed (GM1-ELISA)	N/A	Affinity chromatography (D-galactose-agarose column)	[108]
Yeast culture	Pichia pastoris	Expression plasmid: (pB) transformation	Affinity for GM1-ganglioside confirmed (GM1-ELISA) and immunogenic in mice	50 mg/L culture	IMAC Purification	[109]
Insect cell culture	B. mori (silkworm larvae)	Baculovirus expression system	Affinity for GM1-ganglioside confirmed (GM1-ELISA) and immunogenic in mice	54.4 mg/L larval hemolymph	Centrifugation	[110]
	Solanum tubersosum (potato)	Transgenic (Agrobacterium-mediated transformation.)	Affinity for GM1-ganglioside confirmed (GM1-ELISA)	0.5% of total soluble protein	Centrifugation	[112]
Plants		Transgenic (Agrobacterium-mediated transformation)	Affinity for GM1-ganglioside confirmed (GM1-ELISA)	0.3% of total soluble protein	Non-purified (edible plant vaccine)	[124]
	Daucus carota (carrot)	Transgenic (Agrobacterium-mediated transformation)	Affinity for GM1-ganglioside confirmed (GM1-ELISA)	0.48% of total soluble protein	Non-purified (edible vaccine)	[113]

Table 1. *Cont.*

System	Expression Host	Functional Evaluation	Mode of Expression	CTB Yield	Purification	Reference
Plants	*Oryza sativa* (rice seed)	Affinity for GM1-ganglioside confirmed	Transgenic (*Agrobacterium*-mediated transformation)	2.1% of total soluble protein	Non-purified (edible vaccine)	[101]
		Detected by anti-CTB antibody (Western Blot)	Transgenic (*Agrobacterium*-mediated transformation)	3.37 mg/g rice seeds	IMAC Purification	[114]
		Affinity for GM1-ganglioside confirmed	Transgenic (*Agrobacterium*-mediated transformation)	2.35 mg/g of seed	Non-purified (edible vaccine)	[115]
		Affinity for GM1-ganglioside confirmed	Transgenic (Expression plasmid biolistic-mediated transformation)	2.1% of total seed	Non-purified (edible vaccine)	[116]
	Latuca sativa (lettuce)	Affinity for GM1-ganglioside confirmed (GM1-ELISA)	Transgenic (*Agrobacterium*-mediated transformation)	0.24% of total soluble protein	Non-purified (edible vaccine)	[117]
	Lycopersicon esculentum (tomato)	Affinity for GM1-ganglioside confirmed (GM1-ELISA)	Transgenic (*Agrobacterium*-mediated transformation)	0.04% of total soluble protein	Non-purified (edible vaccine)	[118]
		Detected by anti-CTB antibody and immunogenic in mice	Transgenic (*Agrobacterium*-mediated transformation)	0.081% of total soluble protein	Non-purified (edible vaccine)	[125]
		Affinity for GM1-ganglioside confirmed (GM1-ELISA) and immunogenic in mice	Transient (plant viral vectors)	1.5 mg/g leaf material or 49.9% of total soluble protein	IMAC Purification, Hydroxyapatite Chromatography (CHT column)	[100]
	Nicotiana benthamiana (a tobacco relative)	Affinity for GM1-ganglioside confirmed (GM1-ELISA)	Transgenic (*Agrobacterium*-mediated transformation)	0.56% of total soluble protein	Centrifugation	[112]
		Affinity for GM1-ganglioside confirmed (GM1-ELISA)	Transgenic (*Agrobacterium*-mediated transformation)	0.095% of total soluble leaf protein	Immunoaffinity column chromatography (anti-CT IgG resin)	[119]
		Affinity for GM1-ganglioside confirmed (GM1-ELISA)	Transient (plant viral vectors)	0.14% of total soluble leaf protein	Centrifugation	[120]
	Nicotiana tabacum (tobacco)	Affinity for GM1-ganglioside confirmed (GM1-ELISA)	Transplastomic (Expression plasmid [pLD-LH-CTB] microprojectile bombardment)	4.1% of total soluble protein	Non-purified crude leaf extract	[121]
	Robusta sp. (banana callus)	Detected by anti-CT antibody (Western Blot)	Transgenic (*Agrobacterium*-mediated transformation)	125 µg/g callus tissue	Non-purified (edible vaccine)	[122]
	Zea mays (maize seed)	Affinity for GM1-ganglioside confirmed and immunogenic in mice	Transgenic (Plasmid microprojectile bombardment)	1.56 µg/g dry seed weight	Non-purified (edible vaccine)	[123]

* Immobilized metal ion affinity chromatography (IMAC).

A second approach is to produce rCTB in non-food or feed plants and isolate the immunogen from the tissue for vaccination. This has been undertaken in several tobacco family plants (*Nicotiana tabacum* and *N. benthamiana*) [100,112,119–121]. Daniell et al expressed rCTB in chloroplasts of transplastomic tobacco plants which enabled a high-level accumulation of glycosylation-free rCTB in leaf tissue. Alternatively, we have recently developed a transient mass production platform for a non-glycosylated variant (Asn4→Ser) of rCTB in *N. benthamiana* using a plant virus vector system [100]. Over 1 g of the rCTB variant was produced in 1 kg of tobacco leaf (corresponding to 1000 doses of Dukoral® vaccine) in 5 days post vector inoculation. The protein was efficiently purified via conventional chromatographical processes and shown to be virtually identical to original CTB in terms of physicochemical stability, GM1-ganglioside binding affinity and oral immunogenicity in mice. A major advantage to this method of production is that it is rapidly scalable based on the need for rCTB production, which could obviate the need for large vaccine stockpiling. Although the requirement of protein purification may reduce a previously conceived advantage offered by plant-based systems, it would in turn provide superior controls to the quality and dosage of vaccines and eliminate potential side effects associated with impurities.

5. Concluding Remarks

While first being recognized for its role in the delivery of the virulence factor of *V. cholerae*, the works highlighted in this paper show CTB's broad utility as a cholera vaccine immunogen, vaccine adjuvant (through co-administration or conjugation), immune modulator and/or anti-inflammatory agent. This has led to the development of various rCTB expression systems in an effort to make the protein more efficient and widely available. Given that CTB appears to provide additional efficacy to killed bacteria-based cholera vaccines, development of alternative rCTB production and delivery methods may significantly contribute to cholera prevention and control. Because of the capacity to induce potent mucosal humoral immune responses, antigen-CTB fusion provides a promising strategy for vaccines against enteric pathogens and mucosally transmitted diseases. On the other hand, the immunotherapeutic potential of CTB in inflammatory diseases warrants further investigations; despite a number of studies demonstrating CTB's anti-inflammatory effects, the underlying mechanism remains to be fully disclosed. This could be partly due to the inconsistent quality of CTB used in those studies and also attributed to different pathways altered by CTB, depending on the route/mode of administration and inflammatory conditions. Since many inflammatory diseases involve chronic and recurring inflammation, long-term immunological and toxicological impacts of repeated CTB administration need to be investigated. Nevertheless, several early-stage clinical trials have paved the way for the development of CTB-based anti-inflammatory agents. In summary, CTB has shown utility in many disease states and may ultimately be a compound with many diverse applications. The works highlighted in this paper show great promise for a single protein having multiple applications and perhaps allowing for an evolution in vaccine development.

Acknowledgments

We thank Adam Husk for critical reading of the manuscript. This manuscript is based on work supported by DoD/USMRAA/TATRC/W81XWH-10-2-0082-CLIN1; W81XWH-10-2-0082-CLIN2 and the Helmsley Charitable Trust Fund. KJB was supported by a T32 Environmental Health Sciences Grant (3 T32 ES 11564-10 S1).

Author Contributions

K.J.B. and N.M. conceived and designed the review idea and contents. J.M.R. and K.T.H. contributed to table and figure generation and revising the paper.

Conflicts of Interest

The author declares no conflict of interest.

References

1. Lutz, C.; Erken, M.; Noorian, P.; Sun, S.; McDougald, D. Environmental reservoirs and mechanisms of persistence of *Vibrio cholerae*. *Front. Microbiol.* **2013**, *4*, 375.
2. Chatterjee, S.; Ghosh, K.; Raychoudhuri, A.; Pan, A.; Bhattacharya, M.K.; Mukhopadhyay, A.K.; Ramamurthy, T.; Bhattacharya, S.K.; Nandy, R.K. Phenotypic and genotypic traits and epidemiological implication of *Vibrio cholerae* O1 and O139 strains in India during 2003. *J. Med. Microbiol.* **2007**, *56*, 824–832.
3. Harris, J.B.; LaRocque, R.C.; Qadri, F.; Ryan, E.T.; Calderwood, S.B. Cholera. *Lancet* **2012**, *379*, 2466–2476.
4. Merrell, D.S.; Butler, S.M.; Qadri, F.; Dolganov, N.A.; Alam, A.; Cohen, M.B.; Calderwood, S.B.; Schoolnik, G.K.; Camilli, A. Host-induced epidemic spread of the cholera bacterium. *Nature* **2002**, *417*, 642–645.
5. Stine, O.C.; Morris, J.G., Jr. Circulation and transmission of clones of *Vibrio cholerae* during cholera outbreaks. *Curr. Top. Microbiol. Immunol.* **2014**, *379*, 181–193.
6. Leung, D.T.; Chowdhury, F.; Calderwood, S.B.; Qadri, F.; Ryan, E.T. Immune responses to cholera in children. *Expert Rev. Anti Infect. Ther.* **2012**, *10*, 435–444.
7. Piarroux, R.; Faucher, B. Cholera epidemics in 2010: Respective roles of environment, strain changes, and human-driven dissemination. *Clin. Microbiol. Infect.* **2012**, *18*, 231–238.
8. Orata, F.D.; Keim, P.S.; Boucher, Y. The 2010 cholera outbreak in Haiti: How science solved a controversy. *PLoS Pathog.* **2014**, *10*, e1003967.
9. WHO. Cholera 2012. *Wkly. Epidemiol. Rec.* **2013**, *88*, 321–336.
10. WHO. Cholera vaccines: WHO position paper. *Wkly. Epidemiol. Rec.* **2010**, *85*, 117–128.
11. Siddique, A.K.; Nair, G.B.; Alam, M.; Sack, D.A.; Huq, A.; Nizam, A.; Longini, I.M., Jr.; Qadri, F.; Faruque, S.M.; Colwell, R.R.; *et al.* El Tor cholera with severe disease: A new threat to Asia and beyond. *Epidemiol. Infect.* **2010**, *138*, 347–352.

12. Wernick, N.L.; Chinnapen, D.J.; Cho, J.A.; Lencer, W.I. Cholera toxin: An intracellular journey into the cytosol by way of the endoplasmic reticulum. *Toxins* **2010**, *2*, 310–325.

13. Lencer, W.I.; Tsai, B. The intracellular voyage of cholera toxin: Going retro. *Trends Biochem. Sci.* **2003**, *28*, 639–645.

14. Chinnapen, D.J.; Chinnapen, H.; Saslowsky, D.; Lencer, W.I. Rafting with cholera toxin: Endocytosis and trafficking from plasma membrane to ER. *FEMS Microbiol. Lett.* **2007**, *266*, 129–137.

15. Sanchez, J.; Holmgren, J. Cholera toxin structure, gene regulation and pathophysiological and immunological aspects. *Cell Mol. Life Sci.* **2008**, *65*, 1347–1360.

16. Zhang, R.G.; Scott, D.L.; Westbrook, M.L.; Nance, S.; Spangler, B.D.; Shipley, G.G.; Westbrook, E.M. The three-dimensional crystal structure of cholera toxin. *J. Mol. Biol.* **1995**, *251*, 563–573.

17. Sanchez, J.; Holmgren, J. Cholera toxin—A foe & A friend. *Indian J. Med. Res.* **2011**, *133*, 153–163.

18. Basu, I.; Mukhopadhyay, C. Insights into Binding of Cholera Toxin to GM1 Containing Membrane. *Langmuir* **2014**, *30*, 15244–15252.

19. Saslowsky, D.E.; te Welscher, Y.M.; Chinnapen, D.J.; Wagner, J.S.; Wan, J.; Kern, E.; Lencer, W.I. Ganglioside GM1-mediated transcytosis of cholera toxin bypasses the retrograde pathway and depends on the structure of the ceramide domain. *J. Biol. Chem.* **2013**, *288*, 25804–25809.

20. Charles, R.C.; Hilaire, I.J.; Mayo-Smith, L.M.; Teng, J.E.; Jerome, J.G.; Franke, M.F.; Saha, A.; Yu, Y.; Kovac, P.; Calderwood, S.B.; *et al.* Immunogenicity of a killed bivalent (O1 and O139) whole cell oral cholera vaccine, Shanchol, in Haiti. *PLoS Negl. Trop. Dis.* **2014**, *8*, e2828.

21. Schaetti, C.; Weiss, M.G.; Ali, S.M.; Chaignat, C.L.; Khatib, A.M.; Reyburn, R.; Duintjer Tebbens, R.J.; Hutubessy, R. Costs of illness due to cholera, costs of immunization and cost-effectiveness of an oral cholera mass vaccination campaign in Zanzibar. *PLoS Negl. Trop. Dis.* **2012**, *6*, e1844.

22. Lajoinie, A.; Henin, E.; Kassai, B.; Terry, D. Solid oral forms availability in children: A cost saving investigation. *Br. J. Clin. Pharmacol.* **2014**, *78*, 1080–1089.

23. Clemens, J.D.; Sack, D.A.; Harris, J.R.; Chakraborty, J.; Khan, M.R.; Stanton, B.F.; Kay, B.A.; Khan, M.U.; Yunus, M.; Atkinson, W.; *et al.* Field trial of oral cholera vaccines in Bangladesh. *Lancet* **1986**, *2*, 124–127.

24. Clemens, J.D.; Sack, D.A.; Harris, J.R.; van Loon, F.; Chakraborty, J.; Ahmed, F.; Rao, M.R.; Khan, M.R.; Yunus, M.; Huda, N.; *et al.* Field trial of oral cholera vaccines in Bangladesh: Results from three-year follow-up. *Lancet* **1990**, *335*, 270–273.

25. Alam, M.M.; Leung, D.T.; Akhtar, M.; Nazim, M.; Akter, S.; Uddin, T.; Khanam, F.; Mahbuba, D.A.; Ahmad, S.M.; Bhuiyan, T.R.; *et al.* Antibody avidity in humoral immune responses in Bangladeshi children and adults following administration of an oral killed cholera vaccine. *Clin. Vaccine Immunol.* **2013**, *20*, 1541–1548.

26. Lucas, M.E.; Deen, J.L.; von Seidlein, L.; Wang, X.Y.; Ampuero, J.; Puri, M.; Ali, M.; Ansaruzzaman, M.; Amos, J.; Macuamule, A.; *et al.* Effectiveness of mass oral cholera vaccination in Beira, Mozambique. *N. Engl. J. Med.* **2005**, *352*, 757–767.

27. Taylor, D.N.; Cardenas, V.; Sanchez, J.L.; Begue, R.E.; Gilman, R.; Bautista, C.; Perez, J.; Puga, R.; Gaillour, A.; Meza, R.; *et al.* Two-year study of the protective efficacy of the oral whole cell plus recombinant B subunit cholera vaccine in Peru. *J. Infect. Dis.* **2000**, *181*, 1667–1673.

28. Clemens, J.D.; Jertborn, M.; Sack, D.; Stanton, B.; Holmgren, J.; Khan, M.R.; Huda, S. Effect of neutralization of gastric acid on immune responses to an oral B subunit, killed whole-cell cholera vaccine. *J. Infect. Dis.* **1986**, *154*, 175–178.

29. Saha, A.; Chowdhury, M.I.; Khanam, F.; Bhuiyan, M.S.; Chowdhury, F.; Khan, A.I.; Khan, I.A.; Clemens, J.; Ali, M.; Cravioto, A.; *et al.* Safety and immunogenicity study of a killed bivalent (O1 and O139) whole-cell oral cholera vaccine Shanchol, in Bangladeshi adults and children as young as 1 year of age. *Vaccine* **2011**, *29*, 8285–8292.

30. Luquero, F.J.; Grout, L.; Ciglenecki, I.; Sakoba, K.; Traore, B.; Heile, M.; Diallo, A.A.; Itama, C.; Page, A.L.; Quilici, M.L.; *et al.* Use of *Vibrio cholerae* vaccine in an outbreak in Guinea. *N. Engl. J. Med.* **2014**, *370*, 2111–2120.

31. Kang, S.S.; Yang, J.S.; Kim, K.W.; Yun, C.H.; Holmgren, J.; Czerkinsky, C.; Han, S.H. Anti-bacterial and anti-toxic immunity induced by a killed whole-cell-cholera toxin B subunit cholera vaccine is essential for protection against lethal bacterial infection in mouse pulmonary cholera model. *Mucosal. Immunol.* **2013**, *6*, 826–837.

32. Holmgren, J.; Adamsson, J.; Anjuere, F.; Clemens, J.; Czerkinsky, C.; Eriksson, K.; Flach, C.F.; George-Chandy, A.; Harandi, A.M.; Lebens, M.; *et al.* Mucosal adjuvants and anti-infection and anti-immunopathology vaccines based on cholera toxin, cholera toxin B subunit and CpG DNA. *Immunol. Lett.* **2005**, *97*, 181–188.

33. Williams, N.A.; Hirst, T.R.; Nashar, T.O. Immune modulation by the cholera-like enterotoxins: From adjuvant to therapeutic. *Immunol. Today* **1999**, *20*, 95–101.

34. Bharati, K.; Ganguly, N.K. Cholera toxin: A paradigm of a multifunctional protein. *Indian J. Med. Res.* **2011**, *133*, 179–187.

35. Holmgren, J.; Lycke, N.; Czerkinsky, C. Cholera toxin and cholera B subunit as oral-mucosal adjuvant and antigen vector systems. *Vaccine* **1993**, *11*, 1179–1184.

36. Bowman, C.C.; Selgrade, M.K. Utility of rodent models for evaluating protein allergenicity. *Regul. Toxicol. Pharmacol.* **2009**, *54*, S58–S61.

37. Oyoshi, M.K.; Oettgen, H.C.; Chatila, T.A.; Geha, R.S.; Bryce, P.J. Food allergy: Insights into etiology, prevention, and treatment provided by murine models. *J. Allergy Clin. Immunol.* **2014**, *133*, 309–317.

38. Elson, C.O.; Ealding, W. Generalized systemic and mucosal immunity in mice after mucosal stimulation with cholera toxin. *J. Immunol.* **1984**, *132*, 2736–2741.

39. Jackson, R.J.; Fujihashi, K.; Xu-Amano, J.; Kiyono, H.; Elson, C.O.; McGhee, J.R. Optimizing oral vaccines: Induction of systemic and mucosal B-cell and antibody responses to tetanus toxoid by use of cholera toxin as an adjuvant. *Infect. Immun.* **1993**, *61*, 4272–4279.

40. Bourguin, I.; Chardes, T.; Bout, D. Oral immunization with Toxoplasma gondii antigens in association with cholera toxin induces enhanced protective and cell-mediated immunity in C57BL/6 mice. *Infect. Immun.* **1993**, *61*, 2082–2088.

41. Holmgren, J.; Czerkinsky, C.; Lycke, N.; Svennerholm, A.M. Strategies for the induction of immune responses at mucosal surfaces making use of cholera toxin B subunit as immunogen, carrier, and adjuvant. *Am. J. Trop. Med. Hyg.* **1994**, *50*, 42–54.

42. Blanchard, T.G.; Lycke, N.; Czinn, S.J.; Nedrud, J.G. Recombinant cholera toxin B subunit is not an effective mucosal adjuvant for oral immunization of mice against *Helicobacter felis*. *Immunology* **1998**, *94*, 22–27.

43. Kubota, E.; Joh, T.; Tanida, S.; Sasaki, M.; Kataoka, H.; Watanabe, K.; Itoh, K.; Oshima, T.; Ogasawara, N.; Togawa, S.; *et al.* Oral vaccination against Helicobacter pylori with recombinant cholera toxin B-subunit. *Helicobacter* **2005**, *10*, 345–352.

44. Mutsch, M.; Zhou, W.; Rhodes, P.; Bopp, M.; Chen, R.T.; Linder, T.; Spyr, C.; Steffen, R. Use of the inactivated intranasal influenza vaccine and the risk of Bell's palsy in Switzerland. *N. Engl. J. Med.* **2004**, *350*, 896–903.

45. Couch, R.B. Nasal vaccination, *Escherichia coli* enterotoxin, and Bell's palsy. *N. Engl. J. Med.* **2004**, *350*, 860–861.

46. Rath, B.; Linder, T.; Cornblath, D.; Hudson, M.; Fernandopulle, R.; Hartmann, K.; Heininger, U.; Izurieta, H.; Killion, L.; Kokotis, P.; *et al.* All that palsies is not Bell's -the need to define Bell's palsy as an adverse event following immunization. *Vaccine* **2007**, *26*, 1–14.

47. Guo, L.; Liu, K.; Xu, G.; Li, X.; Tu, J.; Tang, F.; Xing, Y.; Xi, T. Prophylactic and therapeutic efficacy of the epitope vaccine CTB-UA against Helicobacter pylori infection in a BALB/c mice model. *Appl. Microbiol. Biotechnol.* **2012**, *95*, 1437–1444.

48. De Geus, E.D.; van Haarlem, D.A.; Poetri, O.N.; de Wit, J.J.; Vervelde, L. A lack of antibody formation against inactivated influenza virus after aerosol vaccination in presence or absence of adjuvantia. *Vet. Immunol. Immunopathol.* **2011**, *143*, 143–147.

49. Boustanshenas, M.; Bakhshi, B.; Ghorbani, M. Investigation into immunological responses against a native recombinant CTB whole-cell *Vibrio cholerae* vaccine in a rabbit model. *J. Appl. Microbiol.* **2013**, *114*, 509–515.

50. Baptista, A.A.; Donato, T.C.; Garcia, K.C.; Goncalves, G.A.; Coppola, M.P.; Okamoto, A.S.; Sequeira, J.L.; Andreatti Filho, R.L. Immune response of broiler chickens immunized orally with the recombinant proteins flagellin and the subunit B of cholera toxin associated with *Lactobacillus spp. Poult. Sci.* **2014**, *93*, 39–45.

51. Guo, L.; Yin, R.; Liu, K.; Lv, X.; Li, Y.; Duan, X.; Chu, Y.; Xi, T.; Xing, Y. Immunological features and efficacy of a multi-epitope vaccine CTB-UE against H. pylori in BALB/c mice model. *Appl. Microbiol. Biotechnol.* **2014**, *98*, 3495–3507.

52. Tinker, J.K.; Yan, J.; Knippel, R.J.; Panayiotou, P.; Cornell, K.A. Immunogenicity of a West Nile virus DIII-cholera toxin A2/B chimera after intranasal delivery. *Toxins* **2014**, *6*, 1397–1418.

53. Czerkinsky, C.; Sun, J.B.; Lebens, M.; Li, B.L.; Rask, C.; Lindblad, M.; Holmgren, J. Cholera toxin B subunit as transmucosal carrier-delivery and immunomodulating system for induction of antiinfective and antipathological immunity. *Ann. N. Y. Acad. Sci.* **1996**, *778*, 185–193.

54. Martell, B.A.; Orson, F.M.; Poling, J.; Mitchell, E.; Rossen, R.D.; Gardner, T.; Kosten, T.R. Cocaine vaccine for the treatment of cocaine dependence in methadone-maintained patients: A randomized, double-blind, placebo-controlled efficacy trial. *Arch. Gen. Psychiatry* **2009**, *66*, 1116–1123.

55. Orson, F.M.; Rossen, R.D.; Shen, X.; Lopez, A.Y.; Wu, Y.; Kosten, T.R. Spontaneous development of IgM anti-cocaine antibodies in habitual cocaine users: Effect on IgG antibody responses to a cocaine cholera toxin B conjugate vaccine. *Am. J. Addict.* **2013**, *22*, 169–174.

56. Lebens, M.; Holmgren, J. Mucosal vaccines based on the use of cholera toxin B subunit as immunogen and antigen carrier. *Dev. Biol. Stand.* **1994**, *82*, 215–227.

57. Sun, J.B.; Czerkinsky, C.; Holmgren, J. Mucosally induced immunological tolerance, regulatory T cells and the adjuvant effect by cholera toxin B subunit. *Scand. J. Immunol.* **2010**, *71*, 1–11.

58. Guo, L.; Li, X.; Tang, F.; He, Y.; Xing, Y.; Deng, X.; Xi, T. Immunological features and the ability of inhibitory effects on enzymatic activity of an epitope vaccine composed of cholera toxin B subunit and B cell epitope from *Helicobacter pylori* urease A subunit. *Appl. Microbiol. Biotechnol.* **2012**, *93*, 1937–1945.

59. Lv, X.; Yang, J.; Song, H.; Li, T.; Guo, L.; Xing, Y.; Xi, T. Therapeutic efficacy of the multi-epitope vaccine CTB-UE against Helicobacter pylori infection in a Mongolian gerbil model and its microRNA-155-associated immuno-protective mechanism. *Vaccine* **2014**, *32*, 5343–5352.

60. Kono, M.; Hotomi, M.; Hollingshead, S.K.; Briles, D.E.; Yamanaka, N. Maternal immunization with pneumococcal surface protein A protects against pneumococcal infections among derived offspring. *PLoS One* **2011**, *6*, e27102.

61. Matoba, N.; Magerus, A.; Geyer, B.C.; Zhang, Y.; Muralidharan, M.; Alfsen, A.; Arntzen, C.J.; Bomsel, M.; Mor, T.S. A mucosally targeted subunit vaccine candidate eliciting HIV-1 transcytosis-blocking Abs. *Proc. Natl. Acad. Sci. USA* **2004**, *101*, 13584–13589.

62. Nowroozalizadeh, S.; Jansson, M.; Adamsson, J.; Lindblad, M.; Fenyo, E.M.; Holmgren, J.; Harandi, A.M. Suppression of HIV replication *in vitro* by CpG and CpG conjugated to the non toxic B subunit of cholera toxin. *Curr. HIV Res.* **2008**, *6*, 230–238.

63. Maeto, C.; Rodriguez, A.M.; Holgado, M.P.; Falivene, J.; Gherardi, M.M. Novel mucosal DNA-MVA HIV vaccination in which DNA-IL-12 plus cholera toxin B subunit (CTB) cooperates to enhance cellular systemic and mucosal genital tract immunity. *PLoS One* **2014**, *9*, e107524.

64. Hou, J.; Liu, Y.; Hsi, J.; Wang, H.; Tao, R.; Shao, Y. Cholera toxin B subunit acts as a potent systemic adjuvant for HIV-1 DNA vaccination intramuscularly in mice. *Hum. Vaccin Immunother.* **2014**, *10*, 1274–1283.

65. Hervouet, C.; Luci, C.; Cuburu, N.; Cremel, M.; Bekri, S.; Vimeux, L.; Maranon, C.; Czerkinsky, C.; Hosmalin, A.; Anjuere, F. Sublingual immunization with an HIV subunit vaccine induces antibodies and cytotoxic T cells in the mouse female genital tract. *Vaccine* **2010**, *28*, 5582–5590.

66. Boberg, A.; Gaunitz, S.; Brave, A.; Wahren, B.; Carlin, N. Enhancement of epitope-specific cellular immune responses by immunization with HIV-1 peptides genetically conjugated to the B-subunit of recombinant cholera toxin. *Vaccine* **2008**, *26*, 5079–5082.

67. Zolla-Pazner, S.; Kong, X.P.; Jiang, X.; Cardozo, T.; Nadas, A.; Cohen, S.; Totrov, M.; Seaman, M.S.; Wang, S.; Lu, S. Cross-clade HIV-1 neutralizing antibodies induced with V3-scaffold protein immunogens following priming with gp120 DNA. *J. Virol.* **2011**, *85*, 9887–9898.

68. Matoba, N.; Kajiura, H.; Cherni, I.; Doran, J.D.; Bomsel, M.; Fujiyama, K.; Mor, T.S. Biochemical and immunological characterization of the plant-derived candidate human immunodeficiency virus type 1 mucosal vaccine CTB-MPR(649–684). *Plant Biotechnol. J.* **2009**, *7*, 129–145.

69. Matoba, N.; Griffin, T.A.; Mittman, M.; Doran, J.D.; Hanson, C.V.; Montefiori, D.; Alfsen, A.; Bomsel, M.; Mor, T.S. Transcytosis-blocking Abs elicited by an oligomeric immunogen based on the membrane proximal region of HIV-1 gp41 target non-neutralizing epitopes. *Curr. HIV Res.* **2008**, *6*, 218–229.

70. Matoba, N.; Geyer, B.C.; Kilbourne, J.; Alfsen, A.; Bomsel, M.; Mor, T.S. Humoral immune responses by prime-boost heterologous route immunizations with CTB-MPR(649–684), a mucosal subunit HIV/AIDS vaccine candidate. *Vaccine* **2006**, *24*, 5047–5055.

71. Matoba, N.; Shah, N.R.; Mor, T.S. Humoral immunogenicity of an HIV-1 envelope residue 649–684 membrane-proximal region peptide fused to the plague antigen F1-V. *Vaccine* **2011**, *29*, 5584–5590.

72. Arulanandam, B.P.; Metzger, D.W. Modulation of mucosal and systemic immunity by intranasal interleukin 12 delivery. *Vaccine* **1999**, *17*, 252–260.

73. Hu, B.; Li, C.; Lu, H.; Zhu, Z.; Du, S.; Ye, M.; Tan, L.; Ren, D.; Han, J.; Kan, S.; *et al.* Immune responses to the oral administration of recombinant *Bacillus subtilis* expressing multi-epitopes of foot-and-mouth disease virus and a cholera toxin B subunit. *J. Virol. Methods* **2011**, *171*, 272–279.

74. Guo, L.; Qiu, Y.; Chen, J.; Zhang, S.; Xu, B.; Gao, Y. Effective transcutaneous immunization against hepatitis B virus by a combined approach of hydrogel patch formulation and microneedle arrays. *Biomed. Microdevices* **2013**, *15*, 1077–1085.

75. Anjuere, F.; George-Chandy, A.; Audant, F.; Rousseau, D.; Holmgren, J.; Czerkinsky, C. Transcutaneous immunization with cholera toxin B subunit adjuvant suppresses IgE antibody responses via selective induction of Th1 immune responses. *J. Immunol.* **2003**, *170*, 1586–1592.

76. Glenn, G.M.; Scharton-Kersten, T.; Vassell, R.; Matyas, G.R.; Alving, C.R. Transcutaneous immunization with bacterial ADP-ribosylating exotoxins as antigens and adjuvants. *Infect. Immun.* **1999**, *67*, 1100–1106.

77. Maheshwari, C.; Pandey, R.S.; Chaurasiya, A.; Kumar, A.; Selvam, D.T.; Prasad, G.B.; Dixit, V.K. Non-ionic surfactant vesicles mediated transcutaneous immunization against hepatitis B. *Int. Immunopharmacol.* **2011**, *11*, 1516–1522.

78. Harakuni, T.; Kohama, H.; Tadano, M.; Uechi, G.; Tsuji, N.; Matsumoto, Y.; Miyata, T.; Tsuboi, T.; Oku, H.; Arakawa, T. Mucosal vaccination approach against mosquito-borne Japanese encephalitis virus. *Jpn. J. Infect. Dis.* **2009**, *62*, 37–45.

79. Harokopakis, E.; Hajishengallis, G.; Greenway, T.E.; Russell, M.W.; Michalek, S.M. Mucosal immunogenicity of a recombinant Salmonella typhimurium-cloned heterologous antigen in the absence or presence of coexpressed cholera toxin A2 and B subunits. *Infect. Immun.* **1997**, *65*, 1445–1454.

80. Martin, M.; Hajishengallis, G.; Metzger, D.J.; Michalek, S.M.; Connell, T.D.; Russell, M.W. Recombinant antigen-enterotoxin A2/B chimeric mucosal immunogens differentially enhance antibody responses and B7-dependent costimulation of CD4(+) T cells. *Infect. Immun.* **2001**, *69*, 252–261.

81. Harokopakis, E.; Childers, N.K.; Michalek, S.M.; Zhang, S.S.; Tomasi, M. Conjugation of cholera toxin or its B subunit to liposomes for targeted delivery of antigens. *J. Immunol. Methods* **1995**, *185*, 31–42.

82. O'Hagan, D.T.; McGee, J.P.; Lindblad, M.; Holmgren, J. Cholera toxin B Subunit (CTB) entrapped in microparticles shows comparable immunogenicity to CTB mixed with whole cholera toxin following oral immunization. *Int. J. Pharm.* **1995**, *119*, 251–255.

83. Rosli, R.; Nograles, N.; Hanafi, A.; Nor Shamsudin, M.; Abdullah, S. Mucosal genetic immunization through microsphere-based oral carriers. *Hum. Vaccin Immunother.* **2013**, *9*, 2222–2227.

84. Smits, H.H.; Gloudemans, A.K.; van Nimwegen, M.; Willart, M.A.; Soullie, T.; Muskens, F.; de Jong, E.C.; Boon, L.; Pilette, C.; Johansen, F.E.; *et al.* Cholera toxin B suppresses allergic inflammation through induction of secretory IgA. *Mucosal. Immunol.* **2009**, *2*, 331–339.

85. Stanford, M.; Whittall, T.; Bergmeier, L.A.; Lindblad, M.; Lundin, S.; Shinnick, T.; Mizushima, Y.; Holmgren, J.; Lehner, T. Oral tolerization with peptide 336–351 linked to cholera toxin B subunit in preventing relapses of uveitis in Behcet's disease. *Clin. Exp. Immunol.* **2004**, *137*, 201–208.

86. Libby, P.; Nathan, D.M.; Abraham, K.; Brunzell, J.D.; Fradkin, J.E.; Haffner, S.M.; Hsueh, W.; Rewers, M.; Roberts, B.T.; Savage, P.J.; *et al.* Report of the National Heart, Lung, and Blood Institute-National Institute of Diabetes and Digestive and Kidney Diseases Working Group on Cardiovascular Complications of Type 1 Diabetes Mellitus. *Circulation* **2005**, *111*, 3489–3493.

87. Odumosu, O.; Nicholas, D.; Payne, K.; Langridge, W. Cholera toxin B subunit linked to glutamic acid decarboxylase suppresses dendritic cell maturation and function. *Vaccine* **2011**, *29*, 8451–8458.

88. Denes, B.; Fodor, I.; Langridge, W.H. Persistent suppression of type 1 diabetes by a multicomponent vaccine containing a cholera toxin B subunit-autoantigen fusion protein and complete Freund's adjuvant. *Clin. Dev. Immunol.* **2013**, *2013*, 578786.

89. Shil, P.K.; Kwon, K.C.; Zhu, P.; Verma, A.; Daniell, H.; Li, Q. Oral delivery of ACE2/Ang-(1–7) bioencapsulated in plant cells protects against experimental uveitis and autoimmune uveoretinitis. *Mol. Ther.* **2014**, *22*, 2069–2082.

90. Xiong, Q.; Li, J.; Jin, L.; Liu, J.; Li, T. Nasal immunization with heat shock protein 65 attenuates atherosclerosis and reduces serum lipids in cholesterol-fed wild-type rabbits probably through different mechanisms. *Immunol. Lett.* **2009**, *125*, 40–45.

91. Klingenberg, R.; Lebens, M.; Hermansson, A.; Fredrikson, G.N.; Strodthoff, D.; Rudling, M.; Ketelhuth, D.F.; Gerdes, N.; Holmgren, J.; Nilsson, J.; *et al.* Intranasal immunization with an apolipoprotein B-100 fusion protein induces antigen-specific regulatory T cells and reduces atherosclerosis. *Arterioscler Thromb. Vasc. Biol.* **2010**, *30*, 946–952.

92. Salazar-Gonzalez, J.A.; Rosales-Mendoza, S.; Romero-Maldonado, A.; Monreal-Escalante, E.; Uresti-Rivera, E.E.; Banuelos-Hernandez, B. Production of a Plant-Derived Immunogenic Protein Targeting ApoB100 and CETP: Toward a Plant-Based Atherosclerosis Vaccine. *Mol. Biotechnol.* **2014**, *56*, 1133–1142.

93. Weiner, H.L. The mucosal milieu creates tolerogenic dendritic cells and T(R)1 and T(H)3 regulatory cells. *Nat. Immunol.* **2001**, *2*, 671–672.

94. Hernandez, H.J.; Rutitzky, L.I.; Lebens, M.; Holmgren, J.; Stadecker, M.J. Diminished immunopathology in Schistosoma mansoni infection following intranasal administration of cholera toxin B-immunodominant peptide conjugate correlates with enhanced transforming growth factor-beta production by CD4 T cells. *Parasite Immunol.* **2002**, *24*, 423–427.

95. Yu, X.; Song, B.; Huang, C.; Xiao, Y.; Fang, M.; Feng, J.; Wang, P.; Zhang, G. Prolonged survival time of allografts by the oral administration of RDP58 linked to the cholera toxin B subunit. *Transpl. Immunol.* **2012**, *27*, 122–127.

96. Boirivant, M.; Fuss, I.J.; Ferroni, L.; de Pascale, M.; Strober, W. Oral administration of recombinant cholera toxin subunit B inhibits IL-12-mediated murine experimental (trinitrobenzene sulfonic acid) colitis. *J. Immunol.* **2001**, *166*, 3522–3532.

97. Coccia, E.M.; Remoli, M.E.; Di Giacinto, C.; Del Zotto, B.; Giacomini, E.; Monteleone, G.; Boirivant, M. Cholera toxin subunit B inhibits IL-12 and IFN-γ production and signaling in experimental colitis and Crohn's disease. *Gut* **2005**, *54*, 1558–1564.

98. Stal, P.; Befrits, R.; Ronnblom, A.; Danielsson, A.; Suhr, O.; Stahlberg, D.; Brinkberg Lapidus, A.; Lofberg, R. Clinical trial: The safety and short-term efficacy of recombinant cholera toxin B subunit in the treatment of active Crohn's disease. *Aliment. Pharmacol. Ther.* **2010**, *31*, 387–395.

99. Tamura, S.; Yamanaka, A.; Shimohara, M.; Tomita, T.; Komase, K.; Tsuda, Y.; Suzuki, Y.; Nagamine, T.; Kawahara, K.; Danbara, H.; *et al.* Synergistic action of cholera toxin B subunit (and *Escherichia coli* heat-labile toxin B subunit) and a trace amount of cholera whole toxin as an adjuvant for nasal influenza vaccine. *Vaccine* **1994**, *12*, 419–426.

100. Hamorsky, K.T.; Kouokam, J.C.; Bennett, L.J.; Baldauf, K.J.; Kajiura, H.; Fujiyama, K.; Matoba, N. Rapid and scalable plant-based production of a cholera toxin B subunit variant to aid in mass vaccination against cholera outbreaks. *PLoS Negl. Trop. Dis.* **2013**, *7*, e2046.

101. Nochi, T.; Takagi, H.; Yuki, Y.; Yang, L.; Masumura, T.; Mejima, M.; Nakanishi, U.; Matsumura, A.; Uozumi, A.; Hiroi, T.; *et al.* Rice-based mucosal vaccine as a global strategy for cold-chain- and needle-free vaccination. *Proc. Natl. Acad. Sci. USA* **2007**, *104*, 10986–10991.

102. Lebens, M.; Johansson, S.; Osek, J.; Lindblad, M.; Holmgren, J. Large-scale production of *Vibrio cholerae* toxin B subunit for use in oral vaccines. *Biotechnology* **1993**, *11*, 1574–1578.

103. Dakterzada, F.; Mobarez, A.M.; Roudkenar, M.H.; Forouzandeh, M. Production of Pentameric Cholera Toxin B Subunit in *Escherichia coli. Avicenna J. Med. Biotechnol.* **2012**, *4*, 89–94.

104. Boustanshenas, M.; Bakhshi, B.; Ghorbani, M.; Norouzian, D. Comparison of two recombinant systems for expression of cholera toxin B subunit from *Vibrio cholerae. Indian J. Med. Microbiol.* **2013**, *31*, 10–14.

105. Slos, P.; Speck, D.; Accart, N.; Kolbe, H.V.; Schubnel, D.; Bouchon, B.; Bischoff, R.; Kieny, M.P. Recombinant cholera toxin B subunit in *Escherichia coli*: High-level secretion, purification, and characterization. *Protein Exp. Purif.* **1994**, *5*, 518–526.

106. Bakhshi, B.; Boustanshenas, M.; Ghorbani, M. A single point mutation within the coding sequence of cholera toxin B subunit will increase its expression yield. *Iran Biomed. J.* **2014**, *18*, 130–135.

107. Okuno, T.; Kashige, N.; Satho, T.; Irie, K.; Hiramatsu, Y.; Sharmin, T.; Fukumitsu, Y.; Uyeda, S.; Yamada, S.; Harakuni, T.; *et al.* Expression and secretion of cholera toxin B subunit in *Lactobacilli. Biol. Pharm. Bull.* **2013**, *36*, 952–958.

108. Yasuda, Y.; Matano, K.; Asai, T.; Tochikubo, K. Affinity purification of recombinant cholera toxin B subunit oligomer expressed in *Bacillus brevis* for potential human use as a mucosal adjuvant. *FEMS Immunol. Med. Microbiol.* **1998**, *20*, 311–318.

109. Miyata, T.; Harakuni, T.; Taira, T.; Matsuzaki, G.; Arakawa, T. Merozoite surface protein-1 of Plasmodium yoelii fused via an oligosaccharide moiety of cholera toxin B subunit glycoprotein expressed in yeast induced protective immunity against lethal malaria infection in mice. *Vaccine* **2012**, *30*, 948–958.

110. Gong, Z.H.; Jin, H.Q.; Jin, Y.F.; Zhang, Y.Z. Expression of cholera toxin B subunit and assembly as functional oligomers in silkworm. *J. Biochem. Mol. Biol.* **2005**, *38*, 717–724.

111. Arakawa, T.; Chong, D.K.; Langridge, W.H. Efficacy of a food plant-based oral cholera toxin B subunit vaccine. *Nat. Biotechnol.* **1998**, *16*, 292–297.

112. Mikschofsky, H.; König, P.; Keil, G.M.; Hammer, M.; Schirrmeier, H.; Broer, I. Cholera toxin B (CTB) is functional as an adjuvant for cytoplasmatic proteins if directed to the endoplasmatic reticulum (ER), but not to the cytoplasm of plants. *Plant Sci.* **2009**, *177*, 35–42.

113. Kim, Y.S.; Kim, M.Y.; Kim, T.G.; Yang, M.S. Expression and assembly of cholera toxin B subunit (CTB) in transgenic carrot (*Daucus carota* L.). *Mol. Biotechnol.* **2009**, *41*, 8–14.

114. Kajiura, H.; Wasai, M.; Kasahara, S.; Takaiwa, F.; Fujiyama, K. *N*-glycosylation and *N*-glycan moieties of CTB expressed in rice seeds. *Mol. Biotechnol.* **2013**, *54*, 784–794.

115. Yuki, Y.; Mejima, M.; Kurokawa, S.; Hiroiwa, T.; Takahashi, Y.; Tokuhara, D.; Nochi, T.; Katakai, Y.; Kuroda, M.; Takeyama, N.; *et al.* Induction of toxin-specific neutralizing immunity by molecularly uniform rice-based oral cholera toxin B subunit vaccine without plant-associated sugar modification. *Plant Biotechnol. J.* **2013**, *11*, 799–808.

116. Oszvald, M.; Kang, T.J.; Tomoskozi, S.; Jenes, B.; Kim, T.G.; Cha, Y.S.; Tamas, L.; Yang, M.S. Expression of cholera toxin B subunit in transgenic rice endosperm. *Mol. Biotechnol.* **2008**, *40*, 261–268.

117. Kim, Y.S.; Kim, B.G.; Kim, T.G.; Kang, T.J.; Yang, M.S. Expression of a cholera toxin B subunit in transgenic lettuce (*Lactuca sativa* L.) using Agrobacterium-mediated transformation system. *Plant Cell Tiss. Organ. Cult.* **2006**, *87*, 203–210.

118. Jani, D.; Meena, L.S.; Rizwan-ul-Haq, Q.M.; Singh, Y.; Sharma, A.K.; Tyagi, A.K. Expression of cholera toxin B subunit in transgenic tomato plants. *Transgenic Res.* **2002**, *11*, 447–454.

119. Wang, X.G.; Zhang, G.H.; Liu, C.X.; Zhang, Y.H.; Xiao, C.Z.; Fang, R.X. Purified cholera toxin B subunit from transgenic tobacco plants possesses authentic antigenicity. *Biotechnol. Bioeng.* **2001**, *72*, 490–494.

120. Rattanapisit, K.; Bhoo, S.H.; Hahn, T.R.; Mason, H.S.; Phoolcharoen, W. Rapid transient expression of cholera toxin B subunit (CTB) in *Nicotiana benthamiana*. *In Vitro Cell Dev. Biol. Plant* **2013**, *49*, 107–113.

121. Daniell, H.; Lee, S.B.; Panchal, T.; Wiebe, P.O. Expression of the native cholera toxin B subunit gene and assembly as functional oligomers in transgenic tobacco chloroplasts. *J. Mol. Biol.* **2001**, *311*, 1001–1009.

122. Renuga, R.S.; Babu Thandapani, A.; Arumugam, K.R. Expression of Cholera toxin B subunit in Banana callus culture. *J. Pharm. Sci. Res.* **2010**, *2*, 26–33.

123. Karaman, S.; Cunnick, J.; Wang, K. Expression of the cholera toxin B subunit (CT-B) in maize seeds and a combined mucosal treatment against cholera and traveler's diarrhea. *Plant Cell Rep.* **2012**, *31*, 527–537.

124. Arakawa, T.; Chong, D.K.; Merritt, J.L.; Langridge, W.H. Expression of cholera toxin B subunit oligomers in transgenic potato plants. *Transgenic Res.* **1997**, *6*, 403–413.

125. Jiang, X.L.; He, Z.M.; Peng, Z.Q.; Qi, Y.; Chen, Q.; Yu, S.Y. Cholera toxin B protein in transgenic tomato fruit induces systemic immune response in mice. *Transgenic Res.* **2007**, *16*, 169–175.

126. Nochi, T.; Yuki, Y.; Katakai, Y.; Shibata, H.; Tokuhara, D.; Mejima, M.; Kurokawa, S.; Takahashi, Y.; Nakanishi, U.; Ono, F.; *et al.* A rice-based oral cholera vaccine induces macaque-specific systemic neutralizing antibodies but does not influence pre-existing intestinal immunity. *J. Immunol.* **2009**, *183*, 6538–6544.

Do the A Subunits Contribute to the Differences in the Toxicity of Shiga Toxin 1 and Shiga Toxin 2?

Debaleena Basu and Nilgun E. Tumer

Abstract: Shiga toxin producing *Escherichia coli* O157:H7 (STEC) is one of the leading causes of food-poisoning around the world. Some STEC strains produce Shiga toxin 1 (Stx1) and/or Shiga toxin 2 (Stx2) or variants of either toxin, which are critical for the development of hemorrhagic colitis (HC) or hemolytic uremic syndrome (HUS). Currently, there are no therapeutic treatments for HC or HUS. *E. coli* O157:H7 strains carrying Stx2 are more virulent and are more frequently associated with HUS, which is the most common cause of renal failure in children in the US. The basis for the increased potency of Stx2 is not fully understood. Shiga toxins belong to the AB$_5$ family of protein toxins with an A subunit, which depurinates a universally conserved adenine residue in the α-sarcin/ricin loop (SRL) of the 28S rRNA and five copies of the B subunit responsible for binding to cellular receptors. Recent studies showed differences in the structure, receptor binding, dependence on ribosomal proteins and pathogenicity of Stx1 and Stx2 and supported a role for the B subunit in differential toxicity. However, the current data do not rule out a potential role for the A$_1$ subunits in the differential toxicity of Stx1 and Stx2. This review highlights the recent progress in understanding the differences in the A$_1$ subunits of Stx1 and Stx2 and their role in defining toxicity.

Reprinted from *Toxins*. Cite as: Basu, D.; Tumer, N.E. Do the A Subunits Contribute to the Differences in the Toxicity of Shiga Toxin 1 and Shiga Toxin 2? *Toxins* **2015**, *7*, 1467-1485.

1. Introduction

Shiga toxin producing *Escherichia coli* (STEC) strains such as *E. coli* O157:H7, as well as other serotypes, are the major causative agents of severe gastroenteritis, which can lead to life-threating complications including hemorrhagic colitis (HC) and hemolytic uremic syndrome (HUS) [1,2]. HUS is the most common cause of renal failure in children in the US [3]. The recent multi-state outbreak of *E. coli* O157:H7 in the US and a HUS outbreak in Germany in 2011 caused by *E. coli* O104:H4 highlight the public health impact of this pathogen [4–7]. STEC strains produce Shiga toxin 1 (Stx1) and/or Shiga toxin 2 (Stx2) or variants of either toxin. *E. coli* strains carrying Stx2 are more virulent and are more frequently associated with HUS [8–10]. However the molecular basis for the higher potency of Stx2 is unknown. Although extensive research is being undertaken to develop effective vaccines and therapeutics to protect against HUS, there are no current therapies available. In order to develop inhibitors against Shiga toxins, there is a need for better understanding of their underlying mechanism of toxicity.

Shiga toxin (Stx) from *Shigella dysenteriae* and Stx1 (Stx1) and 2 (Stx2) from Shiga toxin-producing *Escherichia coli* (STEC) are a family of structurally and functionally related proteins [5,11]. Stx, Stx1 and Stx2 are ribosome inactivating proteins (RIPs), a class of proteins that irreversibly damage the ribosome catalytically by modifying the large rRNA and inhibiting protein synthesis [12–16]. RIPs are present throughout the plant kingdom and are also found in bacteria [12–14]. RIPs are

N-glycosidases that remove a specific adenine from the highly conserved α-sarcin/ricin loop (SRL) in the 28S rRNA of the large ribosomal subunit [12–14]. Irreversible modification of the target adenine blocks elongation factor (EF)-1- and EF-2-dependent GTPase activity and renders the ribosome unable to bind EF-2, thereby blocking translation [17,18]. The RIPs are divided into three types based on their physical properties. Type 1 RIPs such as pokeweed antiviral protein (PAP), trichosanthin (TCS) and saporin are single chain, highly basic monomeric enzymes, approximately 30 kDa in size [19–21] Type 2 RIPs consist of an A chain and variable number of B chains. The A chain is the active chain, while the B chain can bind receptors on the surface of eukaryotic cells and mediate retrograde transport of the A-chain to the cytosol. Potent toxins like ricin, abrin and Shiga toxins fall into this category. Type 3 RIPs are synthesized as inactive precursors (proRIPs) that require proteolytic processing to occur between the amino acids involved in formation of the active site [13,15]. Because of their potent and selective toxicity, RIPs have been exploited as potential agents of bioterrorism and have garnered interest for use in antiviral and anticancer therapy [14,22,23]. Shiga toxin and its B subunit have been investigated as novel therapeutic agents against pancreatic cancer and colon cancer [24,25].

2. Structure

Stx derives its name from the dysentery causing bacteria, *Shigella dysenteriae*, which was first described by Kiyoshi Shiga in 1898. While Stx from *S. dysenteriae* differs from Stx1 by one amino acid [26,27], Stx1 and Stx2 have only 56% amino acid similarity [28] and are antigenically distinct [28–30]. STEC can produce either one type of toxin or a combination of variants of one or both types of toxin [31]. Stx1 and Stx2, which are also referred to as Stx1a and Stx2a [32], are type II RIPs, which consist of a catalytically active A chain associated with a pentamer of B subunits responsible for the binding of the Shiga toxins to their common cellular receptor, globotriaosylceramide (Gb3) [33,34]. The B subunits (7.7 kDa each) form a central pore which harbors the *C*-termini of the A subunit [35]. The crystal structure of the Stx1 B subunit pentamer, bound with Gb3 shows that each B monomer contains three distinct binding sites for the glycan component of Gb3, referred to as P^k trisaccharide, α-D-Gal*p*-(1-4)-β-D-Gal*p*-(1-4)-β-D-Glc*p*-(1-O) for a total of 15 sites [36]. Of these three binding sites (labeled 1–3), site 2 has the highest occupancy of electron density defining the position of the trisaccharide, while site 1 has the lowest [36,37]. The only known exception to this Gb3-dependence of Shiga toxins is for the Stx2 variant Stx2e, which exhibits specific affinity for globotetraosylceramide (Gb4) [38,39], although Stx1 and Stx2 can bind to Gb4 weakly [40]. Stx2e has recently been shown to bind to the Forssman glycolipid, which makes this subtype unique among the Stx subtypes [41]. Recently, the crystal structure of Stx2 bound to a P^k derivative has been published [42]. This structure showed that only two of three previously identified binding sites on the B_5 pentamer was functional in Stx2, indicating that there are differences in receptor binding between Stx1 and Stx2a [42].

The A subunit of Shiga toxin consists of A_1 and A_2 chains which are bound together by a disulfide bond between C242 and C261 forming a loop [43]. The X-ray crystal structures of *Shigella* Stx and Stx2 are highly similar [34,35]. However, structural differences have been identified between Stx1 and Stx2 [34,35]. In Stx1, part of the active site is blocked by the A_2 chain, while it is accessible in

Stx2 [35]. The active site of Stx2 is accessible to the adenine substrate and Stx2 cleaves the adenine when it is crystallized in the presence of adenosine [44]. In the crystal structure, the A subunit in Stx2 is in a different orientation with respect to the B subunit, which may affect receptor affinity of Stx2 [35]. The *C*-terminus of Stx2 extends through the pore formed by the B pentamer, which is thought to interfere with receptor binding [35]. However, it is not known if the A subunits contribute to the interaction of the holotoxins with the receptor and whether the A subunit interferes with Gb3 binding. Stx1 and ricin have been shown to interact with human neutrophils, which do not express Gb3 or Gb4, through their A subunit without inducing their internalization [45]. TLR4 has recently been identified as the receptor that recognizes the A subunits of Stx1 and Stx2 in human neutrophils [46].

Once the toxins bind the globotriaosylceramide (Gb3) receptor, they are endocytosed by a clathrin-dependent or independent pathway [11,47]. They then undergo retrograde transport to the Golgi apparatus and then to the endoplasmic reticulum (ER). The active A_1 subunit is cleaved enzymatically from the A_2-B5 complex [43]. The cleavage occurs between R251 and M252 in Shiga toxin and Stx1 and between R250 and A251 in Stx2 by the furin protease. After cleavage, the A_1 fragment remains bound to the A_2 fragment through the disulfide bond. The A_1 chain is then released from the A_2-B5 complex by reduction of the disulfide bond in the ER and undergoes retrotranslocation from the ER into the cytosol where it depurinates the ribosome and inhibits protein synthesis [11,47].

3. Catalytic Activity and Cytotoxicity of Stx1 and Stx2

3.1. Differences in Cytotoxicity

An extensive review on the pathophysiology of Stx-related disease in different animal models can be found in [48] and is briefly described here. Epidemiological studies suggest that majority of the HUS-associated fatalities are caused by *E. coli* O157:H7 strains carrying Stx2 [8–10]. Previous studies using Shiga toxins have shown that while Stx2 is more potent in animal models, Stx1 is more toxic to Vero cells [49,50]. The 50% lethal dose for purified Stx2 was 400-fold lower than for Stx1 in a mouse model, and only Stx2-treated mice developed renal complications and death [49,51]. However, animal models have limitations compared with the observations from humans and do not replicate the disease in humans. Nonhuman primate models (Baboon) showed renal damage consistent with HUS upon intravenous injection of the toxins. Treatment of non-human primates with four doses of 25 ng/kg Stx2 caused HUS, while an equal dose of Stx1 had no effect [50]. In another study comparison of the effects of the two toxins showed interesting differences, including different proinflammatory responses and different timings with delayed organ injury after Stx2 challenge [52]. Baboons treated with Stx1 developed HUS within two to three days, while those with Stx2 took longer (3–5 days), indicating the role of other factors in producing delayed renal injury upon challenge by Stx2. Furthermore, Stx1 incited a stronger proinflammatory response earlier, while the proinflammatory response induced by Stx2 was gradual and delayed by several days [52]. A subsequent study using baboon models showed that both Stx1 and Stx2 can affect kidney function. Although Stx2 was found to cause more severe damage to the kidney than Stx1, the damage inflicted on the kidney by Stx1 was significant [53].

In comparison to animal models studies in Vero cells suggested that the cytotoxicity of Stx1 is 10-fold greater than Stx2 [49]. The basis for the differential toxicity of Stx1 and Stx2 in animal models *versus* mammalian cell lines is unknown. Shiga toxins trigger endothelial damage in kidney and brain by targeting Gb3. However, differences have been observed in the sensitivity of endothelial cells to Stx1 and Stx2. The current knowledge of endothelial cell damage caused by Stx1 and Stx2 is reviewed in [54]. Stx2 had a higher potency for human renal microvascular endothelial cells (HRMEC) than to human umbilical vein endothelial cells (HUVEC), where toxicity of Stx1 and Stx2 was similar [55]. These results indicated selective sensitivity of renal endothelial cells to Stx2 although the renal endothelial cells possessed fewer Gb3 receptor binding sites for Stx2 than Stx1 [55]. The Stx receptor distribution in the different renal cell populations and the sensitivity of the different kidney cell types to Stx1 and Stx2 is reviewed in [11]. Comparison of cellular injury induced by Stx1 and Stx2 in human brain microvascular endothelial cells (HBMEC) and HUVEC derived EA.hy 926 macrovascular endothelial cells indicated that these cell lines had differential susceptibility to the toxins. HBMEC cells were over 1000-fold more susceptible to Stx2, while EA.hy 926 cells were around 10-fold more susceptible to Stx1 [56]. Stx1 caused both necrosis and apoptosis, while Stx2 induced mainly apoptosis in both cell lines [56]. The basis for the differential susceptibility of endothelial cells to Stx1 and Stx2 is not well understood. Holotoxin stability, enzymatic activity and receptor affinity were proposed as potential factors contributing to the differential toxicity. In addition, the cytotoxicity comparisons between Stx1 and Stx2 in animals and cells are critically dependent on the specific batches of toxin used and can vary accordingly.

The B subunits of Stx1 and Stx2 have been hypothesized to play an important role in mediating the differences in potency. The B subunits of Stx1 and Stx2 display differences in receptor recognition, as well as in the number of potential binding sites [57,58]. Studies in Vero cells demonstrated that Stx1 has a higher affinity for the Gb3 receptor [49,59–61]. Using purified Gb3, it was shown that Stx1 has a 10-fold higher affinity for Gb3 compared to Stx2 [59]. It has been suggested that Stx1 might bind to Gb3 variants in the lung, preventing it from reaching more susceptible organs, such as the kidneys, whereas Stx2 binds preferentially to Gb3 variants in kidney. As a result, Stx1 shows decreased binding to kidney cells, which are the main targets for lethality in mice [62]. Analysis of binding kinetics to the glycolipid receptor analog using surface plasmon resonance (SPR) showed that Stx1 bound to the receptor analog better than Stx2 and had faster association and dissociation rates [40]. These results suggest that the differences in binding kinetics and affinity of the B subunits for the Gb3 receptor may be responsible for the greater toxicity of Stx1 to Vero cells.

The B subunits of Stx1 and Stx2 also display differences in structural stability [63]. The B pentamer of Stx1 was more stable than the B pentamer of Stx2 and bound the receptor with higher affinity than the B pentamer of Stx2 [63]. Moreover, while Stx1B subunits were able to bind glycolipids only as a stable pentamer, Stx2B subunits bound to glycolipids in lower oligomeric states [63]. These results suggested that differences in receptor affinity and receptor binding preferences may contribute to the differential toxicity of Stx1 and Stx2 by affecting their targeting to susceptible tissues.

Stx A/B subunit chimeras, where the A and B subunits of the two toxins have been interchanged, were used to study the contribution of the individual A and B subunits to toxicity [59]. The holotoxin as well as the chimeric toxins were used in mouse and in Vero cells to differentiate the roles of the subunits in toxicity [49,59]. However, the chimeric toxins were usually found to be less stable than the holotoxins due to incorrect folding [59] or showed equivalent cytotoxicity [64]. Chimeric toxins, created by operon fusions displayed cytotoxicity intermediate to Stx1 and Stx2 [37] or did not produce a functional chimera [65]. Therefore, clear conclusions regarding the role of each subunit in toxicity could not be deduced from these studies. A recent study used the A_2 subunit along with the B subunit to increase the stability of the chimeric toxin [66]. The binding of the chimeric toxins to the Gb3 receptor and their translocation through the monolayers of the polarized HCT-8 cells were dependent on the origin of the B subunit, and the chimeric toxin with the Stx1B subunit had a higher affinity for the receptor than the Stx2B chimera. The toxicity of the chimeric toxins to Vero and HCT-8 cells indicated the importance of the origin of the B subunit although the B subunit accounted for less than 50% of the differential toxicity for Vero cells [66]. Perhaps, due to the instability of the chimeric toxins at pH 3, the oral administration of the chimeric toxin where the A subunit was from Stx1 in mice required at least 10 times more toxin as compared to native Stx2, while Stx1 or the chimeric toxin where the A subunit was from Stx2 failed to show any mortality in mice, even at a very high concentration. This study highlighted the importance of the B subunit in the differential toxicity of Stx1 and 2. The differential lethality in mouse was thought to take place at the level of toxicity to the kidney [66]. However, although *in vivo* results indicate that the B subunits are involved in differences in the severity of the intoxication, they do not rule out a potential role for the A subunits in the differential toxicity of Stx1 and Stx2. The critical question regarding why Stx2 is more potent than Stx1 *in vivo* still remains unanswered.

3.2. Differences in Catalytic Activity

The A subunits of Shiga toxins and ricin play a critical role in the toxicity of each toxin. They have the same catalytic activity and show conservation in amino acids at the active site. Mutagenesis studies identified Glu167, Arg170, Tyr77, Tyr114, Trp203 and Arg205, which are critical for the catalytic activity and are conserved between Stx1 and Stx2 [67–72]. Current knowledge about the mode of action of the A subunits is obtained from studies with either cultured cell lines or *in vitro* systems. *In vivo* studies at the molecular level or at the level of the whole organism are limited due to the extreme cytotoxicity of these toxins and the lack of available model systems. Using the yeast, *Saccharomyces cerevisiae* as a model, we identified the amino acids critical for the cytotoxicity of Stx1A and Stx2A and showed that the activity of the A subunits can be differentiated [70]. The results showed that Asn75 and Tyr77 were more critical for the depurination activity of Stx2A, while Arg176 was more critical for the depurination activity of Stx1A. Analysis of solvent accessible surface areas indicated that Asn75 and Tyr77 were more exposed in Stx2A, while Arg176 was more exposed in Stx1A [70]. Arg176 was subsequently shown to be critical for ribosome binding of Stx1A_1 [73], suggesting that there may be differences in the ribosome binding of Stx1A_1 and Stx2A_1.

Several studies used cell-free translation inhibition assays to compare the enzymatic activity of the A subunits of Stx1 and Stx2 [49,50]. The A subunits displayed similar translation inhibitory activities [49,59,74], leading to the conclusion that the enzymatic activities of the A subunits are not responsible for the toxicity differences between Stx1 and Stx2. As a result the role of the A subunit in the differential potency of Stx1 and Stx2 has not received much attention. Since translation inhibition is a downstream effect of depurination, these assays did not directly compare the catalytic activity of Stx1A1 and Stx2A1 on the ribosome. Further, while in some studies the holotoxin was used [49], in others the holotoxin was activated by digestion with trypsin to release the A1 chain from the A2-B5 complex and/or by chemical treatment with DTT to break the disulfide bond between the A1 and the A2 chains [50,75]. These methods frequently yield variable amounts of activated protein and can cause degradation, preventing comparison of enzymatic activity directly [75]. Therefore, due to technical limitations, the role of the A1 subunit in increased potency of Stx2 has not been fully investigated and a direct comparison of the catalytic activity has not been carried out.

The unanswered questions regarding the relative catalytic activity of RIPs highlight the importance of quantitative assays, which allow direct comparisons of the depurination activity. Our group developed a quantitative real-time PCR (qRT-PCR) assay that can directly measure the catalytic activity of the RIPs on ribosomes *in vitro* or in yeast and in mammalian cells *in vivo* [76,77]. The qRT-PCR assay exhibited a much wider dynamic range than the previously used primer extension assay and increased sensitivity [76]. Sturm and Schramm described a quantitative enzyme coupled luminescence assay to examine the kinetics of depurination by RIPs [78]. In this assay, adenine released by depurination is converted to AMP by adenine phosphoribosyl transferase (APRTase) and then to ATP by pyruvate orthophosphate dikinase (PPDK). The light generated by ATP via firefly luciferase is detected using a luminometer [78]. The qRT-PCR and the enzyme coupled luminescence assay have been used to examine the activity of the ricin toxin A chain (RTA) and its mutants [79]. A highly sensitive and quantitative assay using SPR was developed by our group to examine the interactions RIPs with ribosomes [16,75,80,81]. The development of these assays will allow direct comparisons of the binding and depurination kinetics of the A1 subunits of Shiga toxins and will help determine whether the A1 subunits of Stx1 and Stx2 have a significant role in their differential toxicity.

4. Ribosome Interactions

Although the SRL is the primary substrate for all RIPs, ribosomal proteins play an important role in of the depurination of intact ribosomes by RIPs. While the K_m of RTA for rat ribosomes and naked 28S rRNA are similar, RTA depurinates ribosomes almost 10^5-fold greater than the naked 28S RNA [82], suggesting that not only the target RNA sequence, but also the structure of the ribosome plays a significant role in the catalytic activity of RIPs.

Previous studies have shown the importance of the ribosomal phosphoproteins (P) of the P-protein stalk for the depurination activity of the RIPs [73,75,80,81,83,84]. Ricin has been shown to crosslink to the stalk protein P0 and the ribosomal protein L9 [85]. Trichosanthin (TCS), which is a type-1 RIP, has been shown to interact with P0, and P1 proteins of the ribosomal stalk using yeast-two hybrid analysis and by *in vitro* binding assays [86]. The last 11 residues of P2, which are conserved in P0, P1 and P2 have been found to be critical for the interaction with trichosanthin (TCS) [87]. The crystal structure of TCS complexed to a peptide corresponding to the *C*-terminal domain (CTD) of human P2, SDDDMGFGLFD, showed that the conserved DDD motif at the *N*-terminal region of this peptide interacts with the positively charged K173, R174, and K177 residues in TCS, while the *C*-terminal region is inserted into a hydrophobic pocket [88]. Using yeast mutants deleted in the stalk proteins (ΔP1 and ΔP2) and highly sensitive SPR and depurination assays, our group provided the first evidence that the ribosomal stalk proteins are essential for the cytotoxicity of RTA *in vivo* and that the ribosomal stalk is the main landing platform for RTA on the ribosome [80]. We subsequently showed that multiple copies of the stalk proteins accelerate the recruitment of RTA to the ribosome for depurination [89].

The ribosomal P-protein stalk is a lateral flexible protuberance of the large ribosomal subunit, which recruits translational factors to the ribosome and participates in the GTPase activation by EF-Tu and EF-G. The eukaryotic P protein stalk consists of 11 kDa P1 and P2 proteins bound to a larger P0 protein. P1 and P2 dimerize via their helical *N*-terminal domains, whereas the highly conserved *C*-terminal tails of P1 and P2 interact with the translational GTPases (tGTPases) [90]. Although the stalk is relatively conserved in eukaryotes there are some notable differences between the stalk structure in mammals and in yeast. The human ribosomal stalk contains two identical heterodimers of P1 and P2 bound to P0 assembled into a pentameric complex [91,92]. In contrast the yeast pentameric stalk consists of four different proteins P1α, P1β, P2α, P2β [93] which form two different heterodimers [94], P1α-P2β and P2α-P1β, bound to P0 [95,96]. The human P1 has 40%–47% sequence identity with P1α and P1β and human P2 has 53%–56% sequence identity with P2α and P2β [97,98]. The prokaryotic equivalent of P1 and P2 are L12 proteins bound to a smaller P0 equivalent L10 [99]. In bacteria, the stalk structure can be a pentamer or heptamer [100], while in archaea the ribosomal stalk is a heptamer [101]. Although the eukaryotic and prokaryotic stalk proteins are analogous in function, there is no sequence homology between these related proteins [91,92].

Using the yeast-two-hybrid and pull-down experiments in HeLa cells, it was demonstrated that the A₁ chain of Stx1 interacts with the P0, P1 and P2 proteins of the P-protein stalk [84]. Removal of the last 17 amino acids of P1 or P2, but not P0 abolished the interaction between Stx1A₁ and the human ribosomal stalk proteins, suggesting that the conserved CTD of P1/P2 proteins allows Stx1 to access the SRL [84]. To determine if Stx1A and Stx2A require the ribosomal stalk for depurination *in vivo*, we examined their depurination activity and cytotoxicity in the yeast P protein deletion mutants [75]. Our results showed that ribosomal stalk is important for both toxins to depurinate the ribosome *in vivo*. Cytoplasmic stalk proteins were critical for Stx1A and Stx2A to access the SRL *in vivo* (Figure 1A). However, Stx1A and Stx2A differed in depurination activity towards ribosomes when P1/P2 binding sites on P0 were deleted. P1/P2 proteins facilitated depurination by Stx1A only if their binding sites on P0 were intact (Figure 1B). In contrast, Stx2A was less dependent on the stalk proteins for activity than Stx1A and could depurinate the ribosomes with a defective stalk better than Stx1A [75]. These results demonstrated that although ribosomal stalk is important for Stx1 and Stx2 to depurinate the ribosome, Stx2 is less dependent on the stalk proteins for depurination activity and suggested that cytosolic P1/P2 proteins deliver the toxins to the ribosome to create a toxin pool near the SRL [75].

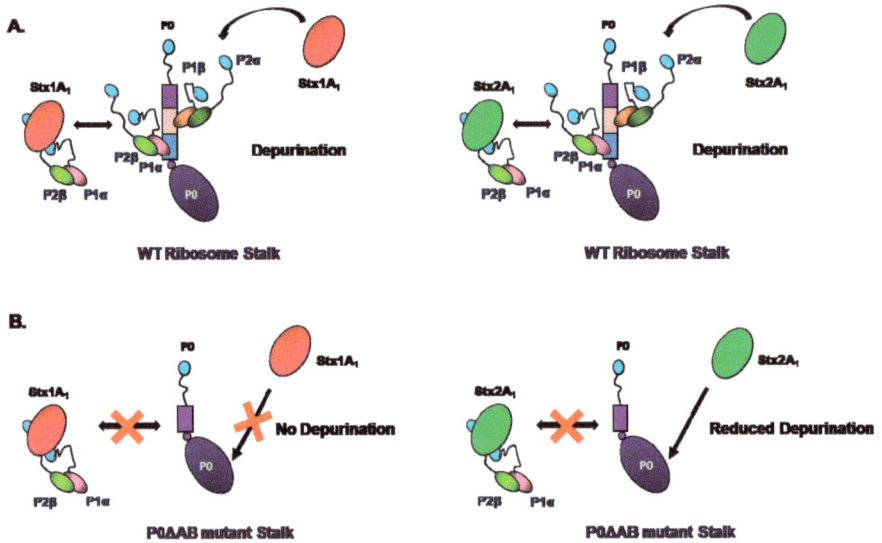

Figure 1. Model illustrating the interaction of Stx1A₁ and Stx2A₁ with the wild type and mutant stalk [75]. (**A**). Both Stx1A₁ and Stx2A₁ are able to interact with free P1α/P2β as well as ribosome bound P1α/P2β to depurinate the ribosome; (**B**). In the P0ΔAB mutant because the binding sites for P1/P2 dimers are deleted, free P1α/P2β proteins are not able to bind to the ribosomal stalk. Stx1A₁ shows almost no depurination activity indicating its dependence on the ribosomal stalk. Stx2A₁ has very little effect on depurination suggesting that it is less dependent on P1/P2 than Stx1A₁.

The A₁ chain of Stx1 was shown to interact with the ribosomal stalk proteins P0, P1, and P2 via the conserved CTD of P2 through hydrophobic and cationic surfaces on the toxin. Point mutations at arginines (Arg172, Arg176, Arg179, and Arg188) on Stx1A₁ perturbed the interaction between the toxin and the P2 peptide [73]. Using a combination of SPR and yeast-two hybrid analysis, these arginines were shown to be critical for the interaction of Stx1A₁ with the P2 peptide. The interactions with the P2 peptide were electrostatic and hydrophobic and took place at a site that was distinct from the active site. Since these residues are conserved between Stx1A₁ and Stx2A₁, it was proposed that Stx2 interacts with the ribosome in a similar manner [73].

The arginine residues, which were critical for binding to the stalk proteins in Stx1A₁ [73] and RTA [79] were on the opposite face of the active site, suggesting that both toxins interact with the ribosome in a similar manner. Analysis of the interaction of RTA with wild type and mutant yeast ribosomes deleted in stalk proteins by SPR showed that this interaction did not fit the 1:1 interaction model [81]. RTA interacted with wild type ribosomes by electrostatic interactions, which followed a two-step binding model. The two-step model is characterized by two different types of interactions with the ribosome, a saturable stalk dependent interaction with rapid association and dissociation rates and a much slower non-saturable stalk independent interaction with slower association and dissociation rates. The faster stalk dependent interaction was stronger than the slower stalk independent interaction. Further, the yeast mutant ribosomes lacking an intact stalk interacted with RTA by a 1:1

interaction model, which mirrored the slower interaction with wild type ribosomes [81]. According to the two-step interaction model shown schematically in Figure 2, in the first step RTA/Stx1A₁ molecules are first concentrated on the surface of the ribosome via slow non-specific electrostatic interactions and are guided to the stalk. In the second step, rapid, more specific electrostatic interactions occur between the stalk binding surface of RTA/Stx1A₁ and the CTD of the stalk proteins. In the third step, the P-protein stalk delivers RTA/Stx1A₁ to the SRL by a conformational change of the flexible hinge region and allows RTA/Stx1A₁ to depurinate the SRL at a very high rate. Consistent with this model, the interaction between RTA and the isolated native pentameric stalk complex from yeast fit well with a single step interaction model [89].

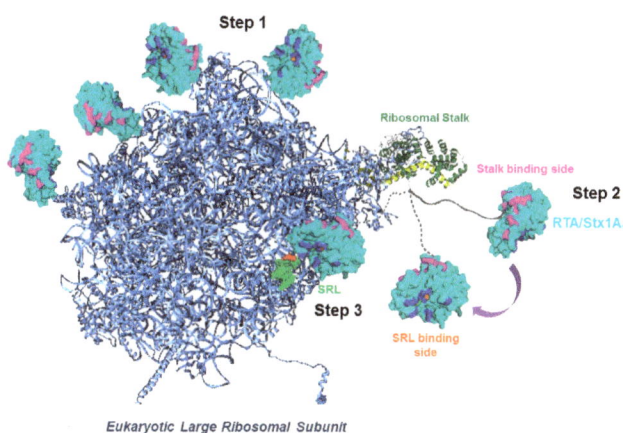

Figure 2. Model of how RTA and Stx1A₁ may access the α-sarcin/ricin loop (SRL) of the large rRNA [79]. Eukaryotic large ribosomal subunit was created using Protein Data Bank (PDB) ID: 3U5I and Protein Data Bank ID: 3U5H (blue) using the PyMOL software (The PyMOL Molecular Graphics System, Version 1.3 Schrödinger, LLC) with the SRL (green). The fitted cartoon structure of P0 fragment complexed with the N-terminal domain of P-proteins (Protein Data Bank ID: 3A1Y) from archaea is depicted as yellow and green, respectively as described [79]. The flexible CTD domain of a P-protein is represented as a gray line. Ricin toxin A chain (RTA) (Protein Data Bank ID: 1RTC) is colored in cyan, its active site is shown in orange, RNA binding site in blue and the stalk binding site is shown in magenta. In Step 1 RTA/Stx1A₁ are concentrated on the ribosome surface by nonspecific electrostatic interactions. In Step 2 RTA/Stx1A₁ interact with the C-terminal domain (CTD) of P-proteins with their ribosome binding surface, which is on the opposite side of the surface that contains the active site. The flexible hinge of P-proteins orients the active site of RTA/Stx1A₁ towards the SRL and in Step3 RTA/ Stx1A₁ establish the specific contacts necessary to hydrolyze a single N-glycosidic bond in the SRL.

The enzyme coupled luminescence assay showed that the K_m values and catalytic rates (k_{cat}) of the ribosome binding mutants of RTA for an SRL mimic RNA were similar to wild type RTA, indicating that their catalytic activity was not altered [79]. However, their K_m was higher and their

k_{cat} was lower towards ribosomes, indicating that the mutations affected ribosome binding and catalytic activity of RTA towards ribosomes without affecting RNA binding or catalytic activity of RTA towards naked RNA [79]. Based on this data, we proposed that arginines located on the opposite face of the active site of RTA bind to the flexible P-proteins of the ribosomal stalk. Stalk binding stimulates the catalysis of depurination by orienting the active site of RTA towards the SRL and thereby allows docking of the target adenine into the active site [79]. This model provided an explanation for why RTA depurinates intact ribosomes much better than free rRNA and how RTA hydrolyzes a single N-glycosidic bond on intact ribosomes from among the 4000 stem-loops in the large rRNA [79].

Subsequent studies showed that the ability to interact with the stalk was conserved in some RIPs, but not all RIPs [102]. PAP, which is a type-1 RIP active against ribosomes from all five kingdoms, interacts with ribosomal protein L3 to depurinate the SRL [103]. Since RTA, TCS and Stx1 were able to interact with the ribosomal stalk, was this ability to interact with the stalk a feature of an ancestral RIP, which has been conserved in some RIPs like ricin, Shiga toxins and TCS and lost in other RIPs like PAP? Phylogenetic analysis suggested that the ability to interact with the CTD of the ribosomal stalk arose independently in different RIPs by convergent evolution [104]. Further, the ability to interact with stalk was considered an adaptive advantage and did not have strong sequence constraints, which made it easy for different proteins to acquire this feature [104]. Based on the wide distribution of RIPs in plants and their presence in some bacteria, it has been postulated that an ancestral RIP domain present in plants was acquired by bacteria by horizontal gene transfer. However, a recent study presented evidence for the presence of RIP genes in Fungi and Metazoa and proposed that the differential loss of paralogous genes accounted for the complex pattern of RIP genes in extant species, rather than horizontal gene transfer [105].

Structural differences were shown between the structures of the CTD of bacterial and eukaryotic stalk proteins. The CTD of bacterial L12 is globular [106]. In contrast, NMR spectroscopy showed that while the N-terminal domain of eukaryotic P1/P2 dimer is structured, the CTD is flexible and can extend away from the dimerization region [107]. It has been suggested that these structural differences in the CTD may facilitate the domain specific recognition of elongation factors [20]. RTA is unable to depurinate intact *E. coli* ribosomes [82]. Similarly, TCS can only depurinate eukaryotic ribosomes, but not bacterial ribosomes. However, TCS was able to depurinate hybrid ribosomes when the bacterial stalk proteins were replaced with the eukaryotic stalk proteins [107]. These results suggested that the CTD and the flexible linker of stalk proteins are responsible for recruiting RIPs to the ribosome [107]. Therefore, RIPs like ricin and TCS that can only depurinate eukaryotic ribosomes may have evolved to bind to the CTD of eukaryotic stalk proteins, thereby hijacking the eukaryotic stalk proteins by binding to their *C*-terminal consensus sequences [20].

However, some critical questions remain. Stx1 can depurinate *E. coli* ribosomes, even though the stalk proteins differ in primary sequence and structure between the prokaryotes and the eukaryotes. Moreover, the conserved CTD of P proteins that can interact with Stx1 *in vitro*, is missing in the *E. coli* stalk proteins. Therefore, it is not clear how Stx1 accesses the SRL on *E. coli* ribosomes. Moreover, although the ribosome binding residues identified in Stx1A₁ are conserved in Stx2A₁, it is not known if they interact similarly with the ribosome. We have shown that there is a difference in the surface exposure of residues between Stx1A and Stx2A [70]. Arg176 is more exposed in Stx1A and is more critical for the depurination activity of Stx1A than Stx2A [70]. Arg176 has been shown to be important for binding of Stx1A$_1$ to the ribosome [73]. It is not known if Arg176 has a similar role in binding of Stx2A$_1$ to the ribosome. Although both Stx1A and Stx2A bind to the stalk, Stx2A is less dependent on the stalk proteins than Stx1A for its depurination activity [75]. These results indicate that there are differences in the ribosome interactions of Stx1 and Stx2, which may lead to differences in their depurination activity.

Evidence for structural differences between Stx1 and Stx2 and their importance in inactivation of the ribosome was obtained when Smith *et al.*, demonstrated that monoclonal antibody (MAb) 11E10, which neutralized both the cytotoxicity and lethality of Stx2, but not Stx1, bound to three specific regions around the active site of Stx2A, but failed to bind to Stx1A [108]. The sequence of the three regions was the most divergent between Stx2 and Stx1, which explained why the antibody specifically recognized Stx2 [108]. MAb 11E10 blocked the enzymatic activity of Stx2 *in vitro* and altered its intracellular trafficking pattern, providing evidence that structural differences lead to differential effects on the catalytic activity and trafficking of Stx1 and Stx2. Another MAb, S2C4, which was able to neutralize Stx2, but not Stx1 [109], was predicted to bind to another region that differed in sequence between Stx2 and Stx1 [82]. This region (residues 176–188) was shown to be important for binding of Stx1A$_1$ to the ribosomal stalk [73], suggesting that structural differences between Stx1 and Stx2 may affect ribosome binding differentially. These results highlight the importance of identifying Stx2 residues, which are important in binding to the ribosome and the role of these residues on ribosome binding and depurination activity of each toxin.

Finally, in order for the toxin to depurinate the SRL specifically, it has to interact with the residues surrounding the SRL. Modeling analysis of the crystal structure of RTA and a 29-mer oligonucleotide hairpin containing the conserved GAGA loop of the SRL identified residues, which may be involved in binding to the 29-mer [110]. The amino acids at the active site are conserved between Stx1 and Stx2 [67–72]. However, there are conformational differences between the active sites of Stx1 and Stx2 [34,35] and the active site residues are more exposed in Stx2 than in Stx1 [70]. Currently it is not known if residues around the active site contribute to the catalytic activity of each toxin similarly. Analysis of depurination kinetics will shed more light on the relative role of these residues in binding to the SRL and in catalytic activity.

5. Conclusions

STEC are a serious cause of morbidity and mortality and a better understanding of their mechanism of virulence is of high significance. *In vivo* data indicate that the B subunits are involved in differential toxicity of Stx1 and Stx2, but do not rule out a potential role for the A subunits, suggesting that steps in addition to receptor binding and trafficking likely contribute to the differential toxicity of Stx1 and Stx2. The role of the A subunits in differential toxicity has not been fully examined. New discoveries indicate that the A subunits of Stx1 and Stx2 differ in their dependence on the ribosomal stalk proteins, suggesting that the role of the A subunits in ribosome binding, depurination activity and cytotoxicity may differ. Structural studies identified conformational differences in the active sites of the A subunits of Stx1 and Stx2. Monoclonal antibodies that selectively bind and neutralize Stx2 indicated important differences in enzymatic action and intracellular trafficking. A better understanding of the interaction of the A_1 subunits of Stx1 and Stx2 with the ribosome and with the SRL, and comparative analyses of the catalysis of ribosome depurination are necessary to fully understand the factors that may contribute to the more pathogenic effects of Stx2 *in vivo*. These studies are relevant because they will have implications for the pathogenesis of HUS and may lead to the identification of novel therapeutic targets for Stx-associated HUS, for which there are no current therapies available.

Acknowledgments

We thank Jennifer Nielsen Kahn and Xiao-Ping Li for helpful comments. This work is supported by National Institutes of Health grants AI092011 and AI072425 to Nilgun E. Tumer.

Author Contributions

D.B. performed the literature review; D.B. and N.E.T. wrote the manuscript.

Conflicts of Interest

The authors declare no conflict of interest.

References

1. Boerlin, P.; McEwen, S.; Boerlin-Petzold, F.; Wilson, J.; Johnson, R.; Gyles, C. Associations between virulence factors of Shiga toxin-producing *Escherichia coli* and disease in humans. *J. Clin. Microbiol.* **1999**, *37*, 497–503.
2. Scallan, E.; Hoekstra, R.M.; Angulo, F.J.; Tauxe, R.V.; Widdowson, M.-A.; Roy, S.L.; Jones, J.L.; Griffin, P.M. Foodborne illness acquired in the United States—Major pathogens. *Emerg. Infect. Dis.* **2011**, *17*, 7.
3. Siegler, R.; Oakes, R. Hemolytic uremic syndrome; pathogenesis, treatment, and outcome. *Curr. Opin. Pediatr.* **2005**, *17*, 200–204.
4. Frank, C.; Werber, D.; Cramer, J.P.; Askar, M.; Faber, M.; an der Heiden, M.; Bernard, H.; Fruth, A.; Prager, R.; Spode, A. Epidemic profile of Shiga-toxin-producing *Escherichia coli* O104: H4 outbreak in Germany. *New Engl. J. Med.* **2011**, *365*, 1771–1780.
5. Kaper, J.; O'Brien, A. Overview and historical perspectives. *Microbiol. Spectr.* **2014**, *2*, doi:10.1128/microbiolspec.EHEC-0028-2014.
6. Karch, H.; Denamur, E.; Dobrindt, U.; Finlay, B.B.; Hengge, R.; Johannes, L.; Ron, E.Z.; Tønjum, T.; Sansonetti, P.J.; Vicente, M. The enemy within us: Lessons from the 2011 European *Escherichia coli* O104: H4 outbreak. *EMBO Mol. Med.* **2012**, *4*, 841–848.
7. Bielaszewska, M.; Mellmann, A.; Zhang, W.; Köck, R.; Fruth, A.; Bauwens, A.; Peters, G.; Karch, H. Characterisation of the *Escherichia coli* strain associated with an outbreak of haemolytic uraemic syndrome in Germany, 2011: A microbiological study. *Lancet Infect. Dis.* **2011**, *11*, 671–676.

8. Nataro, J.P.; Kaper, J.B. Diarrheagenic *Escherichia coli*. *Clin. Microbiol. Rev.* **1998**, *11*, 142–201.

9. Pickering, L.; Obrig, T.; Stapleton, F. Hemolytic-uremic syndrome and enterohemorrhagic *Escherichia coli*. *Pediatr. Infect. Dis. J.* **1994**, *13*, 459.

10. Manning, S.D.; Motiwala, A.S.; Springman, A.C.; Qi, W.; Lacher, D.W.; Ouellette, L.M.; Mladonicky, J.M.; Somsel, P.; Rudrik, J.T.; Dietrich, S.E. Variation in virulence among clades of *Escherichia coli* O157: H7 associated with disease outbreaks. *Proc. Natl. Acad. Sci. USA* **2008**, *105*, 4868–4873.

11. Bergan, J.; Lingelem, A.B.D.; Simm, R.; Skotland, T.; Sandvig, K. Shiga toxins. *Toxicon* **2012**, *60*, 1085–1107.

12. Stirpe, F. Ribosome-inactivating proteins: From toxins to useful proteins. *Toxicon* **2013**, *67*, 12–16.

13. Nielsen, K.; Boston, R.S. Ribosome-inactivating proteins: A plant perspective. *Annu. Rev. Plant Biol.* **2001**, *52*, 785–816.

14. Zhabokritsky, A.; Kutky, M.; Burns, L.; Karran, R.; Hudak, K. RNA toxins: Mediators of stress adaptation and pathogen defense. *Wiley Interdiscip. Rev. RNA* **2010**, *2*, 890–903.

15. Stirpe, F. Ribosome-inactivating proteins. *Toxicon* **2004**, *44*, 371–383.

16. May, K.L.; Yan, Q.; Tumer, N.E. Targeting ricin to the ribosome. *Toxicon* **2013**, *69*, 143–151.

17. Clementi, N.; Chirkova, A.; Puffer, B.; Micura, R.; Polacek, N. Atomic mutagenesis reveals A2660 of 23S ribosomal RNA as key to EF-G GTPase activation. *Nat. Chem. Biol.* **2010**, *6*, 344–351.

18. Shi, X.; Khade, P.K.; Sanbonmatsu, K.Y.; Joseph, S. Functional role of the sarcin-ricin loop of the 23S rRNA in the elongation cycle of protein synthesis. *J. Mol. Biol.* **2012**, *419*, 125–138.

19. Domashevskiy, A.V.; Goss, D.J. Pokeweed antiviral protein, a ribosome inactivating protein: Activity, inhibition and prospects. *Toxins* **2015**, *7*, 274–298.

20. Choi, A.K.; Wong, E.C.; Lee, K.-M.; Wong, K.-B. Structures of eukaryotic ribosomal stalk proteins and its complex with trichosanthin, and their implications in recruiting ribosome-inactivating proteins to the ribosomes. *Toxins* **2015**, *7*, 638–647.

21. Di, R.; Tumer, N.E. Pokeweed antiviral protein: Its cytotoxicity mechanism and applications in plant disease resistance. *Toxins* **2015**, *7*, 755–772.

22. Walsh, M.J.; Dodd, J.E.; Hautbergue, G.M. Ribosome-inactivating proteins: Potent poisons and molecular tools. *Virulence* **2013**, *4*, 774–784.

23. Parikh, B.; Tumer, N. Antiviral activity of ribosome inactivating proteins in medicine. *Mini Rev. Med. Chem.* **2004**, *4*, 523.

24. Distler, U.; Souady, J.; Hülsewig, M.; Drmić-Hofman, I.; Haier, J.; Friedrich, A.W.; Karch, H.; Senninger, N.; Dreisewerd, K.; Berkenkamp, S. Shiga toxin receptor Gb3Cer/CD77: Tumor-association and promising therapeutic target in pancreas and colon cancer. *PLoS ONE* **2009**, *4*, e6813.

25. Maak, M.; Nitsche, U.; Keller, L.; Wolf, P.; Sarr, M.; Thiebaud, M.; Rosenberg, R.; Langer, R.; Kleeff, J.; Friess, H. Tumor-specific targeting of pancreatic cancer with Shiga toxin B-subunit. *Mol. Cancer Ther.* **2011**, *10*, 1918–1928.

26. Johannes, L.; Römer, W. Shiga toxins—From cell biology to biomedical applications. *Nat. Rev. Microbiol.* **2010**, *8*, 105–116.

27. Strockbine, N.A.; Jackson, M.; Sung, L.; Holmes, R.; O'Brien, A.D. Cloning and sequencing of the genes for Shiga toxin from *Shigella dysenteriae* type 1. *J. Bacteriol.* **1988**, *170*, 1116–1122.

28. Jackson, M.P.; Neill, R.J.; O'Brien, A.D.; Holmes, R.K.; Newland, J.W. Nucleotide sequence analysis and comparison of the structural genes for Shiga-like toxin I and Shiga-like toxin II encoded by bacteriophages from *Escherichia coli* 933. *FEMS Microbiol. Lett.* **1987**, *44*, 109–114.

29. Strockbine, N.A.; Marques, L.; Newland, J.W.; Smith, H.W.; Holmes, R.K.; O'brien, A.D. Two toxin-converting phages from *Escherichia coli* O157: H7 strain 933 encode antigenically distinct toxins with similar biologic activities. *Infect. Immun.* **1986**, *53*, 135–140.

30. Calderwood, S.B.; Auclair, F.; Donohue-Rolfe, A.; Keusch, G.T.; Mekalanos, J.J. Nucleotide sequence of the Shiga-like toxin genes of *Escherichia coli*. *Proc. Natl. Acad. Sci. USA* **1987**, *84*, 4364–4368.

31. Karch, H.; Tarr, P.; Bielaszewska, M. Enterohaemorrhagic *Escherichia coli* in human medicine. *Int. J. Med. Microbiol.* **2005**, *295*, 405–418.

32. Scheutz, F.; Teel, L.D.; Beutin, L.; Piérard, D.; Buvens, G.; Karch, H.; Mellmann, A.; Caprioli, A.; Tozzoli, R.; Morabito, S. Multicenter evaluation of a sequence-based protocol for subtyping Shiga toxins and standardizing Stx nomenclature. *J. Clin. Microbiol.* **2012**, *50*, 2951–2963.

33. Stein, P.E.; Boodhoo, A.; Tyrrell, G.J.; Brunton, J.L.; Read, R.J. Crystal structure of the cell-binding B oligomer of Verotoxin-1 from *E. coli*. *Nature* **1992**, *355*, 748–750.

34. Fraser, M.; Chernaia, M.; Kozlov, Y.; James, M. Crystal structure of the holotoxin from *Shigella dysenteriae* at 2.5 A resolution. *Nat. Struct. Biol.* **1994**, *1*, 59.

35. Fraser, M.E.; Fujinaga, M.; Cherney, M.M.; Melton-Celsa, A.R.; Twiddy, E.M.; O'Brien, A.D.; James, M.N. Structure of Shiga toxin type 2 (Stx2) from *Escherichia coli* O157: H7. *J. Biol. Chem.* **2004**, *279*, 27511–27517.

36. Ling, H.; Boodhoo, A.; Hazes, B.; Cummings, M.D.; Armstrong, G.D.; Brunton, J.L.; Read, R.J. Structure of the Shiga-like toxin I B-pentamer complexed with an analogue of its receptor Gb3. *Biochemistry* **1998**, *37*, 1777–1788.

37. Shimizu, T.; Sato, T.; Kawakami, S.; Ohta, T.; Noda, M.; Hamabata, T. Receptor affinity, stability and binding mode of Shiga toxins are determinants of toxicity. *Microb. Pathog.* **2007**, *43*, 88–95.

38. Samuel, J.; Perera, L.; Ward, S.; O'brien, A.; Ginsburg, V.; Krivan, H. Comparison of the glycolipid receptor specificities of Shiga-like toxin type II and Shiga-like toxin type II variants. *Infect. Immun.* **1990**, *58*, 611–618.

39. DeGrandis, S.; Law, H.; Brunton, J.; Gyles, C.; Lingwood, C. Globotetraosylceramide is recognized by the pig edema disease toxin. *J. Biol. Chem.* **1989**, *264*, 12520–12525.

40. Nakajima, H.; Kiyokawa, N.; Katagiri, Y.U.; Taguchi, T.; Suzuki, T.; Sekino, T.; Mimori, K.; Ebata, T.; Saito, M.; Nakao, H. Kinetic analysis of binding between Shiga toxin and receptor glycolipid Gb3Cer by surface plasmon resonance. *J. Biol. Chem.* **2001**, *276*, 42915–42922.

41. Müthing, J.; Meisen, I.; Zhang, W.; Bielaszewska, M.; Mormann, M.; Bauerfeind, R.; Schmidt, M.A.; Friedrich, A.W.; Karch, H. Promiscuous Shiga toxin 2e and its intimate relationship to Forssman. *Glycobiology* **2012**, *22*, 849–862.

42. Jacobson, J.M.; Yin, J.; Kitov, P.I.; Mulvey, G.; Griener, T.P.; James, M.N.; Armstrong, G.; Bundle, D.R. The crystal structure of Shiga toxin type 2 with bound disaccharide guides the design of a heterobifunctional toxin inhibitor. *J. Biol. Chem.* **2014**, *289*, 885–894.

43. Van Deurs, B.; Sandvig, K. Furin-induced cleavage and activation of Shiga toxin. *J. Biol. Chem.* **1995**, *270*, 10817–10821.

44. Fraser, M.E.; Cherney, M.M.; Marcato, P.; Mulvey, G.L.; Armstrong, G.D.; James, M.N. Binding of adenine to Stx2, the protein toxin from *Escherichia coli* O157: H7. *Acta Crystallogr. Sect. F Struct. Biol. Cryst. Commun.* **2006**, *62*, 627–630.

45. Arfilli, V.; Carnicelli, D.; Rocchi, L.; Ricci, F.; Pagliaro, P.; Tazzari, P.; Brigotti, M. Shiga toxin 1 and ricin A chain bind to human polymorphonuclear leucocytes through a common receptor. *Biochem. J.* **2010**, *432*, 173–180.

46. Brigotti, M.; Carnicelli, D.; Arfilli, V.; Tamassia, N.; Borsetti, F.; Fabbri, E.; Tazzari, P.L.; Ricci, F.; Pagliaro, P.; Spisni, E. Identification of TLR4 as the receptor that recognizes Shiga toxins in human neutrophils. *J. Immunol.* **2013**, *191*, 4748–4758.

47. Sandvig, K.; van Deurs, B. Delivery into cells: Lessons learned from plant and bacterial toxins. *Gene Ther.* **2005**, *12*, 865–872.

48. Mayer, C.L.; Leibowitz, C.S.; Kurosawa, S.; Stearns-Kurosawa, D.J. Shiga Toxins and the pathophysiology of hemolytic uremic syndrome in humans and animals. *Toxins* **2012**, *4*, 1261.

49. Tesh, V.L.; Burris, J.; Owens, J.; Gordon, V.; Wadolkowski, E.; O'brien, A.; Samuel, J. Comparison of the relative toxicities of Shiga-like toxins type I and type II for mice. *Infect. Immun.* **1993**, *61*, 3392–3402.

50. Siegler, R.L.; Obrig, T.G.; Pysher, T.J.; Tesh, V.L.; Denkers, N.D.; Taylor, F.B. Response to Shiga toxin 1 and 2 in a baboon model of hemolytic uremic syndrome. *Pediatr. Nephrol.* **2003**, *18*, 92–96.

51. Wadolkowski, E.; Sung, L.; Burris, J.; Samuel, J.; O'brien, A. Acute renal tubular necrosis and death of mice orally infected with *Escherichia coli* strains that produce Shiga-like toxin type II. *Infect. Immun.* **1990**, *58*, 3959–3965.

52. Stearns-Kurosawa, D.; Collins, V.; Freeman, S.; Tesh, V.L.; Kurosawa, S. Distinct physiologic and inflammatory responses elicited in baboons after challenge with Shiga Toxin type 1 or 2 from enterohemorrhagic *Escherichia coli*. *Infect. Immun.* **2010**, *78*, 2497.

53. Stearns-Kurosawa, D.J.; Oh, S.-Y.; Cherla, R.P.; Lee, M.-S.; Tesh, V.L.; Papin, J.; Henderson, J.; Kurosawa, S. Distinct renal pathology and a chemotactic phenotype after enterohemorrhagic *Escherichia coli* Shiga toxins in non-human primate models of hemolytic uremic syndrome. *Am. J. Pathol.* **2013**, *182*, 1227.

54. Bauwens, A.; Betz, J.; Meisen, I.; Kemper, B.; Karch, H.; Müthing, J. Facing glycosphingolipid-Shiga toxin interaction: Dire straits for endothelial cells of the human vasculature. *Cell. Mol. Life Sci.* **2013**, *70*, 425–457.

55. Louise, C.B.; Obrig, T.G. Specific interaction of *Escherichia coli* O157: H7-derived Shiga-like toxin II with human renal endothelial cells. *J. Infect. Dis.* **1995**, *172*, 1397–1401.

56. Bauwens, A.; Bielaszewska, M.; Kemper, B.; Langehanenberg, P.; von Bally, G.; Reichelt, R.; Mulac, D.; Humpf, H.-U.; Friedrich, A.W.; Kim, K.S. Differential cytotoxic actions of Shiga toxin 1 and Shiga toxin 2 on microvascular and macrovascular endothelial cells. *Thromb. Haemost.* **2011**, *105*, 515–528.

57. Fuller, C.A.; Pellino, C.A.; Flagler, M.J.; Strasser, J.E.; Weiss, A.A. Shiga toxin subtypes display dramatic differences in potency. *Infect. Immun.* **2011**, *79*, 1329–1337.

58. Flagler, M.J.; Mahajan, S.S.; Kulkarni, A.A.; Iyer, S.S.; Weiss, A.A. Comparison of binding platforms yields insights into receptor binding differences between Shiga toxins 1 and 2. *Biochemistry* **2010**, *49*, 1649–1657.

59. Head, S.; Karmali, M.; Lingwood, C. Preparation of VT1 and VT2 hybrid toxins from their purified dissociated subunits. Evidence for B subunit modulation of a subunit function. *J. Biol. Chem.* **1991**, *266*, 3617–3621.

60. Lingwood, C.A. Role of Verotoxin receptors in pathogenesis. *Trends Microbiol.* **1996**, *4*, 147–153.

61. Zumbrun, S.D.; Hanson, L.; Sinclair, J.F.; Freedy, J.; Melton-Celsa, A.R.; Rodriguez-Canales, J.; Hanson, J.C.; O'Brien, A.D. Human intestinal tissue and cultured colonic cells contain globotriaosylceramide synthase mRNA and the alternate Shiga toxin receptor globotetraosylceramide. *Infect. Immun.* **2010**, *78*, 4488–4499.

62. Rutjes, N.W.; Binnington, B.A.; Smith, C.R.; Maloney, M.D.; Lingwood, C.A. Differential tissue targeting and pathogenesis of Verotoxins 1 and 2 in the mouse animal model. *Kidney Int.* **2002**, *62*, 832–845.

63. Karve, S.S.; Weiss, A.A. Glycolipid binding preferences of Shiga toxin variants. *PLoS One* **2014**, *9*, e101173.

64. Ito, H.; Yutsudo, T.; Hirayama, T.; Takeda, Y. Isolation and some properties of A and B subunits of Vero toxin 2 and *in vitro* formation of hybrid toxins between subunits of Vero toxin 1 and Vero toxin 2 from *Escherichia coli* O157: H7. *Microb. Pathog.* **1988**, *5*, 189–195.

65. Weinstein, D.L.; Jackson, M.P.; Perera, L.P.; Holmes, R.K.; O'Brien, A.D. *In vivo* formation of hybrid toxins comprising Shiga toxin and the Shiga-like toxins and role of the B subunit in localization and cytotoxic activity. *Infect. Immun.* **1989**, *57*, 3743–3750.

66. Russo, L.M.; Melton-Celsa, A.R.; Smith, M.J.; O'Brien, A.D. Comparisons of native Shiga Toxins (Stxs) Type 1 and 2 with chimeric toxins indicate that the source of the binding subunit dictates degree of toxicity. *PLoS ONE* **2014**, *9*, e93463.

67. Hovde, C.; Calderwood, S.; Mekalanos, J.; Collier, R. Evidence that glutamic acid 167 is an active-site residue of Shiga-like toxin I. *Proc. Natl. Acad. Sci. USA* **1988**, *85*, 2568.

68. Jackson, M.; Deresiewicz, R.; Calderwood, S. Mutational analysis of the Shiga toxin and Shiga-like toxin II enzymatic subunits. *J. Bacteriol.* **1990**, *172*, 3346.

69. Yamasaki, S.; Furutani, M.; Ito, K.; Igarashi, K.; Nishibuchi, M.; Takeda, Y. Importance of arginine at position 170 of the A subunit of Verotoxin 1 produced by enterohemorrhagic *Escherichia coli* for toxin activity. *Microb. Pathog.* **1991**, *11*, 1–9.

70. Di, R.; Kyu, E.; Shete, V.; Saidasan, H.; Kahn, P.; Tumer, N. Identification of amino acids critical for the cytotoxicity of Shiga toxin 1 and 2 in *Saccharomyces cerevisiae*. *Toxicon* **2011**, *57*, 525–539.

71. Deresiewicz, R.L.; Austin, P.R.; Hovde, C.J. The role of tyrosine-114 in the enzymatic activity of the Shiga-like toxin I A-chain. *Mol. Genet. Genomics* **1993**, *241*, 467–473.

72. Skinner, L.; Jackson, M. Investigation of ribosome binding by the Shiga toxin A1 subunit, using competition and site-directed mutagenesis. *J. Bacteriol.* **1997**, *179*, 1368–1374.

73. McCluskey, A.; Bolewska-Pedyczak, E.; Jarvik, N.; Chen, G.; Sidhu, S.; Johannes, L. Charged and hydrophobic surfaces on the A chain of Shiga-Like Toxin 1 recognize the *C*-terminal domain of ribosomal stalk proteins. *PLoS ONE* **2012**, *7*, e31191.

74. Brigotti, M.; Carnicelli, D.; Alvergna, P.; Mazzaracchio, R.; Sperti, S.; Montanaro, L. The RNA-N-glycosidase activity of Shiga-like toxin I: Kinetic parameters of the native and activated toxin. *Toxicon* **1997**, *35*, 1431–1437.

75. Chiou, J.-C.; Li, X.-P.; Remacha, M.; Ballesta, J.P.; Tumer, N.E. Shiga toxin 1 is more dependent on the P proteins of the ribosomal stalk for depurination activity than Shiga toxin 2. *Int. J. Biochem. Cell. Biol.* **2011**, *43*, 1792–1801.

76. Pierce, M.; Kahn, J.N.; Chiou, J.; Tumer, N.E. Development of a quantitative RT-PCR assay to examine the kinetics of ribosome depurination by ribosome inactivating proteins using *Saccharomyces cerevisiae* as a model. *RNA* **2011**, *17*, 201–210.

77. May, K.L.; Li, X.P.; Martinez-Azorin, F.; Ballesta, J.P.; Grela, P.; Tchorzewski, M.; Tumer, N.E. The P1/P2 proteins of the human ribosomal stalk are required for ribosome binding and depurination by ricin in human cells. *FEBS J.* **2012**, *279*, 3925–3936.

78. Sturm, M.B.; Schramm, V.L. Detecting ricin: Sensitive luminescent assay for ricin A-chain ribosome depurination kinetics. *Anal. Chem.* **2009**, *81*, 2847–2853.

79. Li, X.-P.; Kahn, P.C.; Kahn, J.N.; Grela, P.; Tumer, N.E. Arginine residues on the opposite side of the active site stimulate the catalysis of ribosome depurination by ricin A chain by interacting with the P-protein stalk. *J. Biol. Chem.* **2013**, *288*, 30270–30284.

80. Chiou, J.C.; Li, X.P.; Remacha, M.; Ballesta, J.P.; Tumer, N.E. The ribosomal stalk is required for ribosome binding, depurination of the rRNA and cytotoxicity of ricin A chain in *Saccharomyces cerevisiae*. *Mol. Microbiol.* **2008**, *70*, 1441–1452.

81. Li, X.-P.; Chiou, J.-C.; Remacha, M.; Ballesta, J.P.; Tumer, N.E. A two-step binding model proposed for the electrostatic interactions of ricin a chain with ribosomes. *Biochemistry* **2009**, *48*, 3853–3863.

82. Endo, Y.; Tsurugi, K. The RNA N-glycosidase activity of ricin A-chain. The characteristics of the enzymatic activity of ricin A-chain with ribosomes and with rRNA. *J. Biol. Chem.* **1988**, *263*, 8735–8739.

83. Endo, Y.; Tsurugi, K.; Yutsudo, T.; Takeda, Y.; Ogasawara, T.; Igarashi, K. Site of action of a Vero toxin (VT2) from *Escherichia coli* O157: H7 and of Shiga toxin on eukaryotic ribosomes. *Eur. J. Biochem.* **1988**, *171*, 45–50.

84. McCluskey, A.J.; Poon, G.M.; Bolewska-Pedyczak, E.; Srikumar, T.; Jeram, S.M.; Raught, B.; Gariépy, J. The catalytic subunit of Shiga-like toxin 1 interacts with ribosomal stalk proteins and is inhibited by their conserved *C*-terminal domain. *J. Mol. Biol.* **2008**, *378*, 375–386.

85. Vater, C.A.; Bartle, L.M.; Leszyk, J.D.; Lambert, J.M.; Goldmacher, V.S. Ricin A chain can be chemically cross-linked to the mammalian ribosomal proteins L9 and L10e. *J. Biol. Chem.* **1995**, *270*, 12933–12940.

86. Chan, S.H.; Hung, F.S.J.; Chan, D.S.B.; Shaw, P.C. Trichosanthin interacts with acidic ribosomal proteins P0 and P1 and mitotic checkpoint protein MAD2B. *Eur. J. Biochem.* **2001**, *268*, 2107–2112.

87. Chan, D.S.; Chu, L.-O.; Lee, K.-M.; Too, P.H.; Ma, K.-W.; Sze, K.-H.; Zhu, G.; Shaw, P.-C.; Wong, K.-B. Interaction between trichosanthin, a ribosome-inactivating protein, and the ribosomal stalk protein P2 by chemical shift perturbation and mutagenesis analyses. *Nucleic Acids Res.* **2007**, *35*, 1660–1672.

88. Too, P.H.-M.; Ma, M.K.-W.; Mak, A.N.-S.; Wong, Y.-T.; Tung, C.K.-C.; Zhu, G.; Au, S.W.-N.; Wong, K.-B.; Shaw, P.-C. The *C*-terminal fragment of the ribosomal P protein complexed to trichosanthin reveals the interaction between the ribosome-inactivating protein and the ribosome. *Nucleic Acids Res.* **2009**, *37*, 602–610.

89. Li, X.-P.; Grela, P.; Krokowski, D.; Tchórzewski, M.; Tumer, N.E. Pentameric organization of the ribosomal stalk accelerates recruitment of ricin a chain to the ribosome for depurination. *J. Biol. Chem.* **2010**, *285*, 41463–41471.

90. Bargis-Surgey, P.; Lavergne, J.-P.; Gonzalo, P.; Vard, C.; Filhol-Cochet, O.; Reboud, J.-P. Interaction of elongation factor eEF-2 with ribosomal P proteins. *Eur. J. Biochem.* **1999**, *262*, 606–611.

91. Grela, P.; Sawa-Makarska, J.; Gordiyenko, Y.; Robinson, C.; Grankowski, N.; Tchórzewski, M. Structural properties of the human acidic ribosomal P proteins forming the P1-P2 heterocomplex. *J. Biochem.* **2008**, *143*, 169.

92. Wool, I.; Chan, Y.; Glück, A.; Suzuki, K. The primary structure of rat ribosomal proteins P0, P1, and P2 and a proposal for a uniform nomenclature for mammalian and yeast ribosomal proteins. *Biochimie* **1991**, *73*, 861.

93. Planta, R.J.; Mager, W.H. The list of cytoplasmic ribosomal proteins of *Saccharomyces cerevisiae*. *Yeast* **1998**, *14*, 471–477.

94. Tchórzewski, M.; Boguszewska, A.; Dukowski, P.; Grankowski, N. Oligomerization properties of the acidic ribosomal P-proteins from *Saccharomyces cerevisiae*: Effect of P1A protein phosphorylation on the formation of the P1A-P2B hetero-complex. *BBA Clin.* **2000**, *1499*, 63.

95. Krokowski, D.; Boguszewska, A.; Abramczyk, D.; Liljas, A.; Tchórzewski, M.; Grankowski, N. Yeast ribosomal P0 protein has two separate binding sites for P1/P2 proteins. *Mol. Microbiol.* **2006**, *60*, 386–400.

96. Hagiya, A.; Naganuma, T.; Maki, Y.; Ohta, J.; Tohkairin, Y.; Shimizu, T.; Nomura, T.; Hachimori, A.; Uchiumi, T. A mode of assembly of P0, P1 and P2 proteins at the GTPase-associated center in animal ribosome: *In vitro* analyses with P0 truncation mutants. *J. Biol. Chem.* **2005**, *280*, 39193–39199.

97. Grela, P.; Krokowski, D.; Gordiyenko, Y.; Krowarsch, D.; Robinson, C.V.; Otlewski, J.; Grankowski, N.; Tchórzewski, M. Biophysical properties of the eukaryotic ribosomal stalk. *Biochemistry* **2010**, *49*, 924–933.

98. Ballesta, J.P.; Remacha, M. The large ribosomal subunit stalk as a regulatory element of the eukaryotic translational machinery. *Prog. Nucleic Acid Res. Mol. Biol.* **1996**, *55*, 157–193.

99. Gonzalo, P.; Reboud, J.P. The puzzling lateral flexible stalk of the ribosome. *Biol. Cell* **2003**, *95*, 179–193.

100. Diaconu, M.; Kothe, U.; Schlünzen, F.; Fischer, N.; Harms, J.M.; Tonevitsky, A.G.; Stark, H.; Rodnina, M.V.; Wahl, M.C. Structural basis for the function of the ribosomal L7/12 stalk in factor binding and GTPase activation. *Cell* **2005**, *121*, 991–1004.

101. Maki, Y.; Hashimoto, T.; Zhou, M.; Naganuma, T.; Ohta, J.; Nomura, T.; Robinson, C.V.; Uchiumi, T. Three binding sites for stalk protein dimers are generally present in ribosomes from archaeal organism. *J. Biol. Chem.* **2007**, *282*, 32827–32833.

102. Ayub, M.J.; Smulski, C.R.; Ma, K.-W.; Levin, M.J.; Shaw, P.-C.; Wong, K.-B. The *C*-terminal end of P proteins mediates ribosome inactivation by trichosanthin but does not affect the pokeweed antiviral protein activity. *Biochem. Biophys. Res. Commun.* **2008**, *369*, 314–319.

103. Hudak, K.A.; Dinman, J.D.; Tumer, N.E. Pokeweed antiviral protein accesses ribosomes by binding to L3. *J. Biol. Chem.* **1999**, *274*, 3859–3864.

104. Lapadula, W.J.; Sanchez-Puerta, M.; Juri Ayub, M. Convergent evolution led ribosome inactivating proteins to interact with ribosomal stalk. *Toxicon* **2012**, *59*, 427–432.

105. Lapadula, W.J.; Puerta, M.V.S.; Ayub, M.J. Revising the taxonomic distribution, origin and evolution of ribosome inactivating protein genes. *PLoS ONE* **2013**, *8*, e72825.

106. Bernado, P.; Modig, K.; Grela, P.; Svergun, D.I.; Tchorzewski, M.; Pons, M.; Akke, M. Structure and dynamics of ribosomal protein L12: An ensemble model based on SAXS and NMR relaxation. *Biophys. J.* **2010**, *98*, 2374–2382.

107. Lee, K.-M.; Yusa, K.; Chu, L.-O.; Yu, C.W.-H.; Oono, M.; Miyoshi, T.; Ito, K.; Shaw, P.-C.; Wong, K.-B.; Uchiumi, T. Solution structure of human P1 P2 heterodimer provides insights into the role of eukaryotic stalk in recruiting the ribosome-inactivating protein trichosanthin to the ribosome. *Nucleic Acids Res.* **2013**, *41*, 8776–8787.

108. Smith, M.J.; Melton-Celsa, A.R.; Sinclair, J.F.; Carvalho, H.M.; Robinson, C.M.; O'Brien, A.D. Monoclonal antibody 11E10, which neutralizes Shiga toxin type 2 (Stx2), recognizes three regions on the Stx2 A subunit, blocks the enzymatic action of the toxin *in vitro*, and alters the overall cellular distribution of the toxin. *Infect. Immun.* **2009**, *77*, 2730–2740.

109. Jiao, Y.-J.; Zeng, X.-Y.; Guo, X.-L.; Shi, Z.-Y.; Feng, Z.-Q.; Wang, H. Monoclonal antibody S2C4 neutralizes the toxicity of Shiga toxin 2 and its variants. *Prog. Biochem. Biophys.* **2009**, *36*, 736–742.

110. Olson, M.A.; Cuff, L. Free energy determinants of binding the rRNA substrate and small ligands to ricin A-chain. *Biophys. J.* **1999**, *76*, 28–39.